Undergraduate Topics in Computer Science

'Undergraduate Topics in Computer Science' (UTiCS) delivers high-quality instructional content for undergraduates studying in all areas of computing and information science. From core foundational and theoretical material to final-year topics and applications, UTiCS books take a fresh, concise, and modern approach and are ideal for self-study or for a one- or two-semester course. The texts are authored by established experts in their fields, reviewed by an international advisory board, and contain numerous examples and problems, many of which include fully worked solutions.

The UTiCS concept centers on high-quality, ideally and generally quite concise books in softback format. For advanced undergraduate textbooks that are likely to be longer and more expository, Springer continues to offer the highly regarded *Texts in Computer Science* series, to which we refer potential authors.

Rod Downey

Computability and Complexity

Foundations and Tools for Pursuing Scientific Applications

 Springer

Rod Downey
School of Mathematics and Statistics
Victoria University
Wellington, New Zealand

ISSN 1863-7310   ISSN 2197-1781 (electronic)
Undergraduate Topics in Computer Science
ISBN 978-3-031-53743-1   ISBN 978-3-031-53744-8 (eBook)
https://doi.org/10.1007/978-3-031-53744-8

This Springer imprint is published by the registered company Springer Nature Switzerland AG
The registered company address is: Gewerbestrasse 11, 6330 Cham, Switzerland

Paper in this product is recyclable.

To Kristin, with love

Preface

One of the great achievements of humanity has been the clarification of the concept of an algorithmic process. Having done this, we then sought to understand the structure of algorithms and their complexity. Whilst this story goes back literally thousands of years, it has only been relatively recently that we have had the mathematical tools to deal with these issues.

This book deals with the amazing intellectual development of the ideas of computability and complexity over the last 150 years, or so. These ideas are deep. They should be known by working mathematicians and computer scientists.

The goal of this book is to give a straightforward introduction to the main ideas in these areas, at least from my own biased point of view. The book has been derived from various courses I have taught at Victoria University of Wellington, University of Madison, Wisconsin, Notre Dame, University of Chicago, Cornell University, Nanyang University of Technology, and elsewhere over the last 35 years. The target audience is somewhere around advanced undergraduate and beginning graduate students. That is, in the British system, final year undergraduate students and Masters Part 1 students, and in the US system, seniors and first year graduate students.

This text could be also used by a professional mathematician to gain a thorough grounding in computability and computational complexity.

The topics covered include basic naive set theory, regular languages and automata, models of computation, undecidability proofs, classical computability theory including the arithmetical hierarchy and the priority method, the basics of computational complexity, hierarchy theorems, NP and PSPACE completeness, structural complexity such as oracle results, parameterized complexity and other methods of coping such as approximation, average case complexity, generic case complexity and smoothed analysis.

There are a number of topics of this book which have never been given in a basic text before. I have included the fascinating series of reductions used by John Conway to show a variation of the "Collatz $3n + 1$ Function" is a universal programming language, something I have found is a great teaching tool for introducing undecidability proofs. I have also included (a detailed sketch of) the proof that Exponential Diophantine sets are undecidable as this is reasonably short and self-contained. This is one of the main building blocks in the solution to Hilbert's 10th Problem, and the proof we give shows the usefulness of coding with register machines. The solution to Hilbert's 10th Problem is one of the great results of the 20th century. I have included some computable structure theory; that is, the application of the computability theory to understand computational processes in algebra, especially linear orderings. Also I have included some computable calculus.

I also have included for the first time in an undergraduate text, the basics of parameterized complexity and a short account of both smoothed analysis and of generic case complexity.

There is too much material to be able to cover by normal students in a single semester. The overall level varies from relatively straightforward to some rather complex material. It is possible to make a good course on computability and undecidability using the first half of the book. You could then follow up with more advanced texts such as Soare [Soa87, Soa16], Rogers [Rog87], Odifreddi [Odi90], or be equipped to work in algorithmic randomness via texts like Nies [Nie09], Li and Vitanyi [LV93], or Downey and Hirschfeldt [DH10], or finally to move into some computable analysis or algebra with texts such as Ash and Knight [AK00], Downey and Melnikov [DMar], and Pour-El and Richards [PER89].

You can also make a nice course in computational complexity, by selecting from the first half of the book for enough computability theory to make the second half accessible. After this, you would be in good shape to progress to advanced texts on computational complexity such as Kozen [Koz06], Arora and Borak [AB09], or Homer and Selman [HS11]. The reader could move to more specialized texts such as Vazirani [Vaz01], or Ausiello et. al. [ACG+99] on approximation algorithms, or to Downey and Fellows [DF13], Flum and Grohe [FG06], Cygan et. al. [CFK+16], Fomin et. al. [FLSZ19] and Niedermeier [Nie06] for parameterized algorithms and complexity.

I believe that someone who masters a good fragment of the book will be equipped with the tools to move on to more advanced graduate texts. With this in mind, I have endeavoured to point at research papers where material in the text has been further developed. Rather than pointing at secondary sources such as books, whenever possible, I have also given source references as citations for the various results[1]. It strikes me that we tend to do our subject a disservice by losing track of the historical development of the ideas.

---

[1] Because I have often used exercises based on my own papers (as I know them well) and also a lot of material from parameterized complexity also based on work of mine, this has the unfortunate and rather embarrassing side-effect that I have a lot of self-references in the bibliography, and the index. Do not take the number of such self-references to be any reflection of importance, only reflecting things I know!

Other areas of Science always cite source papers[2]. (Of course this can only be done within reason, as we would otherwise endlessly be citing Euclid!) I sincerely hope that the rather large list of references should serve as a resource for the reader.

The theory of computation is now a huge subject. It is a great intellectual achievement. It is also a bit daunting to those who enter. I hope this text will serve as a guide to help you on your first steps.

---

[2] It is a striking fact that up to around 2012, the centenary of Turing's birth, Turing's most cited paper was not the fundamental work he did on laying the foundations of the theory of computation, but one in biology called "The Chemical Basis for Morphogenesis" [Tur52]. This work describes how patterns in nature, such as stripes (e.g. in Zebras) and spirals, can arise naturally from a homogeneous, uniform state. It was also one of the earliest simulations of nonlinear partial differential equations.

Acknowledgements

The author wishes to thank Springer-Verlag and particularly Ronan Nugent and Wayne Wheeler for encouraging me to see this project through.

Thanks to Matthew Askes and Brendan Harding who helped with the diagrams.

The author wishes to thank many people who have provided feedback on the courses this book was based upon. Detailed feedback on various drafts on this book was provided by Eric Allender, Thomas Forster, Bill Gasarch, Carl Jockusch, and Paul Shafer. Of course, the many remaining errors are all mine.

Introduction

This is a book about *computation*, something which is ubiquitous in the modern world. More precisely, we'll be looking at *computability theory* and *computational complexity theory*.

- Computability theory is the part of mathematics and computer science which seeks to clarify what we mean by computation or algorithm. When is there a computational solution possible to some question? How can we show that none is possible? How computationally hard is the question we are concerned with? Arguably, this area lead to the development of digital computers.

- (Computational) complexity theory is an intellectual heir of computability theory. Complexity theory is concerned with understanding what resources are needed for computation, where typically we would measure the resources in terms of time and space. Can we perform some task in a feasible number of steps? Can we perform some algorithm with only a limited memory? Does randomness help? Are there standard approaches to overcoming computational difficulty?

Computation has been at heart of mathematics since its dawn. To the author's knowledge, the first recorded example of a number known by *approximation* is $\sqrt{2}$. An approximation of $\sqrt{2}$ to six decimal places is contained on a preserved Babylonian clay tablet from around 1700 BC. The cuneiform tablet[3] called YBC 7289, states $\sqrt{2}$ is

$$1 + \frac{24}{60} + \frac{51}{60^2} + \frac{10}{60^3} \approx 1.414213.$$

---

[3] https://en.wikipedia.org/wiki/YBC_7289.

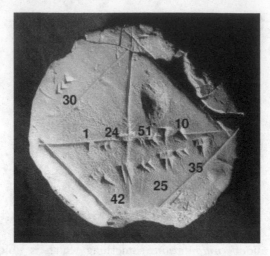

It is reasonable to assume that Babylonian mathematicians had an algorithm for approximating $\sqrt{2}$ to an arbitrary precision [FR98].[4] In India, the text Shatapatha Brahmana used an approximation of $\frac{339}{108}$ for π, correct to 4 decimal places. This work is dated (if you can believe it) *either* 600, 700 or 800 BC. Again, there must have been some algorithm for this approximation.

At the heart of our understanding of computability in mathematics and computer science is the notion of an *algorithm*. The etymology of this word goes back to Al-Khwārizmī, a Persian astronomer and mathematician who wrote a treatise in 825 AD, *On Calculation with Hindu Numerals,* together with an error in the Latin translation of his name.

Likely the first algorithm which we might be taught in an elementary algebra course at university is Euclid's algorithm devised around 300 BC. Euclid, or perhaps *Team Euclid*[5], devised this algorithm for determining the greatest common divisor of of two numbers. Here is an example of Euclid's Algorithm.

Euclid's Algorithm

- To find gcd(1001,357).
- $1001 = 357 \cdot 2 + 287$
- $357 = 287 \cdot 1 + 70$
- $287 = 70 \cdot 4 + 7$
- $70 = 7 \cdot 10$
- $7 =$gcd(1001,357).

---

[4] Fowler and Robson use their understanding of Old Babylonian mathematics to give an algorithm that they may have used. Possibly it was something called "Hero's Method", which begins with a guess and refines it with iterative methods.

[5] In those days, people such as Euclid would have had a number of disciples working for him, so it is not completely clear who was responsible for such results.

What is going on here? What we have specified is some kind of *mechanical* process which, beginning with some primitives we regard as computable, when followed yields the solution from those primitives. In Euclid's algorithm we are taking as primitives division, multiplication, etc and combining them in reasonable ways sequentially. Euclid's algorithm remains the fastest algorithm for computing the GCD and is still used today. (We discuss what we mean by "fastest" in Chapter 7.)

But what is an algorithm? This is really quite a modern question, at least in terms of the history of mathematics. Virtually all of mathematics up to the last quarter of the 19th century could be viewed as algorithmic. It was only with the work of Cantor on set theory and algebraists introducing non-computational techniques in the early 20th century that much of mathematics became non-algorithmic. We refer the reader to a nice article by Metakides and Nerode [MN82] which discusses the history of the introduction of "non-constructive" and "non-algorithmic" methods into mathematics. Mathematicians were quite aware of these issues at the time. For example, David Hilbert [Hil90] proved *Hilbert's Basis Theorem*, solving something called *Gordan's Problem*. Hilbert showed that certain polynomial rings always had finite bases, but without giving any clue how to find them. It was a *pure existence proof.* At the time it is claimed[6] that Gordan rejected the approach saying

"Das ist nicht Mathematik. Das ist Theologie[7]." (This is not Mathematics. This is Theology.)

The invariants guaranteed by Hilbert's Basis Theorem are important for applications and hence we need to know them. Is there an algorithm which *generates* them? And, if so, how *quickly* can we compute them? Actually finding these invariants, rather than simply proving that they exist, requires significant algorithmic insight. This quest lead to a new subject called *Gröbner Basis Theory*, an area actually due to Buchberger [Buc65].

Hilbert was the leading mathematician of his day. In a famous list of problems for the new century (i.e. the 20th Century, published in [Hil12]), in the 1900 International Congress of Mathematicians, Hilbert asked what is called the *Entscheidungsproblem*. (The decidability of first order logic.) More precisely, we know that there is a way of deciding if a formula of propositional logic is a tautology: you draw up a truth table, and see if all the lines are true. That is, propositional logic has a *decision procedure* for validity[8]. Suppose we enrich our language and add quantifiers and predicates to form first order

---

[6] This is likely a myth as the first reference to this "quote" was 25 years after Hilbert's paper and after Gordan's death. But is one of those things that should be true even if it is not!

[7] Another version says "Zauberei" (sorcery) in place of "Theologie".

[8] Not a very efficient method since a propositional formula with 12,000 many variables would have truth table with $2^{12,000}$ many lines. In Chapter 7, we will examine the question of whether there is there is a shorter decision method for this logic.

(predicate) logic. The Entscheidungsproblem asks if there is a similar *decision procedure* for first order logic[9].

If there is such a decision procedure, then we would give one and nod sagely saying "yes, that works". But suppose that we can't find such a procedure?

How can we show there is no decision procedure?

This apparently obscure question from mathematical logic, leads to a very deep question: What do we mean by a decision procedure anyway? Presumably any decision procedure would be what we now call an algorithm; so the question is "what do we mean by *an algorithm?*"

Unless we give a mathematical definition of an algorithm, then surely we would have no hope of showing that none exists. A definition was given in the mid-1930's. The reader should note, however, that an intuitive understanding of computability existed well before the 1930's. Moreover, the proofs given during this time were mostly constructive. For example, in analysis, Borel [Bor12] was seeking algorithmic way of approximating real numbers, and stated that a real x is 'computable' if, given any natural number n, we can obtain a rational q within $\frac{1}{n}$ of x [Bor12][10]. For example the number $e = \sum_{n=0}^{\infty} \frac{1}{n!}$ certainly fits the bill as this series converges *very quickly*. What Borel means by 'computable' is uncertain, particularly since it would be another 20 years before any formal notion of computation emerged. It is unclear what Borel intended when he spoke of 'obtaining' a rational close to x. However, in a footnote Borel writes;

> "I intentionally leave aside the practical length of operations, which can be shorter or longer; the essential point is that each operation can be executed in finite time with a safe method that is unambiguous."

Other examples, such as the work of Kronecker, Hermann, Dehn and Von Mises [Kro82], [Her26], [Deh11] show that phrases such as 'by finite means' or 'by constructive measures' were relatively standard, but lacked any precise definition.

It was the work of a collection of logicians in the 1930's that the notion of algorithm was given a mathematical definition. The culmination of this work was by Alan Turing with the famous paper of 1936 [Tur36]. This definition used a model now called a *Turing Machine*. Turing gave a detailed conceptual analysis of human thought, and used this as the basis for the justification of his model for "decision procedure". The point being that in the 1930's, *a decision procedure would be something a human could do.* Turing's model was immediately accepted as a true model of computation, and allowed him

---

[9] Strictly speaking, Hilbert asked for mathematicians to give such a decision procedure. See §4.3.6

[10] Quotes and comments from Borel's paper [Bor12] are based on a translation (French to English) by Avigad and Brattka [AB12].

to give a proof that first order logic was *undecidable*. There is no decision procedure.

We will look at Turing's work and other models of computation in Chapter 3. These include partial recursive functions and register machines. The other models all are equivalent to Turing's model, and hence support what is called the *Church-Turing Thesis*. This states that Turing machines capture the intuitive model of computation. (We will discuss this further in Chapter 3.)

These models are also useful in establishing other undecidability results, such as ones from "normal mathematics" such as in group theory, algebra, analysis, etc. Typical undecidability results work as follows: We have some problem we suspect is algorithmically undecidable. To establish undecidability, we use a method called "the method of reductions". First we will establish that a certain problem is algorithmically unsolvable, namely the *Halting problem for Turing Machines*. Then, given a problem, we show that if the given problem was decidable, *then* we could solve the halting problem. That is, we "code" the halting problem into the problem at hand.

Because the models are all equivalent, we don't need to start with Turing machines, but could code any undecidable problem for the given model. As an illustration, we will look at a seemingly tame problem generalizing the famous Collatz Conjecture: The Collatz problem takes a positive integer n, and defines a function:

$$f(n) = \begin{cases} \frac{n}{2} & \text{if } n \text{ is even} \\ 3n+1 & \text{if } n \text{ is odd} \end{cases}$$

we then look at the sequence of iterates $f(n), f(f(n)), \ldots$. The Collatz Conjecture that we always get back to 1 no matter what n we start with. In [Con72], John Conway defined a variation of this where we replace even and odd by a sequence of congruences mod d, and asks what can we predict algorithmically about the sequence of iterates we get. That is, we give a set of rationals $\{r_0, \ldots, r_k, d_1, \ldots d_{p-1}\}$ and define $g(n) = r_i n + d_i$ if $n \equiv i \mod p$. In the original Collatz conjecture, $p = 2$ we have $r_0 = \frac{1}{2}, d_0 = 0; r_1 = 3, d_1 = 1$. Amazingly, for suitably chosen rationals r_i, and $d_i = 0$ *for all i*, and integer p, $g = g_{\langle r_1, \ldots, r_k \rangle}$ can simulate a universal model of computation equivalent to the Turing machine! Hence we can't predict anything about their behaviour!

We prove Conway's result in Chapter 4. We do so by coding Register Machines into Conway/Collatz functions. In Chapter 4 we will also prove several other natural undecidability results-undecidability results concerning problems which appear in "normal mathematics", such as in algebra or analysis. These include word problems in semigroups and we will sketch the undecidability of the Entscheidungsproblem. To complete the chapter, we will look at the Exponential case of Hilbert's 10th Problem. This algorithmic problem asks for an algorithm to determine zeroes of multivariable polynomials with rational coefficients. Again, we will show that the relevant problem is algorithmically undecidable; and again we will be coding Register Machines into the problems.

In Chapter 5, we'll deepen the development of classical computability theory. We'll look at important results such as Rice's Theorem, Recursion Theorem, and the like, as well as developing the arithmetical hierarchy, used for calibrating the difficulty of problems.

Computational Complexity. As with computability, people had an intuitive understanding of what it meant to be a *computationally difficult task*. That is, one that could be solved algorithmically, but seemed very hard to do. Classical ciphers were based on the belief that decrypting the cipher would take more time than is available. Indeed, modern public key encryption methods are all based on problems we think are computationally infeasible. For example, we believe that the public key cryptosystem called RSA is secure (more or less) because we believe that it is computationally infeasible to factorize a large integer. The infeasibility of factorization is challenged if we ever build a quantum computer.

What do we mean by infeasibility anyway?

We have seen that we could prove *undecidability* once we had a mathematical definition of being *computable*. Analogously, can we prove *infeasibility* once we have an acceptable definition of what this concept means and a core problem which we regarded as infeasible, in the same way that the halting problem is undecidable?

Historically, mathematicians noticed that certain tasks seemed to be hard to compute, and others seemed easy. A wide class of such problems seemed to require us to "try all possibilities" and, if the best algorithm we could hope for is to try all possibilities, then the problem *must* be infeasible for large inputs. The well-known games of Sodoku and of Go, but for arbitrarily large boards, seemed to fit this; and many computer games were derived from problems of this character. They are believed to be challenging because of this "complexity" issue. Russian mathematicians from the 1950's gave problems of this general kind, apparently needing complete search, the name *perebor* (see Trakhtenbrot [Tra84]), meaning "brute force".

One well-known problem from graph theory is the following:

HAMILTON CYCLE
Input: A graph G.
Question: Is there a cycle through the *vertices* of G going through every vertex exactly once?

Essentially the only known way to solve this is to "try all possibilities". For a graph with $50,000,000$ vertices would involve looking at around $50,000,000!$

(*factorial* 50,000,000) many possibilities. This number is beyond the estimated number of particles in the universe. On the other hand, consider the following problem:

EULER CYCLE
Input: A graph G.
Question: Does G have a cycle through all the *edges* of the graph exactly once?

Long ago, Euler [Eul36] proved[11] that G has an Euler cycle iff G is connected and has only vertices of even degree. For a graph with $50,000,000$ vertices deciding this would take around $50,000,000^2$ many steps, well within the range of modern computers, as it involves simply looking at all pairs of vertices, first seeing if the graph is connected (which uses an efficient algorithm called network flow running in around n^2 many steps) and then seeing what the degree of each vertex of G is, (taking about n many steps where n is the number of vertices).

Why is one problem, HAMILTON CYCLE, seemingly impossible to solve efficiently and the other, EULER CYCLE, "fairly quick" to solve. After all, all we have done is replaced "all vertices" by "all edges". In the case of HAMILTON CYCLE, can we show that we cannot do better than trying all possibilities? That would certainly show that the problem is infeasible, whatever we mean by infeasible.

Is it that we are not smart enough to find some efficient algorithm? What do we mean by "efficient" anyway? What is our definition?

It was in the 1960's that authors such as Hartmanis and Sterns [HS65] gave a formal framework for the investigation of the complexity of algorithms[12]. Their framework was based on putting time and space constraints on the model used for computability theory. They used their model to calibrate complexity into time and space bounded classes. At more or less the same time, Edmonds [Edm65] also suggested asymptotic analysis of the behaviour of algorithms as a basis for classification of their difficulty, also suggesting counting computation steps and using polynomial time was a reasonable approximation for *feasibility*.

The major advance was the identification of the class NP. This is the class of problem which can be solved in *nondeterministic* polynomial time. Nondeterminism was discovered decades earlier in an area called automata theory. However, the introduction of nondeterminism into complexity analyses was found to give great insight into the understanding the intrinsic difficulty of

---

[11] Actually, in 1736, Euler only proved the necessary condition, and it was only later, in 1871, that Hierholzer [Hie73] established that the conditions were sufficient, although this was stated by Euler.

[12] There are many other contributors to the birth of computational complexity such as Pocklington (namely [Poc12]), Cobham, Rabin, and others. There is also a famous letter by Gödel where he describes the class NP well before it was defined. The reader should read the preface of Demaine, Gasarch and Hajiaghayi [DGH24].

problems. As we will see in Chapter 2.2, nondeterminism had been around since the 1950's, but it was Stephen Cook [Coo71], and Leonid Levin [Lev73] who showed that there were problems complete for a computational complexity class called NP. In a way that we will make precise, being complete means that the problem lies in this class, and are as *computationally difficult as any problem* in the class, at least as measured by worst case asymptotic behaviour. The Cook-Levin Theorem was proven by miniaturizing ideas from computability and the proof of the undecidability of the halting problem. Soon after, Richard Karp showed that there were *many* natural problems which are NP-complete. For example, our friend HAMILTON CYCLE is NP-complete. The point of all of this, is that showing that *any* of the problems had feasible solutions would prove that *all* of them do. There is something intrinsic happening. It is fair to say that the NP-completeness phenomenon remains something we still do not completely understand. Garey and Johnson's classic text [GJ79] gave a beautiful and accessible introduction to the widespread nature of apparent intractability. There are several other more recent books; and one successor to Garey and Johnson is [DGH24], which should appear in 2024.

In Chapter 7, we will look at this material: We classify tasks which *can* be computed, and ask questions about their intrinsic difficulty. That is, we establish the basic ideas of complexity theory. By analogy with the material on computability, we will introduce the notions of complexity classes, and examine some of the main characters in the story. They include P, NP, PSPACE and others. These concepts will all be explained in detail in that chapter.

The reader should note that proving a problem undecidable means that for any proposed algorithm A claiming to solve the problem, there are *instances* where the problem is *cannot be solved by A*. Showing that a problem is NP-complete means that, assuming that P $\neq$ NP, given any algorithm A running in polynomial time, there are *instances* where we think the problem cannot be solved by A; that is, the problem is *computationally infeasible*.

Sometimes, NP-complete and even harder problems can be efficiently solved for instances where the combinatorial explosion is caused by an identifiable parameter which can be restricted using *parameterized complexity*. With this parameter fixed, or restricted to a small range, the problem can well becomes feasible. There is a growing literature around this area, which allows for an involved dialog with the problems to discover how to make it feasible, and methods to show it likely infeasible. There is no undergraduate level text treating this area, in spite of the fact that there are many long graduate texts devoted to the area such as Downey and Fellows [DF13], Flum and Grohe [FG06], Cygan et. al. [CFK+16], Fomin et. al. [FLSZ19] and Niedermeier [Nie06]. In Chapter 9, we give a basic treatment of the positive toolkit and completeness theories for this area, an area which has seen significant development in the last 30 years[13].

---

[13] As one of the founders of this area, I also feel "honour bound" to give a treatment of the area!

In the last chapter, Chapter 10, we will rather briefly look at other methods of dealing with the murky universe of intractability. This will include approximation algorithms (where we give up on an exact solution and seek once which is "close" by some approximation measure, e.g. Ausiello et. al. [ACG⁺99]); and average case complexity (where we seek to understand how algorithms behave "on a typical input", Levin [Lev86]). We will also give, for the first time in a text, a treatment of the ideas of *generic case complexity* (Kapovich et. al. [KMSS03]) which is a variation of average case complexity, within which sometimes even undecidable problems become feasible. In the last section of the chapter, we will look at the idea behind Spielman and Teng's *smoothed analysis* ([ST01]), which is a relatively recent approach to understanding why well-known algorithms such as the SIMPLEX METHOD for linear programming actually work much more quickly in practice than we would expect; given the worst case performance is known to be exponential time. In a basic textbook such as this one, it is impossible to give a complete treatment of these approaches to dealing with intractability, as each of these topics would deserve a book to itself (and has already in most cases!). But I believe that it is important for the reader to see some of the horizons and hopefully be inspired to pursue further. This is a starred section which is not essential for the remainder of the book, and more of a *tour de horizon*.

The only chapters we have not yet discussed are the first two. The first two chapters are concerned with, respectively, naive set theory and automata theory.

Naive Set Theory gives a platform for the material to follow, and introduces a number of key ideas, such as coding and diagonalization, more easily understood in a situation where there is no computation happening.

The second chapter gives a general introduction to computation via a primitive computation device called an *automaton*. It also introduces the notion of *nondeterminism*. Aside from the use for the later sections of the book, regular languages and automata theory are a thriving and important area of computer science in their own right.

Starred Material. Some material is *starred*. This means that it can easily be skipped, especially on a first reading. Starred material is not necessary for the rest of the book.

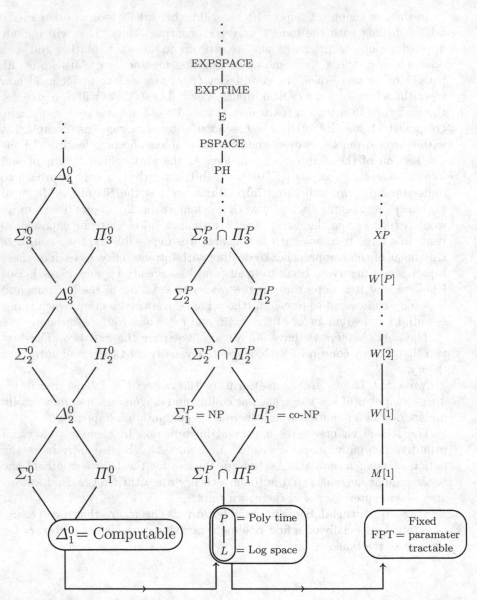

Some computational classes considered in this book

Contents

Part I
Background

Chapter 1
Some Naive Set Theory

Abstract This chapter gives meaning to the notion of size (cardinality) for infinite sets. We define countable and uncountable sets, and introduce Gödel numbering, coding, and diagonalization arguments. These ideas will be recycled throughout the book.

Key words: Coding, cardinality, Gödel numbering, countable, uncountable, diagonalization

1.1 Introduction

It may seem strange that we begin our journey towards understanding the basics of the theory of computation with something much older: naive set theory. This goes back to famous work of Georg Cantor [Can74, Can78, Can79] in the late 19th century. This work attempts to give meaning to $|A|$, the "size" of a set A for A *infinite*. The reason we begin with set theory is that the ideas and techniques such as *diagonalization* and *coding* will have reflections in much of what follows, so it is useful to see them in the pure situation of not having the overlay of computation.

We will be looking at "naive" set theory, which takes as an undefined term "set", and allows definitions of sets as $S = \{x : R(x) \text{ holds}\}$, without caring about what properties R we allow. Russell's Paradox from 1901 points out that this naive approach leads to problems when, for example, we consider $S = \{x : x \notin x\}$, the collection of sets which are not members of themselves. If we then ask "Is $S \in S$?" either answer leads to a contradiction. There are methods to circumvent such paradoxes, and these belong to an area of mathematical logic called *axiomatic* set theory, and the reader is urged to pursue these issues (see, for example Kunen [Kun11]). For us, we'll be concerned with sets commonly found in mathematics such as $\mathbb{N}$, $\mathbb{Z}$, $\mathbb{R}$ and the like, and such issues don't generally arise; at least for the issues we will be concerned with.

1.1.1 Basic definitions

How to give meaning to the size of sets? If we have two bags of candy and ask children who can't count who has the most, they would likely pair the candies off one at a time, and the one who had some left would know they had more, or if they finished together, they would have the same size. This is the basic idea behind relative cardinality.

Definition 1.1.1. Let A and B be sets. We say that the *cardinality* of A is *less than or equal to that of* B, written $|A| \leqslant |B|$ if $A = \emptyset$ or there is a 1-1 function $f : A \to B$.

The following result actually needs an axiom called the *axiom of choice*, but we will take it as a theorem for our purposes.

Theorem 1.1.1. $|A| \leqslant |B|$ *iff* $A = \emptyset$ *or there is a function* g *mapping* B *onto* A. *(Recall* g *being onto means that for all* $a \in A$ *there is a b in B with* $g(a) =.$)

Remark 1.1.1 (No Special Cases for Empty sets). Unless otherwise specified, henceforth we will assume sets are nonempty, as things which are obviously true for empty sets will be obviously true for empty sets, without treating them as a special case.

Proof. Suppose that $|A| \leqslant |B|$. Let $f : A \to B$ be 1-1. The obvious thing to do is to invert f, and then map any of B left over to a fixed member $a \in A$. Thus g is defined as $g(b) = f^{-1}(b)$ if $b \in \mathrm{ra}f$, and $g(b) = a$ if $b \notin \mathrm{ra}f$.

Conversely[1], suppose that $g : B \to A$ is onto. We'd like to invert this map, but it could be that many elements of B go to the same element of A and we need to choose one to make the inversion 1-1. We say that $b_1 \equiv b_2$ if $g(b_1) = g(b_2)$. For each equivalence class $[b] = \{\widehat{b} \in B : \widehat{b} \equiv b\}$ choose one representative. The to define $f : A \to B$, for each $a \in A$, define $f(a) = b'$, where b' is the representative of $[\{b : g(b) = a\}]$. The f is well-defined since for all $a \in A$ there is a $b \in B$ with $g(b) = a$. And it is 1-1 since we are choosing one b' for each a. $\square$

---

[1] It is here we are using the *Axiom of Choice*. This axiom says that if $\{H_i : i \in D\}$ is a family of subsets of a set A, there there is a "choice function" f, such that, for all $i \in D$, $f(i) \in H_i$. In the present proof the subsets are the pre-images and the choices are the representatives of the equivalence classes.

The following is easy by composition of functions.

Proposition 1.1.1. $|A| \leqslant |B|$ *and* $|B| \leqslant |C|$ *implies* $|A| \leqslant |C|$.

Proof. If g maps C onto B and f maps B onto A, then $f \circ g$ maps C onto A. □

We can define $|A| = |B|$ iff $|A| \leqslant |B|$ and $|B| \leqslant |A|$. But the candy matching motivation would lead us to expect that $|A| \leqslant |B|$ and $|B| \leqslant |A|$ would mean the there is a bijection, that is a 1-1 correspondence, between A and B; not simply that each could be injected into the other. This intuition is indeed correct, there is such a bijection, but proving this based on the assumption that $|A| \leqslant |B|$ and $|B| \leqslant |A|$ is surprisingly difficult to prove. Do not be concerned if you find this proof tricky; *it is*. Also we really only will be using the *fact* that the result is true after this, so could well be skipped. I am including this as it is good for the soul to see such a pretty proof.

Theorem 1.1.2 (Cantor-Schröder-Bernstein). [2]
$|A| = |B|$ *iff there is a bijection* $h : A \to B$.

Proof. * If there is a bijection h then obviously $|A| \leqslant |B|$ (by h) and $|B| \leqslant |A|$ (by h^{-1}).

Conversely, suppose that $|A| \leqslant |B|$ and $|B| \leqslant |A|$, via injective functions $f : A \to B$ and $g : B \to A$. Let $P(A)$ denote the power set of A, the set of all subsets of A. Let $\varphi : P(A) \to P(A)$ be defined as follows. For $E \in P(A)$, that is $E \subseteq A$, let

$$\varphi(E) = (A \setminus (g(B \setminus f(E)))).$$

That is, $\varphi(E) = \overline{g(\overline{f(E)})}$.

Claim. If $E \subset F$ then $\varphi(E) \subseteq \varphi(F)$.

Proof. Since $E \subseteq F$, and f is 1-1, $F(E) \subseteq f(F)$. Therefore, $(B \setminus f(E)) \supseteq (B \setminus f(F))$. But g is 1-1 also. Thus, $g(B \setminus f(E)) \supseteq g(B \setminus f(F))$. But then $A \setminus g(B \setminus f(E)) \subseteq A \setminus g(B \setminus f(F))$. That is, $\varphi(E) \subseteq \varphi(F)$, as required. □

Let $R = \{E \in P(A) : E \subseteq \varphi(E)\}$, and let

$$D = \cup_{E \in R} E = \cup R.$$

Notice $E \subseteq D$, for all $E \in R$. Now that $E \subseteq D$, and $E \in R$. Then $E \subseteq \varphi(E)$ by definition of R, and $\varphi(E) \subseteq \varphi(D)$, by Claim 1.1.1. Also $D \subseteq \varphi(D)$, since $a \in D$ implies $a \in E$ for some $E \in R$, and hence $a \in \varphi(E)$. By Claim 1.1.1,

---

[2] This theorem has a curious history, first stated by Cantor as a consequence of the axiom of choice, but later proven by, for example, Bernstein who was a graduate student and gave a proof in a seminar at age 19. Schröder presented a proof also, but that contained a flaw. The proof here, or some variation of it, is apparently due to König.

again, $\varphi(D) \subseteq \varphi(\varphi(D))$. Therefore $\varphi(D) \in R$, and hence $\varphi(D) \subseteq D$. We showed that $D \subseteq \varphi(D)$. Thus

$$\varphi(D) = D.$$

That is

$$\varphi(D) = A \setminus g(B \setminus f(D)) = D$$

Hence, $g(B \setminus f(D)) = A \setminus D$. Therefore $B \setminus f(D) = g^{-1}(B \setminus D)$.

But now we are finished: f maps D 1-1 to $f(D)$, and g^{-1} maps $A \setminus D$ 1-1 to $B \setminus f(D)$.. There fore we can define

$$h(a) = \begin{cases} f(a) & \text{if } a \in D \\ g^{-1}(a) & \text{if } a \in A \setminus D \end{cases}$$

Then as f and g^{-1} are 1-1 and onto for these restricted domains, h is a bijection. □

1.1.2 Countable sets

As we see later in Section 1.3, Cantor's insight is that using this definition of relative size some infinite sets cannot be put into 1-1 correspondence with each other. We will start this analysis with the simplest infinite sets.

Definition 1.1.2. We say that A is *countable* if $|A| \leqslant |\mathbb{N}|$.

The name comes from the intuition that if we had $g : \mathbb{N} \to A$, being onto, as we "counted" $0, 1, 2, \ldots$ we are implicitly counting $A = \{g(0), g(1), g(2), \ldots\}$. Clearly every $B \subseteq \mathbb{N}$ is countable, as we can map B into $\mathbb{N}$ in a 1-1 way by mapping $b \mapsto b$ for $b \in b$. Are there any other interesting examples of countable sets? Definitely yes. We will look at several examples, and then introduce some generic *coding* techniques for establishing countability. These techniques will be recycled when we look at computable functions in Chapter 5, and the undecidability of the Halting Problem in Chapter 3.

Example 1.1.1. $\mathbb{Z}$ is countable. There is a simple bijection from $\mathbb{Z}$ to $\mathbb{N}$ defined via $f(0) = 0$ and, for $n \in \mathbb{N}$, $f(-n) = 2n$ and $f(n) = 2n + 1$. The negative integers are mapped to the even numbers and the positive ones to the odd numbers.

Example 1.1.2 (Cantor [Can74, Can78]). $\mathbb{Q}$ is countable. We will look at Cantor's original technique which uses a "pairing" function (See Exercise 1.2.5). Using the trick for $\mathbb{Z}$ is surely enough to show that $\mathbb{Q}^+$, the positive integers, is a countable set. This technique uses an ingenious method of visualizing the rationals as a 2×2 array and counting them by "zig-zagging".

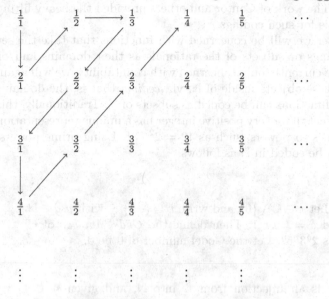

The method of Example 1.1.2 can be used to other situations where such a 2×2 array could be visualized.

Theorem 1.1.3 (Cantor [Can74, Can78]).

1. *If A and B are countable, so is $A \times B$.*
2. *The countable union of countable sets if countable[3]. That is, if $h(\mathbb{N}) \to P(\mathbb{N})$ is a 1-1 function, such that $h(i)$ is a countable set A_i for each i, then $\cup_i h(i) = \{x : x \in A_i \text{ for some } i\}$, is also countable.*

Proof. 1. Suppose that $f(\mathbb{N}) \to A$ and $g(\mathbb{N}) \to B$ are surjective functions. Then we can replace the array we used for $\mathbb{Q}^+$ by the array whose i-th line is $\langle f(i), g(0) \rangle, \langle f(i), g(1) \rangle, \langle f(i), g(2) \rangle, \ldots$. The array would include all the members of $A \times B$, and the same counting method would work.

2. This is more or less the same, and is left to the reader. □

1.2 Gödel numberings and other coding techniques

The methods of the previous section are valuable, and should be internalized. However, today we think of data as being digitized, and this involves coding the data by some representation. Typically this would be base 2. Implicitly,

---

[3] Again choice is being used here, as we did with Theorem 1.1.1.

we are providing a counting of such objects in a certain "algorithmic" way, which will be essential in later sections. In fact historically things worked the other way around. The work of Cantor and others provided the heavy lifting of a conceptual basis for such codings.

For example, later we will be concerned with functions that take the set of finite binary strings or subsets of the rationals as their domains and/or ranges. Since we are currently only concerned with countability, we will again seek to represent these objects inside of $\mathbb{N}$, via *coding*; that is, the domains and ranges of such functions can be coded as subsets of $\mathbb{N}$. Traditionally, this was done using the fact that every positive integer has a unique representation as products of primes to powers, such as $12 = 2^2 \cdot 3^1$. Using prime powers, the rationals $\mathbb{Q}$ can be coded in $\mathbb{N}$ as follows.

Definition 1.2.1. Let $r \in \mathbb{Q} \setminus \{0\}$ and write $r = (-1)^\delta \frac{p}{q}$ with $p, q \in \mathbb{N}$ in lowest terms and $\delta = 0$ or 1. Then define the *Gödel number* of r, denoted by $\#(r)$, as $2^\delta 3^p 5^q$. Let the Gödel number of 0 be 0.

The function $\#$ is an injection from $\mathbb{Q}$ into $\mathbb{N}$, and given $n \in \mathbb{N}$ we can decide exactly which $r \in \mathbb{Q}$, if any, has $\#(r) = n$. Similarly, if σ is a finite binary string, say $\sigma = a_1 a_2 \ldots a_n$, then we can define $\#(\sigma) = 2^{a_1+1} 3^{a_2+1} \ldots (p_n)^{a_n+1}$, where p_n denotes the nth prime.

There are myriad other codings possible, of course. For instance, one could code the string σ as the binary number 1σ, so that, for example, the string 01001 would correspond to 101001 in binary and this identifies a unique member of $\mathbb{N}$. Coding methods such as these are called "effective codings", since they include algorithms for deciding the resulting injections, in the sense discussed above for the Gödel numbering of the rationals.

One classic result here is:

Example 1.2.1. The collection of all finite subsets of $\mathbb{N}$ is a countable set. To see this, if F is a finite set of nonnegative integers, then arrange $F = \{n_1 < \cdots < n_k\}$, and define $\#(F) = 2^{n_1+1} \cdot \cdots \cdot (p_k)^{n_k+1}$. Then each finite set is sent to a unique $m \in \mathbb{N}$. Again, there are lots of other codings. For example we could represent F by $n_1 + 1$ ones followed by 0 followed by $n_2 + 1$ ones, etc.

We know that there is a bijection from $\mathbb{N} \times \mathbb{N} \to \mathbb{N}$, given by any of the techniques above, or by Exercise 1.2.5 below, which we will refer to as Gödel numbering. We will adopt the following useful convention:

Convention 1.2.1 Henceforth we will identify the ordered pair (x, y) with its Gödel number so that when we write $\langle x, y \rangle$, we will simultaneously consider it to be some $n \in \mathbb{N}$ with $\#((x, y)) = n$, and the pair itself. It will be easy to figure out which we mean by context.

1.2.1 *Exercises*

Exercise 1.2.2 A *string on an alphabet* $\Sigma = \{a_1, \ldots, a_n\}$ is a finite concatenation of symbols from Σ. For example, if $\Sigma = \{1, 2, 3\}$ then 221331133 is a string in this alphabet. We let Σ^* denote the collection of all strings obtainable from Σ.

1. Show that if Σ is a finite alphabet, then Σ^* is countable.
2. We can also extend the definition of strings to the case where Σ is infinite. Show that if Σ is a countable set, then Σ^* is countable.
3. An "infinite string" is usually referred to as a *sequence*. A binary sequence $\alpha = a_0 a_1 \ldots$ is called *periodic* if there exists a k such that $\alpha = a_0 \ldots a_k a_0 \ldots a_k a_0 \ldots a_k \ldots$. Show that the collection of periodic binary sequences is countable.
4. A sequence $\beta = b_0 b_1 \ldots$ is called *eventually* periodic, if there is a finite string $b_0 \ldots b_j$ such that $\beta = b_0 \ldots b_j \alpha$ where α is periodic. Show that the collection of eventually periodic binary sequences is a countable.

Exercise 1.2.3 A set $A \subseteq \mathbb{N}$ is called *cofinite* if $\mathbb{N} \setminus A$ is finite. Show that the collection of cofinite sets is countable.

Exercise 1.2.4 A polynomial is an expression of the form $a_0 + a_1 x + \cdots + a_k x^k$, and this would be called a *rational* polynomial if $a_i \in \mathbb{Q}$ for $i \in \{0, \ldots, k\}$. Here k is arbitrary.

1. Let C be the collection of rational polynomials. Show that C is a countable set.
2. A complex number z is called *algebraic* if it is a solution (root) to an equation of the form $P(x) = 0$, where $P(x)$ is a rational polynomial. Show that the collection of algebraic numbers is countable. You may assume that a polynomial of degree k has at most k roots.

Exercise 1.2.5 (Cantor) Show that the *pairing* function

$$p(x, y) = \frac{(x+y)(x+y+1)}{2} + y$$

is a bijection from $\mathbb{N} \times \mathbb{N} \to \mathbb{N}$.
This is not at all easy. A hint is that from basic arithmetic:

$$\sum_{k=0}^{x+y} k = \frac{(x+y)(x+y+1)}{2}.$$

1.3 Diagonalization and Uncountable Sets

So are all sets countable? Set theory would certainly be simpler if that was true, but Cantor had the intuition that $\mathbb{N}$ *cannot* be put into 1-1 correspondence with $\mathbb{R}$, or even $\mathbb{R} \cap (0,1)$. Thus there are different cardinalities of infinite sets. Cantor's method was an ingenious new method called *Diagonalization* which has under-pinned a lot of modern mathematics and computer science. In the proof below we will add a diagram which makes it clear why the technique is called diagonalization.

Theorem 1.3.1 (Cantor [Can79]). $|\mathbb{N}| < |\mathbb{R} \cap (0,1)|$.

Proof. Certainly $|\mathbb{N}| \leqslant |\mathbb{R} \cap (0,1)|$ via a map such as $n \mapsto \cdot 0n$ where we write n in unary (base 1). (Thus, for instance $2 \mapsto \cdot 011$.) Thus we need to show that $\mathbb{R} \cap (0,1)$ is not countable. Suppose not. Then, by the Cantor-Schröder-Bernstein Theorem, there is a bijection $f : \mathbb{N} \to \mathbb{R} \cap (0,1)$. That is $\{f(0), f(1), \dots\}$ lists all of the reals between 0 and 1.

Now we think of a real between 0 and 1 via its decimal expansion. Thus $r = .a_1 a_2 \dots$ with $a_i \in \mathbb{N}$. This is not the greatest representation in the world, since we can't really decide which to chose of, say, $\cdot 010000\dots$ and $\cdot 009999999\dots$. But the method below will dispose of both such representations simultaneously.

We can think of $f(i) = \cdot a_{i,1} a_{i,2} a_{i,3} \dots$.

We define a real r which

1. Should be on the list as it is in $(0,1)$, and
2. *Can't* be on the list, because of the way we construct it, *to diagonalize* against the list.

We define $r = \cdot r_1 r_2 \dots$ where we specify $r_i = a_{i,i} + 5 \mod 10$. So if $a_{i,i} = 6$ then $r_i = 1$, for example. Now we can imagine this using the diagram below:

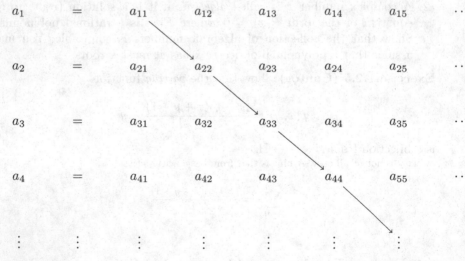

You can see we are working down the diagonal, and making r_i, the i-th decimal place of r significantly different from $f(i)$ its its i-th place $a_{i,i}$. Thus the distance between r and $f(i)$ is *at least* 4×10^{-i}. Thus, $r \neq f(i)$ for all i, each $f(i)$ being "diagonalized" at decimal place i.

Now if $\mathbb{R} \cap (0,1)$ is countable and is counted by f, then as $r \in (0,1)$, there must be some n with $r = f(n)$ as $f(\mathbb{N})$ lists *all* of $\mathbb{R} \cap (0,1)$. But this is a contradiction, $r_n = a_{n,n} + 5 \mod 10$. Hence $r \neq a_n = f(n)$. The contradiction is obtained from the assumption that f exists, and hence there is no such f. That is $\mathbb{R} \cap (0,1)$ is uncountable. $\square$

Essentially the same proof will show the following.

Corollary 1.3.1. *The collection of infinite binary sequences is uncountable.*

Note that we can think of a subset of $\mathbb{N}$ as an infinite binary sequence, via what is called the *characteristic function*:

$$\chi_A(x) = \begin{cases} 1 \text{ if } x \in A, \text{ and} \\ 0 \text{ if } x \notin A \end{cases}$$

Corollary 1.3.2. $P(\mathbb{N})$ *is uncountable.*

The reader should note that we did not state this as $|\mathbb{N}| < |P(\mathbb{N})|$ although this is easily seen to be true since $a \mapsto \{a\}$ is a suitable injection from $\mathbb{N}$ into $P(\mathbb{N})$. But for any infinite set X, if it is not countable then $|\mathbb{N}| < |X|$. This fact is somewhat a consequence of the following "Trichotomy" result, which is also equivalent to what is called the Axiom of Choice. We will take the result as a *Black Box*, since delving into the niceties of axioms of set theory will distract from our main mission.

Theorem 1.3.2 (Trichotomy). *For any sets A, B exactly one of the following holds: $|A| < |B|, |A| = |B|,$ or $|B| < |A|$.*

We remark that $|\mathbb{N}| < |P(\mathbb{N})|$ is a special case of a more general result which uses an abstract form of diagonalization, saying that for any set A, $|A| < |P(A)|$ as we see below. We will recycle the diagonalization method when we prove that the halting problem for Turing Machines is algorithmically undecidable in Theorem 4.1.4, and later the Hierarchy Theorem (Theorem 6.1.5) in §6, two of the core theorems of this book.

Theorem 1.3.3 (Cantor [Can79]). *For any set A, $|A| < |P(A)|$.*

Proof. Even without Trichotomy, it is clear that $|A| \leqslant |P(A)|$, because we can map $x \mapsto \{x\}$. Now, following the template, suppose that $|A| \geqslant |P(A)|$. Let $h : A \to P(A)$ be a surjection. Let $B = \{x : x \notin h(x)\}$. Then $B \subseteq A$ and hence $B \in P(A)$. Since h is onto, there is some $a \in A$ with $h(a) = B$. But if $a \in B$ then by definition of B, $a \notin h(a) = B$, but if $a \notin B$, then $a \notin h(a)$, and hence $a \in B$. Either case is a contradiction, and hence no such h can exist. Thus $|P(A)| \nleqslant |A|$. $\square$

1.3.1 Exercises

Exercise 1.3.4 A *power series* is an "infinite polynomial" of the form $\sum_{n=0}^{\infty} a_n X^n$. It is called *binary* if $a_n \in \{0, 1\}$ for all n. Two binary power series $a(X) = \sum_{n=0}^{\infty} a_n X^n$ and $b(X) = \sum_{n=0}^{\infty} b_n X^n$ are equal iff $a_n = b_n$ for all n. Show that the collection of binary power series forms an uncountable set.

Exercise 1.3.5 Let $X = \{x_1, x_2, \dots\}$ be a countable set. Let $\{Y_n : n \in \mathbb{N}\}$ be a countable listing of all finite subsets of X. Show that the set

$$S = \{n : x_n \notin Y_n\}$$

is not finite.

Exercise 1.3.6 Suppose that A is countable and B is uncountable. Show that $|A| \times |B| = |B|$.

Exercise 1.3.7 Show that $|\mathbb{R} \times \mathbb{R}| = |\mathbb{R}|$.

1.4 Set Theory in Mathematics

It is hard for modern mathematicians to really appreciate the impact of the work of Cantor and his contemporaries on mathematics. In fact, at the time Cantor introduced set theory, it was very controversial. One widely reproduced comment, attributed to Kronecker, a leading mathematician of the time was

> "I don't know what predominates in Cantor's theory philosophy or theology, but I am sure that there is no mathematics there."

It is hard to find a source of this comment and it might be apocryphal, but certainly at the time many mathematicians did not accept the idea of a completed infinity.

But the impact on mathematics of Cantor's work is immense. For example, today we are taught the point-set definition of a function as a collection of ordered pairs $\{\langle x, f(x)\rangle : x \in \operatorname{dom} f\}$, whereas historically a function was essentially some kind of *rule* or *formula*. When we defined the counting of $\mathbb{N} \times \mathbb{N}$ we used a "zig-zag" counting rather than the formula used in Exercise 1.2.5. The original definition of a function in terms of ordered pairs is usually attributed to Lobachevsky [Lob51] and Dirichlet [Dir89] in the early 19th Century but was not widely used until the work of Cantor. Classically, when mathematicians gave examples of functions which were, for example, nowhere differentiable and yet continuous, they had to dream up some formula which did this. One notable example of this phenomenon was the 19th century belief that if a function satisfied the intermediate value theorem (namely for all $x < y$ and $z \in [f(x), f(y)]$, there is a $q \in [x, y]$ such that $f(q) = z$), then it must be continuous. This is easily seen to be false by using the point-set definition of function, and in fact it is possible to have a function which assumed all real values in every interval, and was everywhere discontinuous! Cantor [Can74] showed the insight gained by the point to point definition of an injective function.

One of the first applications of Cantor's work concerned transcendental numbers. Transcendental numbers are reals which are not algebraic. We met algebraic numbers in Exercise 1.2.4; they being defined as those real numbers which are roots of polynomials with rational coefficients. Heroic efforts in the 19th century with Louiville showing in 1844 that transcendental numbers do exist. In 1851, Louiville showed that the "explicit number"

$$\sum_{n=1}^{\infty} 10^{-n!}$$

is transcendental. In the mid-19th century the familiar numbers e and π were also shown to be transcendental. Since Cantor proved that the collection of reals is uncountable, this means that "most" reals are transcendental, as there are only countably many algebraic numbers. Of course, Cantor's method is pure existence: it gives us no inkling as to how to show a given number such as π is transcendental.

Cantor's and Gödel's ideas of coding and diagonalization have had repercussions throughout the 20th century to this day. We will see how coding formed an essential part of the ideas behind the theory of computation, and, by logical extension, the digitization of the modern world. Diagonalization is essentially the only technique for separating computational classes we meet later. It is also used in analysis with the work of Lebesgue and others. For example, if we begin with the continuous functions (from $\mathbb{R}$ to $\mathbb{R}$) as a certain basic class B_0, and then make the next class pointwise limits of a sequence of continuous functions (that is $f \in B_1$ iff there exist $\{f_i : i \in \mathbb{N}\}$ with $f_i \in B_0$ (i.e. f_i continuous) and for each x, $f(x) = \lim_i f_i(x)$.), and construct the the next class B_2 the pointwise limit of elements of elements of B_1, etc, we get to what is called the Baire hierarchy of functions. To prove that this is a proper hierarchy we show use diagonalization. Cantor's ideas underpin the association and use of various invariants (based on Cantor's ordinals and cardinals) to algebraic objects such as ranking functions such as the Ulm invariants for Abelian groups. Finally Cantor's ideas the deep work in logic called *forcing*, a mainstay of proof theory.

1.4.1 The Cardinality of a Set and the Continuum Hypothesis*

The reader might note that we have skirted the issue of what the actual size of a set is. In view of Cantor-Schröder-Bernstein, we can think of relative sizes as be subdivided into equivalence classes where $A \equiv B$ means $|A| = |B|$. These equivalence classes correspond to the sizes of certain well-orderings called *ordinals*, a discussion of which would take us on a beautiful trip, too expensive for us to travel in this (relatively) short book. Now, for the finite A's we know the names of $[A]$'s, these ordinals are simply the finite numbers. For $|\mathbb{N}|$ the ordinal is ω which is $\mathbb{N}$ considered as an ordering $\omega = 0 < 1 < 2 \dots$. This sequence is extended by adding a new element in the ordering beyond ω so bigger than all the finite ordinals and call this $\omega + 1$, thence $\omega + 2$, and so on. Ordinals have many other uses in mathematics beyond set theory

The equivalence class of sets which have the same cardinality as $\mathbb{N}$, or more precisely the order-type of this ordering ω, is called $\aleph_0$. This is referred to as a *cardinal*. The next cardinal is called $\aleph_1$, and it is again associated with a particular ordinal, a well ordering, akin to ω and is called ω_1. It is the first uncountable ordinal. Thus, the order type of ω_1 is the first uncountable cardinal. The cardinality of $\mathbb{R}$ is the cardinality of the set of functions from $\mathbb{N}$ to $\{0, 1\}$, that is, the number of subsets of $\mathbb{N}$. In the arithmetic of cardinals, this is $2^{\aleph_0}$.

The *continuum hypothesis,* says that $\aleph_1 = 2^{\aleph_0}$. That is, *there is no set A with*

$$|\mathbb{N}| < |A| < |\mathbb{R}|.$$

This remarkable hypothesis was at the heart of set theory for over 60 years; and is really part of *axiomatic* set theory. It turns out that relative to the accepted axioms of set theory, it is both consistent that the continuum hypothesis is true, and that the continuum hypothesis is false. The consistency of it being true was proven by Gödel [Goe40]. Cohen [Coh63, Coh64] proved the consistency of it being false. Both proofs involved truly significant new ideas (constructibility in Gödel's case, and forcing in Cohen's case) the ramifications of which are still being felt today. We refer the reader to Kunen [Kun11] if they want to pursue this fascinating topic.

Part II
Computability Theory

Chapter 2
Regular Languages and Finite Automata

Abstract We introduce the notion of a regular language, and show that regular languages are precisely those that are accepted by deterministic finite automata. We introduce nondeterminism, and prove that for automata, nondeterministic and deterministic machines have the same power, the trade-off being an exponential increase in the number of states. We finish with the Myhill-Nerode Theorem which shows how finite state is that same as having finite index for a certain canonical equivalence relation.

Key words: Finite state, automata, regular, Myhill-Nerode Theorem, nondeterminism, formal language

2.1 Introduction

In this chapter we will begin by look at yet another apparently unrelated area: *formal language theory*. But we will find that formal languages go hand-in-hand with a primitive notion of computation called an *automaton*. Whilst automata post-date the Turing machines we will soon meet in Chapter 3, studying them first makes the material from Chapter 3 more easily digested. To do this, we will first introduce a syntactic notion[1] called regularity, a very important notion in an area called formal language theory. As we see, regularity coincides with being accepted by an automaton. This fact shows that things which seem far from being concerned with computation can be intertwined with computation theory.

Even now, the area of formal language theory, and its widespread applications in computing, remain significant areas of research interest. In particular, many automata-based algorithms are used in compilers, operating systems (e.g. avoiding deadlock), and security (they are used to model security). More

---

[1] That is one based on symbols and rules for manipulating them.

widely, even algorithm design for some aspects in graph theory (such as for parse-based graph families like those of bounded path and treewidth) are reliant on automata-theoretical methods. Many basic algorithms in computer science, such as in software engineering, rely heavily on such methods. Thus, beyond motivation for later material, this material is something "very good to know". The principal proof technique for this section is mathematical induction, applied to both the number or length of an object, or its structure.

2.2 Regular Languages

2.2.1 Formal languages

In this section, we will be considering subsets of the set of all strings obtained from an alphabet $\Sigma = \{a, \ldots, a_n\}$, which we will always be finite in this Chapter. Recall from Exercise 1.2.2 that a *string* or *word* is a finite concatenation of symbols from Σ. For example, if $\Sigma = \{a, b, c\}$ then *bbaccaacc* is a string in this alphabet Σ. We sometimes write w^j for w concatenated with itself j times. Thus, this previous expression could be written as $b^2ac^2a^2c^2$. The reader should note that if $\Sigma = \{00, 11\}$, then 1100 would be a string, but 01 would not be. We let Σ^* denote the collection of all strings obtainable from Σ. Notice that this set could be obtained using an inductive definition: λ, the empty string, is a string. If $\sigma \in \Sigma^*$, then $\sigma a \in \Sigma^*$ for each $a \in \Sigma$. Σ^* is called the *Kleene Star* of Σ, named in honour of the American mathematician Stephen Cole Kleene [Kle56], the first to realize the significance of the class of languages we will study, the regular languages.

Definition 2.2.1. We will refer to subsets $L \subseteq \Sigma^*$ as *languages*. Languages are also important when we later look at complexity theory in Chapter 7, where we are concerned with, for example, running times, and how we represent data becomes important.

For example, if Σ denotes the letters of the English alphabet, then strings would be potential English words. Only some of these strings are English words, and we could form

$$L = \{\sigma : \sigma \in \Sigma^* \text{ and } \sigma \text{ is an English word}\}.$$

In this section, we'll be looking at mathematical properties P, and studying certain kinds of languages

$$L = \{\sigma : \sigma \in \Sigma^* \text{ and } \sigma \text{ has property } P\}.$$

It is easy to motivate such languages. For example, Σ might be an alphabet containing $\{(,)\}$, and P might be the property that every "(" is balanced by a ")". This property would be something checked in some initial algorithm compiling some program. Now we develop some notations and terminology for dealing with strings. We will call σ a *substring* of a string τ, if there exist strings u, v (either of which might be λ) such that $\tau = u\sigma v$. We say that σ is an *initial segment* of ρ, written $\sigma \preccurlyeq \rho$ if there exists a string v (which might be λ), with $\rho = \sigma v$. If $v \neq \lambda$ here, then σ is a *proper* initial segment of ρ and we would written $\sigma \prec \rho$.

Definition 2.2.2 (Length). For a fixed alphabet Σ, we define the length of a string $\sigma \in \Sigma^*$, $|\sigma|$, as follows.

1. λ, the empty string has length 0. That is $|\lambda| = 0$.
2. If $|\sigma| = n$ and $a \in \Sigma$, then $|\sigma a| = |\sigma| + 1$.

Thus, $|\sigma|$ is the number of symbols in σ, using Σ as a basis. Note that $|\sigma$ should not be confused with $|A|$, the cardinality of a set A from the last chapter. Sometimes $|\sigma|$ might be denoted by $\ell h(\sigma)$, but we would only do this when the meaning is not clear from context. In some proofs, we might have concatenations of symbols which would be considered strings for different alphabets, and hence would have two possible lengths. But again, when this happens we will be quite clear which alphabet we are using, at the time.

2.2.2 Regular expressions and languages

In this subsection, we look at a certain class of expressions used to generalize one of the basic kinds of languages. Thus we will be defining languages *syntactically*, meaning that we use a grammar with certain rules to define our objects.

Definition 2.2.3 (Regular Expression). A *regular expression* over Σ is an expression in the alphabet $\Sigma, \cup, (,)$, and λ defined by induction on logical complexity as follows:

1. λ and each member of Σ is a regular expression.
2. If α and β are regular expressions, so are $(\alpha\beta)$ and $(\alpha \cup \beta)$.
3. If α is a regular expression, so is $(\alpha)^*$.
4. Nothing else.

A formal proof that a string is a regular expression would be the following showing that, for $\Sigma = \{a, b, c, d\}$, $(((a \cup b)c) \cup c^*)$ is a regular expression. a, b, c, d are each regular expressions by (i), so $(a \cup b)$ is regular by (ii), and $((a \cup b)c)$ is regular, also by (ii); c^* is regular by (iii); and then $(((a \cup b)c) \cup c^*)$ is regular by (ii); again. However, we will eschew such formal proofs, since regularity will be clear for a given expression. For example, if $\Sigma = \{a, b, c\}$, then λ, $((a(ab))c)$, $(bc \cup a) \cup a^*$, and (a^*b^*) are all evidently regular expressions.

Naturally, we drop the brackets where the meaning is clear. So we'd write a^*b^* for (a^*b^*). Below we define a language $L(\alpha)$ associated with each regular expression α, and reflecting the intentional meaning.. When we define $L(\alpha)$ below for regular α, we'll see that $L((\gamma \cup \rho) \cup \delta) = L(\gamma \cup (\rho \cup \delta))$, so we will drop brackets for union and concatenation when the meaning is clear.

Definition 2.2.4 (Regular Language).
We associate a language $L(\alpha)$ with a regular expression α as follows.

(i) If $\alpha = \lambda$, then $L(\alpha) = \emptyset$.
(ii) If $\alpha = a$ with $a \in \Sigma$, then $L(\alpha) = \{a\}$.
(iii) If $\alpha = \gamma\beta$, then $L(\alpha) = L(\gamma)L(\beta)$ [$=_{\text{def}} \{xy : x \in L(\gamma)$ and $y \in L(\beta)\}$].
(iv) If $\alpha = \gamma \cup \beta$, then $L(\alpha) = L(\gamma) \cup L(\beta)$.
(v) If $\alpha = \beta^*$, then $L(\alpha) = (L(\beta))^*$ [$=_{\text{def}}$ the set of all strings obtainable from $L(\beta)$.]

We call a language L *regular* if $L = L(\alpha)$ for some regular expression α.

Example 2.2.1. Let $\Sigma = \{0, 1\}$.
$$L(0(0 \cup 1)^*) = L(0)L((0 \cup 1)^*) \text{ by (iii)}$$
$$= \{0\}(L(0 \cup 1))^* \text{ by (ii) and (v)}$$
$$= \{0\}(L(0) \cup L(1))^* \text{ by (iv)}$$
$$= \{0\}(\{0\} \cup \{1\})^* \text{ by (ii)}$$
$$= \{0\}\{0, 1\}^*$$
$$= \{x \in \{0, 1\}^* : x \text{ starts with a } 0\}.$$

Fix an alphabet Σ. Is every language regular? The answer is definitely no, since regular languages correspond to regular expressions. Therefore if $R = \{L : L \subseteq \Sigma^* \wedge L \text{ is regular}\}$, and E is the set of regular languages over Σ, then $|L| \leqslant |E|$. But a regular expression is a finite string, so E is countable. (See Exercise 1.2.2 or use Gödel numbering or coding.) But the number of subsets of Σ^* is uncountable, so "almost all" languages are not regular.

2.2.3 The Pumping Lemma

The fact that most languages are not regular does not give any method to
to decide if some simple-looking language which could be regular is regular.
The following result gives a *necessary condition* for regularity. Note that all
finite languages are regular, since we could use an expression which simply
lists the members. For example if $L = \{a, aa, b, cc\}$ over $\Sigma = \{a, b, c\}$ then
$L = L(\alpha)$ for $\alpha = a \cup aa \cup b \cup cc$. Thus the theorem below is concerned with
necessary conditions for *infinite* languages.

**Theorem 2.2.1 (Pumping Lemma[2], Bar-Hillel, Perles and Shamir
[BHPS61]).** *Suppose that L is regular and infinite. Then there exist strings
x, y, and z such that*

  1. *$y \neq \lambda$ (it could be that x or z are λ), and*
  2. *for all $n \geqslant 1$, $xy^n z \in L$.*

Proof. Let $L = L(\alpha)$ for α a regular expression over $\Sigma = \{a_1, \ldots, a_n\}$. We
prove the theorem by induction on $|\alpha|$.

The base case is that $|\alpha| = 2$. If $|\alpha| < 2$ then $|\alpha| = 1$, and hence $\alpha \in \Sigma \cup \{\emptyset\}$.
But then $L(\alpha)$ is finite. If $|\alpha| = 2$ and $L(\alpha)$ is infinite, then it must be that
$\alpha = a^*$, for some $a \in \Sigma$. Otherwise $\alpha = ab$ for some $a, b \in \Sigma$. Then we can
choose $x = z = \lambda$, and $y = a$.

Now suppose the result for all expressions β of length ℓ with $2 \leqslant \ell \leqslant n$.
Suppose that α has length $n + 1$. Then one of the following cases below
pertains.

Case 1. $\alpha = \beta \cup \rho$ for some nonempty β, ρ of length $\leqslant n$. Then $L(\alpha) = L(\beta) \cup L(\rho)$, and since $L(\alpha)$ is infinite, at least one of $L(\beta)$ or $L(\rho)$ must
be infinite. Suppose that this is $L(\beta)$. Then by induction, $L(\beta)$ contains the
relevant $xy^n z$ for all $n \geqslant 1$, and by definition of $\cup$, this is in $L(\alpha)$.

Case 2. $\alpha = \beta\rho$ for some nonempty β, ρ of length $\leqslant n$. Note that if both
$L(\beta)$ and $L(\rho)$ are finite then $L(\beta)L(\rho) = L(\alpha)$ is also finite, and so at least
one of $L(\beta)$ or $L(\rho)$ must be infinite. Suppose that this is $L(\beta)$. Then by
induction, $L(\beta)$ contains the relevant $xy^n z$ for all $n \geqslant 1$, and hence $L(\alpha)$
contains $xy^n(zr)$ for all $n \geqslant 1$ where $r \in L(\rho)$.

Case 3. $\alpha = (\beta)^*$ for some β of length n. Then choose any string $y \in L(\beta)$,
and $y^n \in L(\alpha)$ for all $n \geqslant 0$.

  □

Inspecting the proof above we obtain the following stronger statement of
the Pumping Lemma.

---

[2] When we prove that regular languages are exactly those accepted by finite state automata,
this result will be clear. However, at this stage, I would like the reader to see proofs based
on structural induction using the inductive definition of regularity.

Corollary 2.2.1 (Pumping Lemma-Restated). *Suppose that L is an infinite regular language. Then there exists a number n_0 such that for all strings w, if $|w| > n_0$, then there exist strings x, y, and z such that $w = xyz$, and such that*

1. $y \neq \lambda$ *(it could be that x or z are λ), and*
2. $|y|, |x|, |z| < n_0$, *and*
3. *for all $n \geqslant 0$, $xy^n z \in L$.*

The Pumping Lemma is a very useful tool for showing that languages are *not* regular.

Example 2.2.2. For $\Sigma = \{a\}$, $L = \{a^p : p \text{ is prime}\}$ is not regular.

Proof. Like many proofs relying on the Pumping Lemma, the proof works by looking at the form of the members of L, and then arguing that the relevant x, y, z cannot exist. In this case, x, y and z must be a^j for fixed j's, and then it would follow that there would be an infinite arithmetical progression in the primes[3]

In more detail, if L were regular, there would need to be $p, q, r \in \mathbb{N}$ such that $x = a^p$, $y = a^q$ and $z = a^r$ with $q \neq 0$, such that for all $n > 0$, $a^p(a^q)^n a^r = a^{p+nq+r} \in L$. But then $p + nq + r$ would need to be prime for all $n > 0$. But let $n = p + 2q + r + 2$. Then $p + nq + r$ is prime and hence $p + (p + 2q + r + 2)q + r$ is prime. But $p + nq + r = p + (p + 2q + r + 2)q + r = (q + 1)(p + 2q + r)$, which is the product of two smaller numbers (as $q \neq \lambda$), and this is a contradiction. $\square$

Similar methods work for the following example.

Example 2.2.3. Let $\Sigma = \{a, b\}$. Then $L = \{a^n b^n : n \in \mathbb{N}\}$ is not regular.

Proof. (Sketch) Suppose not. Then there are suitable x, y, z, with $xy^n z \in L$ for all $n > 0$. Then the cases would be to consider what y might be. It could be only a's or only b', or it could be $y = a^i b^j$. In the first case you would unbalance the a's in higher powers of y, or unbalance the b's. In the second, you'll get a b before an a. Details are left to Exercise 2.2.10. $\square$

2.2.4 Exercises

Exercise 2.2.2 Write regular expressions for the following languages $L \subseteq \Sigma^* = \{a, b\}^*$.

1. All strings in Σ^* with no more than three a's.

---

[3] But there are *arbitrarily long* arithmetical progressions in the primes, this being a remarkable result proven in 2004 by Green and Tao [GT04].

2. All strings in Σ^* with an even number of b's.

Exercise 2.2.3 Show that if L is an infinite regular language, then L has a proper subset which is a regular language.

Exercise 2.2.4 Recall that an arithmetical progression is a set of integers of the form $\{n + t \cdot m : t \in \{1, \ldots, k\}\}$ for some fixed n and m in $\mathbb{N}$. We can also have an infinite arithmetical progression where $t \in \mathbb{N}^+$.

Suppose that $L \subset \{a\}^*$, and $\{q : a^q \in L\}$ forms an arithmetical progression. Show that L is regular.

Exercise 2.2.5 Prove or disprove: For $L \subseteq \Sigma^* = \{x, y\}^*$, if $L = \{xyx^R : x, y \in \Sigma\}$ then L is regular.

Exercise 2.2.6 Rewrite each of the following regular expressions as a simpler one representing the same language. You should prove by induction on the length of words that your simpler expression equals the given one.

1. $(a^*b^*)^*(b^*a^*)^*$.
2. $(a^*b^*)^* \cup (b^*a^*)^*$.

Exercise 2.2.7 Let $\Sigma^* = \{a, b\}^*$. prove that the following languages are regular by giving regular expressions for them.

1. $L_1 = \{\sigma \in \Sigma^* : \sigma \text{ contains exactly 4 } a\text{'s}\}$.
2. $L_2 = \{\sigma \in \Sigma^* : \sigma \text{ has } abb \text{ as a substring}\}$.

Exercise 2.2.8 Fix Σ. Define the *reversal* of a string w, w^R by induction on $|w|$ as follows:

- $\lambda^R = \lambda$.
- If $w = \sigma a$ for $a \in \Sigma$, then $w^R = a(w^R)$.

Prove the following.

1. If $w = uv$ then $w^R = v^R w^R$.
2. Prove that for all $j \geqslant 1$, that $(w^R)^j = (w^j)^R$.
3. If $L = L(\alpha)$ is regular, then so is $L(\alpha^R)$.

Exercise 2.2.9 The *star height* of a regular expression α is defined by induction as follows. $h(\emptyset) = 0, h(a) = 0$ for $a \in \Sigma$, $h((\alpha \cup \beta)) = \max\{h(\alpha), h(\beta)\}$, and $h(\alpha^*) = h(\alpha) + 1$. For example, if $\alpha = (((ab)^* \cup b)^* \cup a^*)$, then $h(\alpha) = 2$. Calculate below regular expressions for the given languages with least star height.

1. $((abc)^*ab)^*$
2. $c(a^*b)^*)^*$.
3. $((a^* \cup b^*) \cup ab)^*$.

Exercise 2.2.10 Complete the details of the proof that $L = \{a^n b^n : n \in \mathbb{N}\}$ is not regular.

2.2.5 Closure properties of regular languages

Given regular languages L_1, L_2 can we make others? It turns out that the following closure properties hold: Regular languages are closed under

1. Union: If L_1 and L_2 are regular, so is $L_1 \cup L_2$.
2. Intersection: If L_1 and L_2 are regular, so is $L_1 \cap L_2$.
3. Concatenation: If L_1 and L_2 are regular, so is $L_1 \cdot L_2 = \{xy : x \in L_1 \wedge y \in L_2\}$.
4. Kleene star: If L is regular, so is L^*.
5. Complementation: If L is regular, so is $\overline{L} = \Sigma^* \setminus L$.

We have the techniques to establish union, concatenation, and Kleene star, as they follow by definition of a regular expression. However, showing closure under intersection and complementation is much more difficult. If we can show closure under complementation, then closure under intersection follows, since $L(\alpha) \cap L(\beta) = \overline{\overline{L(\alpha)} \cup \overline{L(\beta)}}$, by De Morgan's Laws. We will delay proving the following result until §2.6 we show that regular languages are exactly those accepted by automata; as the result becomes *obviously true*.

Theorem 2.2.11. *Regular languages are closed under complementation.*

2.3 Finite State Automata

2.3.1 Deterministic Finite Automata

This book is about computation, and we have not yet seen any computational devices. In this section, we will introduce our first model of computation. It is not a very general one, but turns out to coincide with the syntactic notion of regularity. The reader should think of the definition below as a computation device with a finite number of *internal states*. The device examines the bits of string in order. According to the symbol being scanned and the internal state, moves to another (perhaps the same) state and moves on to the next state. We end when we run out of symbols to be scanned; so that the machine has eaten the whole string and is now in a state q_i reading the empty string λ. If we are in a "yes" (accept) state at the end we put the string in the language, and if not we leave it out. More formally we have the following.

Definition 2.3.1 (Finite Automaton). A *deterministic finite automaton* is a quintuple $M = \langle K, \Sigma, \delta, S, F \rangle$, where:

K	is a finite set called the set of states,
$S \subseteq K$	is called the set of *initial* states,
$F \subseteq K$	is called the set of *accepting* states,
Σ	is the finite underlying set, and

$\delta : K \times \Sigma \mapsto K$ is a (partial) function called the *transition* function.

We will usually denote states by $\{q_i : i \in G\}$ for some set G and have $S = \{q_0\}$. The automata M as starts on the leftmost symbol of a string $\sigma \in \Sigma^*$, in state q_0. The transition function δ induces a rule

"If M is in state q_i reading a ($\in \Sigma$), move one symbol right and change the internal state of M to $\delta(q_i, a) = q_j$ for some $q_j \in K$."

For each $\gamma \in \Sigma^*$, we may write this action as

$$\langle q_i, a\gamma \rangle \vdash \langle \delta(q_i, a), \gamma \rangle = \langle q_j, \gamma \rangle.$$

Here, the symbol $\vdash$ is read as "yields in one step."

We say that M *accepts* a string σ if, after starting M on input σ in state q_0 and then applying $\vdash |\sigma|$ times, M is in one of the states in F. More formally, one can define M to accept σ if, for some $f \in F$,

$$\langle q_0, \sigma \rangle \vdash^* \langle f, \lambda \rangle,$$

where $\vdash^*$ denotes the transitive closure of $\vdash$. That is, $v \vdash^* y$ if $v = y$ or there exist $a_0, \ldots, a_n \in \Sigma$ such that $v = a_0 \ldots a_n y$ and states q_0, q_{i_j} for $j = 1, \ldots, n$ such that $\delta(q_0, a_0) \vdash q_{i_1}$ and more generally, $\delta(q_{i_j}, a_i) \vdash q_{i_{j+1}}$.

We let

$$L(M) = \{\sigma \in \Sigma^* : M \text{ accepts } \sigma\}.$$

Convention For convenience, we will also regard the empty language and the language consisting only of λ as finite state.

Example 2.3.1. Let $K = \{q_0, q_1\}$, $\Sigma = \{0, 1\}$, $S = \{q_0\}$, and $F = \{q_0\}$.

state(q)	symbol(a)	$\delta(q, a)$
q_0	0	q_0
q_0	1	q_1
q_1	1	q_0
q_1	0	q_1

$$\langle q_0, 00110 \rangle \vdash \langle q_0, 0110 \rangle$$
$$\vdash \langle q_0, 110 \rangle$$
$$\vdash \langle q_1, 10 \rangle$$
$$\vdash \langle q_0, 0 \rangle$$
$$\vdash \langle q_0, \lambda \rangle.$$

Now as $q_0 \in F$, M accepts 00110 and hence, $00110 \in L(M)$. On the other hand, one can easily see that $\langle q_0, 0010 \rangle \vdash^* \langle q_1, \lambda \rangle$ and, since $q_1 \notin F$, $0010 \notin L(M)$.

The traditional way of representing automata is via *transition (or state) diagrams*. Here, there is one vertex for each of the states of M and arrows from one state to another labeled by the possible transitions given by δ. Accept states are usually represented by a "double circle" at the node. For instance, the machine M would be represented by the diagram in Figure 2.1. In this simple machine, q_0 has a double circle around it since it is an accept state, whereas q_1 has only a single circle around it since it is not an accept state.

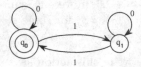

Fig. 2.1: Transition Diagram for Example 2.3.1

Historical Notes

The study of what we now call finite automata essentially began with Mc-Culloch and Pitts [MP43]. Kleene [Kle56] introduced the model we now use. Many others had similar ideas and introduced various models. Hopcroft and Ullman [HU79] is a standard text for finite state automata.

2.4 Exercises

Exercise 2.4.1 Prove that for the machine M of Example 2.3.1,

$$L(M) = \{\sigma : \sigma \text{ contains an even number of ones}\}.$$

Exercise 2.4.2 Construct automata to accept the following languages $L_i \subseteq \{0, 1\}^*$:

1. $L_1 = \{\sigma : \sigma \text{ contains at most 5 ones}\}$.
2. $L_2 = \{\sigma : \sigma \text{ contains no 1 followed by a 0}\}$.
3. $L_3 = \{\sigma : \text{the number of ones in } \sigma \text{ is a multiple of 3}\}$.

Exercise 2.4.3 Prove that if L_1 and L_2 are languages accepted by deterministic finite automata, so is $L_1 \cup L_2$.

2.4.1 Nondeterministic Finite Automata

The basic idea behind nondeterminism is that from any position, there are a number of possible computation paths. For automata nondeterminism is manifested as the generalization the transition relation δ from a function to a multi-function. Thus, δ now becomes a relation so that from a given ⟨state, symbol⟩ there may be several possibilities for δ(state, symbol). Formally, we have the following definition.

Definition 2.4.1 (Nondeterministic Finite Automaton). A *nondeterministic finite automaton* is a quintuple $M = \langle K, \Sigma, \Delta, S, F \rangle$, where K, S, F, and Σ are the same as for a deterministic finite automaton, but $\Delta \subseteq K \times \Sigma \times K$ is a relation called the *transition relation*.

Now, we can interpret the action of a machine M being in state q_i reading a symbol a, as "being able to move to any one of the states q_k with $\langle q_i, a, q_k \rangle \in \Delta$." Abusing notation slightly, we write

$$\langle q_i, a\gamma \rangle \vdash \langle q_k, \gamma \rangle,$$

but now we will interpret $\vdash$ to mean "can *possibly* yield." Again, we will let $\vdash^*$ denote the reflexive transitive closure of $\vdash$ and declare that

$$\sigma \in L(M) \text{ iff } \langle q_0, \sigma \rangle \vdash^* \langle f, \lambda \rangle,$$

for some $f \in F$. The interpretation of the definition of $\vdash^*$ is that we will accept σ provided that there is some way to finish in an accepting state; that is, *we will accept σ provided that there is some computation path of M accepting σ*. Of course, there may be many computation paths in input σ. Perhaps only one leads to acceptance. Nevertheless, we would accept σ.

Example 2.4.1. We specify M by the following sets:
$K = \{q_0, q_1, q_2, q_3, q_4, q_5\}$, $\Sigma = \{a, b\}$, $S = \{q_0\}$, $F = \{q_4\}$, and
$\Delta = \{\langle q_0, a, q_0 \rangle, \langle q_0, b, q_0 \rangle, \langle q_0, b, q_5 \rangle, \langle q_5, b, q_0 \rangle, \langle q_5, a, q_1 \rangle, \langle q_1, b, q_2 \rangle, \langle q_1, a, q_3 \rangle,$
$\langle q_1, b, q_4 \rangle, \langle q_2, a, q_4 \rangle, \langle q_2, b, q_4 \rangle, \langle q_3, b, q_4 \rangle, \langle q_3, a, q_3 \rangle, \langle q_4, a, q_4 \rangle, \langle q_4, b, q_4 \rangle\}$.

Note that upon reading, say, *bab*, M could finish in states q_0, q_2, q_4, or q_5. The reader should verify that

$$L(M) = \{\sigma \in \{a, b\}^* : bab \text{ or } ba(a \ldots a)ab \text{ is a substring of } \sigma\}.$$

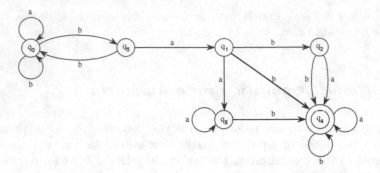

Fig. 2.2: Transition Diagram for Example 2.4.1

2.4.2 The Rabin-Scott Theorem

Nondeterministic automata seem to be more powerful than deterministic ones. But this is not case. This is the gist of classic theorem of Rabin and Scott below.

Definition 2.4.2. We say that two automata M_1 and M_2 are *equivalent* if $L(M_1) = L(M_2)$.

Theorem 2.4.4 (Rabin and Scott's Theorem [RS69]). *Every nonde-terministic finite automaton is equivalent to a deterministic one*[4].

Proof. The construction to follow is often referred to as the *subset construction*. Let $M = \langle K, \Sigma, S, F, \Delta \rangle$ be the nondeterministic automaton. The idea is quite simple. We will consider the states of the new machine *we* create from M as corresponding to *sets* of states from M. We will assume that $S = \{q_0\}$. Now, on a given symbol $a \in \Sigma^*$, M can go to any of the states q_k with $\langle q_0, a, q_k \rangle \in \Delta$. Let $\{q_{k_1}, \ldots, q_{k_p}\}$ list the states with $\langle q_0, a, q_{k_j} \rangle \in \Delta$. The idea is that we will create a state $Q_1 = \{q_{k_1}, \ldots, q_{k_p}\}$, consisting of the *set of states of M reachable from q_0 on input a*. Consider Q_1. For each symbol $b \in \Sigma^*$, and for each $q_{k_j} \in Q_1$, compute the set D_{k_j} of states reachable from q_{k_j} via b, and we let Q_2 denote $\cup_{q_{k_j} \in Q_1} D_{k_j}$. Formally,

---

[4] Furthermore, if M is a (nondeterministic) machine with n states, then we may compute in $O(2^n)$ steps a deterministic machine M' with at most 2^n states. Here, the O only depends on $|\Sigma|$.

$$Q_0 = S = \{q_0\},$$
$$Q_1 = \{q \in K : \langle q_0, a \rangle \vdash \langle q, \lambda \rangle\},$$

and

$$Q_2 = \cup_{q \in Q_1} \{r \in K : \langle q, b \rangle \vdash \langle r, \lambda \rangle\}.$$

The idea continues inductively as we run through all the members of Σ and longer and longer input strings. We continue until no new states are generated. Notice that path lengths are bounded by $n = |K|$, and that the process will halt after $O(2^n)$ steps. The accept states of the machine so generated consists of those subsets of K that we construct and contain at least one accept state of M. It is routine to prove that this construction works. (See Exercise 2.5.3.) □

The algorithm implicit in the proof above is often called *Thompson's construction* [Tho68]. Notice that we may not necessarily construct all the 2^n subsets of K as states. The point here is that many classical proofs of Rabin and Scott's theorem simply say to generate all the 2^n subsets of K. Then, let the initial state of M' be the set of start states of M, the accept states of M' are the subsets of K containing at least one accept state of M, and, if $F \subseteq K$ and $a \in \Sigma$, define

$$\delta(F, a) = \cup_{q \in F} \{r \in K : \langle q, a \rangle \vdash \langle r, \lambda \rangle\}.$$

Example 2.4.2. We apply Thompson's [Tho68] construction to the automaton from Example 2.4.1.

The initial state of Example 2.4.1 is $Q_0 = \{q_0\}$. We obtain the transition diagram of Figure 2.3.

Notice that the deterministic automaton generated by Thompson's construction has only 8 states, whereas the worst possible case would be $2^5 = 32$ states. We remark that is is possible for Thompson's construction and, indeed *any* construction to generate *necessarily* more or less (2^n) states. This is because there exist examples of n state nondeterministic automata where any deterministic automata accepting the same strings necessarily need essentially many (2^n) states. We give one example of this phenomenon in the next section (Exercise 2.4.6).

2.4.3 Exercises

Exercise 2.4.5 Convert the following nondeterministic finite automata into a deterministic one. $\Sigma = \{a, b\}, F = \{q_0\}, s = q_0, K = \{q_0, \ldots, q_4\}$, and $\Delta = \{\langle q_0, a, q_0 \rangle, \langle q_0, a, q_1 \rangle, \langle q_0, b, q_3 \rangle, \langle q_1, a, q_3 \rangle, \langle q_1, a, q_2 \rangle,$
$\langle q_1, b, q_1 \rangle, \langle q_2, a, q_2 \rangle, \langle q_2, b, q_2 \rangle, \langle q_3, a, q_0 \rangle, \langle q_3, b, q_3 \rangle\}.$

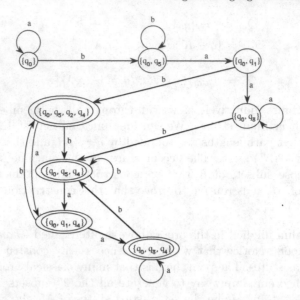

Fig. 2.3: The Deterministic Automaton for Example 2.4.1

Exercise 2.4.6 (Leiss [Lei81]) (This is not easy.) Let $L = L(\alpha)$ for $\alpha = 2(0\cup 1)^*12^*((0\cup 1)2^*)^n$. Prove that L can be represented by a nondeterministic automaton with $n+2$ many states, but any deterministic automaton accepting L must have at least $2^{n+2} + 1$ many states.

2.4.4 Nondeterministic automata with λ-moves

It is very convenient for us to extend the model of a nondeterministic automaton to *nondeterministic finite automata with λ moves*. Here, one allows "spontaneous" movement from one state to another with no head movement.

Example 2.4.3. Consider the automaton M given by Figure 2.4.

For this automaton, on input 110 it is possible for the machine to follow a computation path to *any* of the four states except q_1. For instance, to get to state q_0, M could read 1 in state q_0, move to state q_1, spontaneously move into state q_2 via the empty transition from q_1 to q_2, travel the 1 transition from q_2 back to q_0, and then finally travel the 0 transition from q_0 to q_0. We invite the reader to check that all the other states are accessible on input 110.

Again, it is easy to show that there is no real increase in computational power for the new class of automata.

Theorem 2.4.7 (McNaughton and Yamada [MY60], Rabin and Scott [RS69]). *Every nondeterministic finite automaton with λ moves is equiva-*

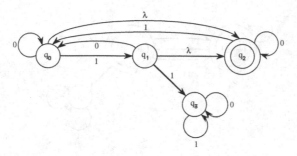

Fig. 2.4: A Nondeterministic Automaton with λ Moves

lent to one without λ moves. Moreover, there is a construction, which when applied to M, a nondeterministic finite automaton with λ moves and n states, yields an equivalent nondeterministic automaton M' with $O(n)$ states in $O(n)$ steps.

Proof. Let M be a given nondeterministic finite automaton with λ moves. We construct an equivalent M' without λ moves. For a state q of M, define the equivalence class of q via

$$E(q) = \{p \in K(M) : \forall w \in \Sigma^*(\langle q, w \rangle \vdash^* \langle p, w \rangle)\};$$

that is, $E(q)$ denotes the collection of states accessible from q with no head movement. In Example 2.4.3, $E(q_0) = \{q_0, q_2\}$. As with the proof of Theorem 2.4.4, for M', we replace states by sets of states of M. This time the set of states are the equivalence classes $E(q)$ for $q \in M$. Of course, the accept states are just those $E(q)$ containing accept states of M, and the new transition on input, a, takes $E(q)$ to $E(r)$, provided that there is some q_i in $E(q)$ and some q_j in $E(r)$ with Δ taking q_i to q_j on input a. It is easy to see that this construction works (Exercise 2.5.4). $\square$

When applied to the automaton of Example 2.4.3, the method used in the proof of Theorem 2.4.7 yields the automaton in Figure 2.5.

Definition 2.4.3. If L is accepted by a finite automaton (over Σ^*), we shall say that L is *finite state* (over Σ^*).

We can use these results to establish a number of closure properties for finite-state languages.

Theorem 2.4.8. *Let $\mathcal{L}$ denote the collection of languages accepted by finite automata. Then $\mathcal{L}$ is closed under the following:*

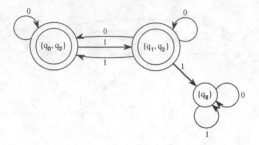

Fig. 2.5: The Automaton Corresponding to Example 2.4.3

1. *union*
2. *intersection*
3. *star (i.e., if $L \in \mathcal{L}$, then so is L^*, the collection of strings obtainable from L)*
4. *complementation*
5. *concatenation (i.e., if $L_1, L_2 \in \mathcal{L}$, then so is $L_1 L_2 = \{xy : x \in L_1 \wedge y \in L_2\}$).*

Proof. Let $M_i = \langle K_i, \Sigma_i, S_i, \Delta_i, F_i \rangle$ for $i = 1, 2$, be automata over Σ. By renaming, we can suppose that $K_1 \cap K_2 = \emptyset$.

For (i), we simply add a new start state S' with a λ move to each of the start states of M_1 and M_2. The accept states for $M_1 \cup M_2$ would be $F_1 \cup F_2$.

For (iv), we need to prove that $\overline{L(M_1)}$ is finite state. This is easy. Let $M' = \langle K_1, \Sigma, S_1, \delta_1, K_1 - F_1 \rangle$. Clearly, $\overline{L(M_1)} = L(M')$.

For (ii), we can use the fact that $L(M_1) \cap L(M_2) = \overline{\overline{L(M_1)} \cup \overline{L(M_2)}}$.

We leave (iii) and (v) for the reader (Exercise 2.5.6). □

Historical Notes

Nondeterministic Finite Automata were introduced by Rabin and Scott [RS69]. Rabin and Scott established the equivalence of nondeterministic finite automata with deterministic finite automata. McNaughton and Yamada consider automata with λ moves. Thompson's construction is taken from Thompson [Tho68].

2.5 Exercises

Exercise 2.5.1 Compute a deterministic finite automaton equivalent to the automaton of Example 2.4.3.

Exercise 2.5.2 Describe the language L accepted by the automaton of Example 2.2.

Exercise 2.5.3 Carefully complete the details of the proof of Theorem 2.4.4 by proving that if $\langle q_0, \sigma \rangle \vdash^*_M \langle r, \lambda \rangle$, then $\langle Q_0, \sigma \rangle \vdash^*_{M'} \langle R, \lambda \rangle$ for some state R of M' containing r.

Exercise 2.5.4 Carefully formalize the proof of Theorem 2.4.7

Exercise 2.5.5 Prove that for any deterministic finite automaton $M = \langle K, \Sigma, S, F, \delta \rangle$ for all $\sigma, \tau \in \Sigma^*$ and $q \in K$,

$$\delta(q, \tau\sigma) = \delta(\delta(q, \tau), \sigma).$$

Exercise 2.5.6 Prove (iii) and (v) of Theorem 2.4.8; that is, prove that the collection of finite-state languages over Σ is closed under star and concatenation.

Exercise 2.5.7 It is possible to generalize the notion of a nondeterministic automaton as follows. Now M is a quintuple $\langle K, \Sigma, \delta, S, F \rangle$ with all as before save that δ is a multi-function from a subset of $K \times \Sigma^* \mapsto K$ giving a rule:

$$\langle q_i, \gamma\sigma \rangle \vdash \langle \delta(q_i, \gamma), \sigma \rangle.$$

The difference is that in the standard definition of an automaton, we can only have $a \in \Sigma$, whereas now we can have $\gamma \in \Sigma^*$. (λ moves can be thought of as a special case of this phenomenon with $\gamma = \lambda$.) We will call such automata nondeterministic automata with *string* moves. Prove that any nondeterministic finite automaton with string moves, M, is equivalent to a standard nondeterministic finite automaton, M'. What is the relationship between the number of states of M and M'?

2.6 Kleene's Theorem: Finite State = Regular

In the last few sections, we have examined regular languages, proving that that determinism and nondeterminism have the same computational power, at least in sofar as accepting languages is concerned. The following beautiful theorem shows that the *syntactic* approach given by regularity and the *semantic* approach using automata coincide. Later we will see an echo of this when we show that partial recursive functions and partial Turing computable functions coincide in Chapter 3.

Theorem 2.6.1 (Kleene [Kle56]). *For a language L, L is finite state iff L is regular. Furthermore, from a regular expression of length n, one can construct a deterministic finite automaton with at most (more or less) (2^n) states accepting L.*

Proof. First we show how to construct automata to accept regular languages. Clearly, λ and $a \in \Sigma$ are fine. Otherwise, a regular language is built inductively from one of lower logical complexity via one of parts (iii)-(v) of Definition 2.2.4. But now the result follows by the closure properties of Theorem 2.4.8. The number of states 2^n comes from looking at the complexity of the inductive definitions of the machines for union, star, and concatenation.

Now the hard part of the proof, turning an automaton into a regular expression. Suppose that L is finite state. Let $L = L(M)$ with $M = \langle K, \Sigma, \delta, S, F \rangle$ a deterministic finite automaton. We need a regular language R with $R = L(M)$. Let $K = \{q_1, \ldots, q_n\}$, and $S = \{q_1\}$. Notice for this proof we are not using a state q_0, and are doing this only for notational convenience in the proof below, specifically for dealing with the definition of $R(i, j, k)$. For $i, j \in \{1, \ldots, n\}$, and $1 \leqslant k \leqslant n + 1$, let $R(i, j, k)$ denote the set of all strings in Σ^* derivable in M from state q_i to state q_j *without using any state q_m for* $m \geqslant k$ in an intermediate step.

That is, we let
$$R(i, j, k) = \{\sigma : \langle q_i, \sigma \rangle \vdash^* \langle q_j, \lambda \rangle \wedge \forall \gamma [\langle q_i, \sigma \rangle \vdash^* \langle q_m, \gamma \rangle \to$$
$$(m < k \vee (\gamma = \lambda \wedge m = j) \vee (\gamma = \sigma \wedge m = i))]\}.$$

For example, $R(i, j, 1)$ are simply those strings which can transition q_i to q_j with no intermediate states; meaning those $a \in \Sigma$ with $\delta(q_i, a) \vdash q_j$. $R(i, j, 2)$ would be $R(i, j, 1)$ together with x's which could transition q_i to q_j using q_1 as an intermediate step. For example, for the right machine, perhaps we might have the transitions $\delta(q_i, a) \vdash q_1$, $\delta(q_1, b) \vdash q_1$, $\delta(q_1, a) \vdash q_j$, and hence aba would transition q_i to q_j only using q_1 as an intermediate, and hence $aba \in R(i, j, 2)$. For this machine, note also that $ab^n a \in R(i, j, 2)$ for all n.

In general, notice that $R(i, j, n + 1) = \{\sigma : \langle q_i, \sigma \rangle \vdash^* \langle q_j, \lambda \rangle\}$. Thus,

$$L(M) = \cup\{R(i, j, n + 1) : q_i \in S \wedge q_j \in F\}.$$

Since the union of a finite number of regular languages is again regular, Theorem 2.6.1 will follow once we verify the lemma below. Theorem 2.6.1 will follow, because the language L accepted by M will be

$$\cup_{j \in F} R(1, j, n + 1),$$

where n is the number of states in M, assuming q_1 is the initial state.

Lemma 2.6.1. *For all i, j, k, $R(i, j, k)$ is regular.*

Proof (Proof of Lemma 2.6.1). We prove Lemma 2.6.1 by induction upon k, the hypothesis being that $R(i, j, p)$ is regular for all $p \leqslant k$.

For $k = 1$,
$$R(i, j, 1) = \{\sigma : \delta(q_i, \sigma) = q_j\}.$$

Then $R(i, j, 1)$ is finite and hence regular. Now, we claim that

$$R(i, j, k + 1) = R(i, j, k) \cup R(i, k, k) R(k, k, k)^* R(k, j, k).$$

To see that the claim holds, note that $R(i,j,k+1)$ is the collection of strings σ with $\langle q_i, \sigma \rangle$ moving to $\langle q_j, \lambda \rangle$ without using states with indices $\geqslant k+1$. Now to go from state q_i to state q_j without involving q_m for $m \geqslant k+1$, we must choose one of the following options.

1. Go from q_i to q_j and only involve states q_p for $p < k$. This is the set $R(i,j,k)$.
2. Otherwise go from q_i to q_j and necessarily involve q_k. Therefore, we must choose one of the three sub-options below.

    a. First go from q_i to q_k using only q_p for $p < k$. [This is $R(i,k,k)$.]
    b. Go from q_k to q_k some number of times, using only states q_p for $p < k$. [This is $R(k,k,k)$ each time and hence gives $R(k,k,k)^*$.]
    c. Finally, go from q_k to q_j only using states q_p for $p < k$. [This is $R(k,j,k)$.]

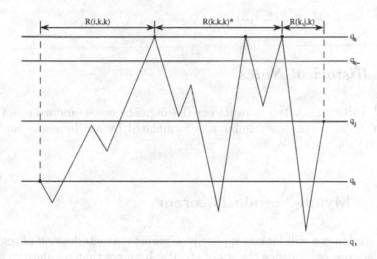

Fig. 2.6: Case (b) of Lemma 2.6.1

Figure 2.6 may be helpful in visualizing (b) above. Thus the contribution of (b) is $R(i,k,k)R(k,k,k)^*R(k,j,k)$. Now we can apply the induction hypothesis to each of the expressions in the decompositions of $R(i,j,k+1)$ of the claim. This observation establishes the claim, and hence the Lemma and finishes the proof of Theorem 2.6.1. □

□

Example 2.6.1. We re-visit Example 2.3.1.

    Let $K = \{q_1, q_2\}$, $\Sigma = \{0,1\}$, $S = \{q_1\}$, and $F = \{q_1\}$. The transitions are

$$\begin{array}{c|c|c}
\text{state}(q) & \text{symbol}(a) & \delta(q,a) \\
\hline
q_1 & 0 & q_1 \\
q_1 & 1 & q_2 \\
q_2 & 1 & q_1 \\
q_2 & 0 & q_2
\end{array}$$

1. $R(1,1,1) = \{0\}, R(1,2,1) = \{1\}, R(2,1,1) = \{1\}, R(2,2,1) = \{0\},$
2. $R(1,1,2) = R(1,1,1) \cup R(1,1,1)R(1,1,1)^*R(1,1,1) = R(1,1,1) \cup \{0\}\{0\}^* \cup \{0\}.$
3. $R(1,2,2) = R(1,2,1) \cup R(1,2,1)R(2,2,1)^*R(2,2,1) = R(1,2,1) \cup \{1\}\{0\}^*\{1\}.$
4. $R(2,1,2) = R(2,1,1) \cup R(2,1,1)R(1,1,1)^*R(1,1,1) = R(2,1,1) \cup \{1\}\{0\}^*\{0\}.$
5. $R(2,2,2) = R(2,2,1) \cup R(2,2,1)R(1,1,1)^*R(1,2,1) = \{1\}\{0\}^*\{0\}.$
6. $R(1,1,3) = R(1,1,2) \cup R(1,1,2)R(2,2,2)^*R(2,1,2).$
7. $R(1,2,3) = R(1,2,2) \cup R(1,2,2)R(2,2,2)^*R(2,2,2).$
8. $R(2,1,3) = R(2,1,2) \cup R(2,1,2)R(2,2,2)^*R(2,1,2).$
9. $R(2,2,3) = R(2,2,2) \cup R(2,2,2)R(2,2,2)^*R(2,2,2).$

Note that $L(M) = R(1,1,3)$.

2.6.1 Historical Notes

Kleene's theorem is taken from Kleene [Kle56]. The now-standard proof that finite-state languages are regular was formulated by McNaughton and Yamada [MY60].

2.7 The Myhill-Nerode Theorem*

In this section, we will look at a purely algebraic characterization of regularity, which uses equivalence classes of strings. It shows that regularity can be thought of as *finiteness*. One of the beauties of this approach is that it allows us to prove languages to be regular purely by showing that a certain equivalence relation has finite index. It is also quite useful for showing languages to be non-regular.

The reader should be warned that it is quite an abstract approach, and requires some sophistication. The Myhill–Nerode Theorem is due to Nerode [Ner58], although Myhill has a similar result in [Myh57]. (See the Historical Notes at the end of the section.)

As we will see when we look at the method of test sets, the Myhill–Nerode Theorem is also of importance in constructing automata for regular languages with only *implicit* computational descriptions, rather than *explicit* ones such as being specified by a regular expression.

The reader should recall that an equivalence relation is a transitive, symmetric, reflexive relation. If $\equiv$ is an equivalence relation on Σ^*, and $x \in \Sigma^*$ then $[x]_\equiv \{y : y \equiv x\}$ is called the equivalence class of x, and the equivalence classes partition Σ^* into a disjoint collection of sets, sometimes called *cells*. We usually drop the $\equiv$ and denote the class by $[x]$ when the meaning is clear.

Definition 2.7.1.

(i) The *index* of an equivalence relation is the number of cells.
(ii) If $\equiv_1$ and $\equiv_2$ are equivalence relations on Σ^*, then $\equiv_1$ is a *refinement* of $\equiv_2$ if the equivalence classes of $\equiv_2$ are unions of equivalence classes of $\equiv_1$

The archetype for this notion is congruence mod k, on the integers. For this equivalence relation there are k cells $\{[0], [1], \ldots, [k-1]\}$. If we consider $k \subset \{2, 4\}$, then congruence mod 4 is a refinement of congruence mod 2.

Definition 2.7.2 (Right Congruence).

1. We call a relation R on a set Σ^* a *right congruence* (with respect to concatenation) iff
   a. R is an equivalence relation on Σ^*
   b. for all $x, y \in \Sigma^*$, xRy iff for all $z \in \Sigma^*$, $xzRyz$.
2. For a language L, the *canonical right congruence* (induced by L) is the relation $\sim_L$ defined by

$$x \sim_L y \text{ iff for all } z \in \Sigma^*, xz \in L \text{ iff } yz \in L.$$

Intuitively, right congruences must agree on all extensions. Similarly, we can define a *left* congruence by replacing "$xzRyx$" in Definition 2.7.2 by "$zxRzy$." Finally, we say that R is a congruence if it is both a left and a right congruence.

Theorem 2.7.1 (The Myhill–Nerode Theorem).

(i) The following are equivalent over Σ^*.

   a. L is finite state.
   b. L is the union of a collection of equivalence classes of a right congruence of finite index over Σ^*.

    c. $\sim_L$ has finite index.

(ii) Furthermore, any right congruence satisfying (ii) is a refinement of $\sim_L$.

Proof. (i) $\rightarrow$ (ii). Suppose that L is regular and is accepted by the deterministic finite automaton $M = \langle K, \Sigma, S, F, \delta \rangle$. Without loss of generality, let $S = \{q_0\}$. We define $R = R_M$ via

$$x R y \text{ iff there is a } q \in K \text{ such that } \langle q_0, x \rangle \vdash^* \langle q, \lambda \rangle \text{ and } \langle q_0, y \rangle \vdash^* \langle q, \lambda \rangle$$

(that is, on input x or y, we finish up in the same state q). It is evident that R_M is a right congruence since, for all $z \in \Sigma^*$,

$$\langle q_0, xz \rangle \vdash^* \langle q, z \rangle$$
$$\langle q_0, yz \rangle \vdash^* \langle q, z \rangle.$$

Also, the index of R_M is finite since there are only finitely many states in K. The theorem follows since $L = \cup \{\sigma \in \Sigma^* : \langle q_0, \sigma \rangle \vdash^* \langle q, \lambda \rangle\}$ with $q \in F$.

(ii) $\rightarrow$ (iii). We prove (b) from which (ii) $\rightarrow$ (iii) follows. Let R be a right congruence of finite index so that L is a union of some of R's equivalence classes. By the fact that R is a right congruence, for all z,

$$xz \in L \text{ iff } yz \in L.$$

Part (b) follows immediately from this.

(iii) $\rightarrow$ (i). This is the hard part of the proof, extracting a machine from the right congruence classes $[x]$ for $x \in \Sigma^*$. Assume that (iii) holds. Then, we have the following implicit description of deterministic finite automaton M accepting L.

1. The states of M are the equivalence classes of $\sim_L$.
2. The initial state is $[\lambda]_{\sim_L}$, the equivalence class of λ.
3. The transition function δ of M is defined by $\delta([a], x) = [ax]$ for $a, x \in \Sigma$.

4. The accepting states of M are those x with $[x] \subseteq L$.

    Notice that item 3 is well defined as $\sim_L$ is a right congruence. The point is that $\sim_L$ being a right congruence makes the definition of δ independent of the particular representative of the equivalence class: If $a \sim_L b$, then

$$\delta([a], x) = [ax] = [bx] = \delta([b], x).$$

We see that $L = L(M)$ and this concludes our proof. $\quad\square$

    It is often quite easy to prove that certain languages are not finite state using the Myhill-Nerode Theorem.

Proposition 2.7.1. $L = \{a^n b^n : n \in \mathbb{N}\}$ *is not regular.*

Proof. Recall that we earlier proved this using the Pumping Lemma. But using the Myhill-Nerode all we need to note is that $\{a^n \not\sim_L a^k : \text{for all } k \neq n \in \mathbb{N}\}$ □

In fact, the classical Pumping Lemma is an easy consequence of the Myhill–Nerode Theorem as we now see.

Corollary 2.7.1 (Pumping Lemma, Bar-Hillel, Perles and Shamir [BHPS61]). *Suppose that L is regular and infinite. Then there exist strings x, y, and z such that for all $n \geqslant 1$,*

$$xy^n z \in L.$$

Proof. Define a collection C of strings inductively. Let $y_0 = \lambda$ and y_{n+1} denote the lexicographically least string extending y_n such that (i) $|y_{n+1}| = n+1$ and (ii) y_{n+1} has infinitely many extensions in L. By the Myhill–Nerode Theorem, there exist $n < m$ such that $y_n \sim_L y_m$. Let $x = y_n$ and $y_m = xy$ and suppose that $y_m z \in L$. Then, by the Myhill–Nerode Theorem, for all k, $xy^k z \in L$. □

We remark that it is possible to use the Myhill–Nerode Theorem directly to establish that certain languages *are* finite state. First, we need a definition.

Definition 2.7.3 (Myhill's Congruence). Define $x \approx_L y$ to mean that for all $u, v \in \Sigma^*$,

$$uxv \in L \text{ iff } uyv \in L.$$

Theorem 2.7.2. *For a formal language L, $\sim_L$ is finite index iff $\approx_L$ has finite index.*

Proof. Suppose that $\approx_L$ is finite index, but $\sim_L$ is not. Let $\{x_i \not\sim_L x_j : i \neq j\}$. But then these also witness $\approx_L$ does not have finite index, a contradiction. Conversely, suppose that $\approx_L$ is not finite index, but $\sim_L$ is. Let $A = \{x_i \not\approx_L x_j : i \neq j\}$. Then for each pair $e = \langle i, j \rangle$ there are y_e, z_e such that

$$y_e x_i z_e \in L \text{ iff } y_e x_j z_e \notin L.$$

But then $y_e x_j \not\sim_L y_e x_i$. So $\sim_L$ is not finite index, after all, a contradiction. □

Example 2.7.1. If L is finite state, so is the reverse L^{rev} of L. To see this, define xRy to hold iff $x^{rev} \approx_L y^{rev}$. As $\approx_L$ is an equivalence relation, so is R. Furthermore, by the Myhill–Nerode Theorem, $\sim_L$ has finite index and hence $\approx_L$ has finite index. This implies that R has finite index. Finally, to see that R is a right congruence, suppose that xRy and $z \in \Sigma^*$. Now $x^{rev} \approx_L$ y^{rev} and since $\approx_L$ is a (left) congruence, we have $z^{rev}x^{rev} \approx_L z^{rev}y^{rev}$ and hence $(xz)^{rev} \approx_L (yz)^{rev}$. This implies that $xzRyz$ and hence R is a right congruence.

Our result will then follow from the Myhill–Nerode Theorem once we can establish that L^{rev} is a union of equivalence classes of R. We need to show that if xRy, then $x \in L^{rev}$ implies $y \in L^{rev}$. If $x \in L^{rev}$, then $x^{rev} \in L$. If xRy, then $x^{rev} \approx_L y^{rev}$ and hence $y^{rev} \in L$, which implies that $y \in L^{rev}$.

In Exercises 2.7.5-2.7.7, we invite the reader to apply the technique of Example 2.7.1 to other languages.

2.7.1 The Method of Test Sets

* The proof of (iii) in Theorem 2.7.1 gives rise to what Downey and Fellows [DF13] call *The Method of Test Sets*.

Suppose that we have a language L which we know is regular but we are not given a regular expression for L, only given a representation via a decision procedure for membership by some algorithm which says, on input x, either *yes* $x \in L$, or *no*, $x \notin L$. Is it possible to deduce an automaton accepting L? Because of Rice's Theorem (which we prove in Chapter 5), we know that we *cannot* deduce a regular expression or automaton for L.

However, given such an algorithm, we *can* deduce the automaton if we additionally know, for instance, some bound on the number of states of such an automaton. The method of test sets allows to solve the following "promise problem."

IMPLICIT BOUNDED REPRESENTATION OF A REGULAR LANGUAGE
Input: A decision procedure saying yes/no for a language L.
Promise: L is regular and there is some M accepting L with at most k states.
Problem: Compute an automaton M' accepting L

Theorem 2.7.3. *The promise problem above is solvable and, furthermore, we can compute M' with at most k states accepting L.*

Proof. The idea is simple. We use the proof (iii) $\to$ (i) of Theorem 2.7.1. The machine is described by items 1 to 4 of that proof. The only fact that we need to check is that we can decide if $x \sim_L y$ for $x, y \in L$. By definition, $x \sim_L y$ iff for all $z \in L$, $xz \in L$ iff $yz \in L$. Suppose that M is any automaton accepting L with n states, and $x \not\sim_L y$. We claim that there some $z \in \Sigma^*$ with $|z| \leqslant n$ and z a witness to $x \not\sim_L y$. The point is that there is some z' such that, say, $xz' \in L$ yet $yz' \notin L$. This means that if we input xz' into L, we will get to an accept state, and if we input yz' into M, we must get to a nonaccept state. But since there are at most n states in M, the longest path from one state to another can have length n. Thus, we can refine z' down to a string of length $\leqslant n$ by deleting all loops.

Thus, there is some z with $|z| \leqslant n$ witnessing that $x \not\sim_L y$. Hence, to decide if $[x] = [y]$, we simply need go through all $z \in \Sigma^*$ with $|z| \leqslant n$ and see if for those z, $xz \in L$ iff $yz \in L$. The same reasoning demonstrates that for all $x \in \Sigma^*$, there is some $x' \in [x]$ with $|x'|, |y'| \leqslant n$. Hence, to implement the method of (iii) $\rightarrow$ (i), we need only look at a table of strings of length $\leqslant n$ and test them with strings of length $\leqslant n$. The fact that $\sim_L$ defines the coarsest congruence means that the number of states of M' we obtain from this technique must be minimal. $\square$

2.7.2 State Minimization*

An important consequence of the proof of the Myhill–Nerode Theorem is the fact that the automaton in (iii) $\rightarrow$ (i) has the least number of states of any deterministic automaton accepting L. Implicitly, the proof provides us with a *state minimization procedure;* that is, we can take an automaton M accepting L, and identify states equivalent under $\sim_L$. Again, since the number of states of L, $|K|$, determines the length of the longest path in L, we only need to look at strings of length $\leqslant |K|$. One minimization algorithm derived from these observations is given below.

Algorithm 2.7.4 (State Minimization Algorithm) Step 1. Make a table of entries $(p, q) \in K \times K$ for $p \neq q$.
Step 2. (*Round 1*) For each (p, q) with $p \in F \wedge q \notin F$, mark (p, q).
Step 3. Repeat the following in rounds until there is no change from one round to the next.
(*Round $i + 1$ for $1 \leqslant i$*) For all (p, q) not yet marked by the end of round i, if for some $a \in \Sigma$, $(\delta(p, a), \delta(q, a))$ is marked, then mark (p, q).
Step 4. Any unmarked squares representing (p, q) represent states that are equivalent and hence can be identified.

In Figure 2.7, we give an application of the minimization algorithm.
We remark that in Round 2, the only unmarked (x, y) are (a, e), (b, c), (b, d), and (d, c). We mark (c, d) since $\delta(c, 0) = d, \delta(d, 0) = e$ and (d, e) was marked in Round 1. We mark (b, d) since $\delta(b, 0) = d$ and $\delta(d, 0) = e$, and, finally, we mark (e, a) since $\delta(e, 1) = e$ and $\delta(a, 1) = b$. In Round 3, there is no change from Round 2 and the algorithm terminates. Notice that the cost of the algorithm is at most $O(n^2)$ if M has n states (for a fixed Σ).

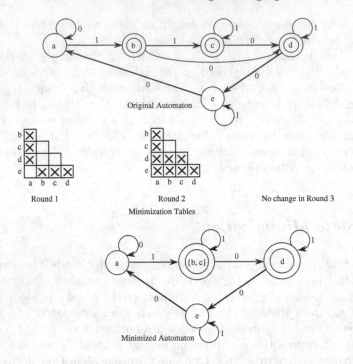

Fig. 2.7: The Minimization Algorithm

2.7.3 Exercises

Exercise 2.7.5 Let L_1 and L_2 be finite state. Define $shuffle(L_1, L_2) = \{x : x \in shuffle(u, v)$ where $u \in L_1$ and $v \in L_2\}$. Here, $shuffle(u, v) = \{y : y$ is obtained by some interleaving of the sequences of letters of u and $v\}$.

Prove that $shuffle(L_1, L_2)$ is finite state using the technique of Example 2.7.1. (Hint: Define R via: xRy iff $(x \sim_{L_1} y \wedge x \sim_{L_2} y \wedge S_x = S_y)$, where $S_z = \{[u]_{\sim_{L_1}}, [v]_{\sim_{L_2}} : z \in shuffle(u, v)\}$.)

Exercise 2.7.6 Let $x, y \in \Sigma^*$. Define the *Hamming distance* between x and y to be

$$d(x, y) = \begin{cases} \text{the number of letters that } x \text{ and } y \text{ differ if } |x| = |y| \\ \infty \text{ otherwise.} \end{cases}$$

Use the method of Example 2.7.1 to prove that for any $r \in \mathbb{N}$, if L is finite state, then so is

$$H_r(L) = \{x \in \Sigma^* : \exists y \in L[d(x, y) \leqslant r]\}.$$

(Hint: For $x \in \Sigma^*$, let $S_{x,i} = \{[u]_{\sim_L} : d(x,u) = i\}$. For $i \leqslant r$, let xR_iy iff $S_{x,i} = S_{y,i}$. Argue via $R = \cap_{0 \leqslant i \leqslant r} R_i$.)

Exercise 2.7.7 Use the method of Example 2.7.1 to prove that if L is finite state, then so is $(1/3)L = \{x \in \Sigma^* : \exists y \in \Sigma^* (xy \in L \wedge |xy| = 3|x|)\}$. (Hint: Define xRy to hold iff $x \approx_L y \wedge S_x = S_y$, where $S_z = \{[u]_{\sim_L} : |zu| = 3|z|\}$.)

2.7.4 Historical Notes

The Myhill–Nerode Theorem is due to Nerode [Ner58]. Myhill [Myh57] proved Theorem 2.7.1 but with the congruence $\approx_L$ of Definition 2.7.3 in place of $\sim_L$ in (iii), and proved that $\approx_L$ is the coarsest congruence (rather than *right* congruence) such that L is a union of equivalence classes of $\approx_L$. The algorithm to minimize the number of states of a finite automaton can be found in Huffmann [Huf54] and Moore [Moo56].

Chapter 3
General Models of Computation

Abstract We give some general models of computation, including Turing Machines, partial recursive functions, and register machines. We discuss the Church-Turing Thesis. We prove the existence of a universal Turing machine and discuss Kleene Normal Form. We also look at primitive recursion.

Key words: Turing machine, Church-Turing Thesis, partial recursive func tions, primitive recursive functions, Kleene Normal Form, universal Turing machine, Minsky machine, register machine

3.1 Introduction

In Chapter 2.2, we gave a method of deciding if some string σ was in $L(\alpha)$ for a regular expression α: namely we would build an automaton M and see if M accepts σ.

Now suppose I ask the question: suppose I wish to show that some process in mathematics is *not* algorithmic, how do I do it? In the case of regular languages we have a definition of computation and a model, an automaton. To show something is not regular we only need to show that it does not obey the definition of being regular. But to show there is *no algorithmic method*, what should we do? Our question is really quite subtle, since we are asking for processes which cannot be solved using *any mechanical process*. What we need is some way to model *any algorithmic process*.

What we plan to do in the next few sections is to offer some models of computation. Then, at the end of the chapter, discuss whether we have answered our question.

3.2 Turing Machines

Today we are surrounded by computing devices and hence a natural model of computation we might choose would be for example, a JAVA program etc. However, we will follow the classical approach and discuss the first universally acceptable model of computation. It was introduced by Alan Turing in the 1930's is what remains a very beautifully readable paper ([Tur36]).

Definition 3.2.1. A **Turing machine** over an alphabet Σ is a finite set of quadruples $\{Q_1, \ldots, Q_n\}$ with each Q_i of the form $\langle q_i, x_j, A_{i,j}, q_{i,j} \rangle$, where

- states $q_i, q_{i,j} \in \{q_0, \ldots, q_m\}$
- action $A_{i,j} \in \Sigma \cup \{R, L\}$
- symbol $x_j \in \Sigma$

We say that the Turing machine is *deterministic* if for each pair q_i, x_j, there is at most one quadruple $\langle q_i, x_j, A_{i,j}, q_{i,j} \rangle$.

In Chapter 2.2, we met automata which chewed through the symbols of a string one at a time. However, we interpret a Turing machine as a device, again with a finite number of internal *states* q_i, scanning one square at a time, on a potentially infinite tape upon which symbols can be written and *overwritten*[1].

The basic actions can be: move the read head one square to the left (L), one square to the right (R), or (over)print the current symbol on the square with one of the symbols from the finite alphabet of symbols. We interpret the quadruple $\langle q_i, x_j, A_{i,j}, q_{i,j} \rangle$ as

"in state q_i reading x_j, so perform action $A_{i,j}$ and move to state $q_{i,j}$".

Some authors use the model of quintuples where one can print a symbol and also move but this clearly makes no essential difference in that one kind of machine can be simulated by the other at the expense of a few extra moves.

All of our machines will have an *initial state* q_0. They will also have a *halt state* q_h. We will use the convention that if the machine gets into a situation where there is no quadruple for an $\langle q_i, S_j, \cdot, \cdot \rangle$ situation, then the machine will halt.

---

[1] One traditional model would have the tape be infinite from the start, but a halting computation will only ever see a finite portion of the tape. The reader should think of the tape as having a tape generating machine at each end, which will manufacture more tape whenever a computation needs more tape.

Non-halting. In contrast with the situation for finite automata, Turing Machine might never halt. For example, consider the machines M_1 and M_2 over the alphabet $\{0\}$ (and the blank square symbol B) which have quadruples $M_1 = \{\langle q_0, 0, L, \langle q_1 \rangle, \langle q_1, B, 0, q_0 \rangle\}$, and $M_2 = \{\langle q_0, 0, L, \langle q_1 \rangle, \langle q_1, B, R, q_0 \rangle\}$. Neither machine will ever halt when started on a tape blank except with a single 0, in state q_0 reading single 0. They do so for different reasons. M_1 never halts but continue left forever. M_2 never halts but moves back and forth from the 0 to the blank on the left. We will use the notation $\uparrow$ to denote a non-halting computation (no matter what the reason), and $\downarrow$ for a halting one.

Now we need to say what we mean by a machine giving a *mechanical method of computing a function*. We restrict our attention to functions on the integers, but as we see this is no real restriction. We will represent $x \in \mathbb{N}$ as a block of x 1's on the tape, meaning that for the present we will be using *unary* notation. When running M on x we begin with an otherwise blank tape on the leftmost 1 of a block of x 1's. For a deterministic machine, we regard the computation as being successful computing y if we finish in the halting state on the leftmost 1 of a block of y 1's of an otherwise blank tape. In this case we would write $M(x) \downarrow$ and $M(x) = y$. For reasons that are not obvious at this stage, we will concern ourselves with *partial functions*.

A *partial function* from $\mathbb{N}$ to $\mathbb{N}$ is a function from a subset of $\mathbb{N}$ to a subset of $\mathbb{N}$. We say that f is *total* if $\operatorname{dom} f = \mathbb{N}$.

Convention 3.2.1 (Equal partial functions) We will use the convention that when we write $f = g$ for partial functions f and g to mean that for all x,

- $x \in \operatorname{dom} f$ iff $x \in \operatorname{dom} g$, and
- $f(x) = g(x)$ for all $x \in \operatorname{dom} f = \operatorname{dom} g$.

Some authors use the notation $\simeq$ in place of $=$ to mean this. Some authors use $f :\subseteq \mathbb{N} \to \mathbb{N}$ to indicate partial functions. However, we feel that things will be clear from context, and don't wish to introduce unnecessary notational burden.

Definition 3.2.2.

(a) A deterministic machine M *simulates* a partial function $f : \mathbb{N} \to \mathbb{N}$ iff
   (i) $\forall x \in \operatorname{dom} f \ (M(x) \downarrow = f(x))$
   (ii) $\forall x \notin \operatorname{dom} f \ (M(x) \uparrow)$

(b) We say that f is partial *Turing computable* if there is some Turing
     machine M which simulates f. If additionally, f is total, we say
     that f is Turing *computable*.

   That is, if x is in the domain of f, the machine will stop on the leftmost
1 of a block of $f(x)$ many 1's, and if x is not in the domain of f, M won't
halt.

Example 3.2.1. $f(x) = 2x$. In the below q_n is the halt state.

$$\langle q_0, B, B, q_n \rangle \qquad \langle q_2, 1, 0, q_1 \rangle$$
$$\langle q_0, 1, 0, q_1 \rangle \qquad \langle q_2, B, L, q_3 \rangle$$
$$\langle q_1, 0, L, q_1 \rangle \qquad \langle q_3, 0, 1, q_3 \rangle$$
$$\langle q_1, B, 0, q_2 \rangle \qquad \langle q_3, 1, L, q_3 \rangle$$
$$\langle q_2, 0, R, q_2 \rangle \qquad \langle q_3, B, R, q_n \rangle$$

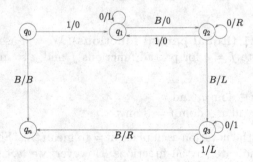

Fig. 3.1: Transition Diagram for $f(x) = 2x$

   We often revert to higher level descriptions of the action of a Turing Machine. For example, we can extend the definition of simulation to include

k-ary functions, by saying that f, when started on the leftmost 1 of a block of x_1 ones, followed by a single blank followed by a block of x_2 ones followed by ... followed by a block of x_k many ones halts if $(x_1, \ldots, x_k) \in \operatorname{dom} f$, etc. With this convention we could simulate $f(x, y) = x \cdot y$ as follows: We know how to compute $2x$, as $2x = x + x$ and this is done by marking off the 1's in x one at a time and copying. Now, for each 1 in y, we can mark them off one at a time and each time, we would add x to the current block. One all the 1's in y are marked off we'd know that we had added x to itself y many times. Similar block marking would also work for division. For example, the function $f(x) = k$ if k is the exponent of 2 in the prime decomposition of x, with $f(x) = \bot$ if $x = 0$. To do this, we would start in state q_0. If it is reading a blank, then output $\bot$. Otherwise, move to state q_1. Now in state q_1 try to mark off two 1's with 0's. If we fail to mark replace two 1's, 11, with 0's, then there must have been only 1, and we erase that one and move to state q_h. Otherwise, we continue marking off lots of 2 ones by 2 0's. If we get to the end either will will be stuck and have one 1 left over (so that x is odd) and we'd again erase the tape and halt on a blank tape, or we would begin a new block with a 3, say, in it. Next, assuming we have done this, we'd go back and replace every 22 by a 1 and begin, more or less, to repeat. Each time we successfully get to the end, we add one more 3 to the fresh block, which we are using to count the number of times we can halve the block. At the end, when we can't divide for the last time, we would erase the block and convert all the 3's back into 1's; and move to the halting state on the first 1 of this block.

The exercises provide the reader with lots of scope for practice.

Nondeterminism We can also extend the notion of nondeterminism we met in Chapter 2.2 (for automata) to Turing Machines. Again, what we need is that from a given $\langle q_i, s_j \rangle$ there might be several quadruples $\langle q_i, S_j, -, - \rangle$. The point is that since there might be more than one possible next move from a configuration q_i, x_j, what we would get are many possible *computation paths*. In the case we want to simulate a partial function f by a nondeterministic machine M, we will mean that $f(x) = y$ iff all paths that lead to halting states result with y, and some path leads to a halting state. However, in terms of Turing computability this leads to nothing new.

Theorem 3.2.2. *A partial function f can be simulated by a nondeterministic machine iff it can be simulated by a deterministic one.*

We will delay the proof of Theorem 3.2.2 until later since to prove it by using explicit quadruples would provide much detail and no insight.

Another model, more or less the same as Turing machines is provided by the following definition.

Definition 3.2.3. A programmable Turing Machine over an alphabet Σ is a finite list of numbered instructions, each instruction being one of the following.

- L
- R
- Go to ⟨line number⟩
- Print x, where $x \in \Sigma$
- If x go to ⟨line number⟩
- Stop

We think of a programmable machine as taking the internal instructions out of a standard Turing machine and putting them on a tape, the line numbers corresponding to states.

Example 3.2.2. $f(x) = 2x$

1. If B go to 14
2. Print 0
3. L
4. If 0 go to 3
5. Print 0
6. R
7. If 0 go to 6
8. If 1 go to 2
9. L
10. Print 1
11. L
12. If 0 go to 10
13. R
14. Stop

3.2.1 Exercises

Exercise 3.2.3 There are other ways a machine may not halt. Using the alphabet $\Sigma = \{0, B\}$, (B for blank) build a machine which, when started on an otherwise blank tape will fill the whole tape with 0's.

Exercise 3.2.4 Show that the function $f(x)$ which is 1 if x is even and $2x$ if x is odd is Turing computable.

Exercise 3.2.5 Construct Turing Machines for the following partial functions $f : \mathbb{N} \to \mathbb{N}$. You may assume that the input is a string of 1's on an otherwise blank tape. Use the alphabet B (for blank), 0, 1.

(i) $f(x) = 2x^2$.

(ii)
$$f(x) = \begin{cases} 0 \text{ if } x = 0 \\ x^x \text{ otherwise.} \end{cases}$$

(iii)
$$f(x) = 1 \begin{cases} 1 \text{ if } x \text{ is prime} \\ 0 \text{ otherwise.} \end{cases}$$

Exercise 3.2.6 *Devise a programmable Turing Machine which computes*

$$f(x) = \begin{cases} x \text{ if } x \text{ is a multiple of 3} \\ 0 \text{ otherwise.} \end{cases}$$

Exercise 3.2.7 Devise quadruples for a Turing Machine which runs on a planar "tape". It should be able to go, say, east, west, up, or down. Devise one which begins on an empty "tape" and covers the whole "tape" with the symbol "A".

Exercise 3.2.8 Devise quadruples for a Turing Machine with two tapes. Each tape should have its own head. The machine should be able to perform an action on one of the two tapes, based on the symbol under each of the two heads. Devise such a Turing Machine to calculate xy, given that the first tape starts with x 1's and the second starts with y 1's. When it finishes the first tape should have xy 1's and the second should be blank.

Exercise 3.2.9 Construct a Turing machine over the alphabet $\{0, 1, 2\}$ which, when started on the leftmost symbol of a block of 0's and 1's of an otherwise blank tape, rearranges them so that all of the 0's are before all of the 1's.

Exercise 3.2.10 Prove that a partial function f can be simulated by a programmable Turing machine iff it can be simulated by a standard one.

3.2.2 Coding, Gödel numbering and other domains beside $\mathbb{N}$

In this subsection, we will introduce a central idea, *coding*, which will be used throughout the book. *This idea should become so ingrained in the reader as to become second nature.*

By way of motivation, looking at $\mathbb{N}$ may seem rather restrictive. For example, later we will be concerned with functions that take the set of finite binary strings or subsets of the rationals as their domains, or other alphabets. We have already met these domains in Chapter 2.2. However, from the point of view of classical computability theory (that is, where resources such as time and memory do not matter, *as long as they are finite!*), our definitions naturally extend to such functions by *coding*. This is a key idea we met in Chapter 1. In Chapter 1, we used coding to show various sets were countable, such as the set of all finite strings from a finite alphabet Σ. Thinking back, the reader will see that the method is intuitively algorithmic. Thus, coding enables us to take other domains and ranges and consider them as natural numbers. This is done all the time in modern computing where files and the like are simply streams of 0's and 1's which can be considered as members of $\mathbb{N}$.

From Chapter 1, recall we met Gödel numbering using prime powers. For example, we would consider rationals as $r = (-1)^\delta \frac{p}{q}$ in lowest terms and use the Gödel number of r, $\#(r) = 2^\delta 3^p 5^q$. This gave an injection of $\mathbb{Q} \to \mathbb{N}$. We could build a Turing Machine which, given input $2^\delta 3^p 5^q$, could figure out δ, p and q. Thus, we can deal with questions about algorithms on $\mathbb{Q}$ purely using $\mathbb{N}$, if perhaps, somewhat inefficiently.

Another illustration is provided using the model of n-ary functions. We *could* define Turing computability for functions $f : \mathbb{N}^n \to \mathbb{N}^k$ taking $(x_1, \ldots, x_n)$ to $(y_1 \ldots, y_k)$ via a machine M starting on the leftmost block of n blocks of x_i 1's separated by a single blank and resulting in the halt state on the leftmost 1 of k blocks of y_i many 1's. But this can also be simulated by considering the input as $2^{x_1+1} 3^{x_2+1} \ldots p_n{}^{x_n+1}$ (where p_i denotes the i-th prime) and the output similarly. Or even putting a 0 between the blocks and considering this as a single input over the alphabet $\{0,1\}^*$.

In a nutshell, *a Turing computable coding* is one that could be performed by a Turing machine, and given a number $d \in \mathbb{N}$, we can figure out whether d corresponds to an object, and if so, what object being coded corresponds to d.

All of the codings we looked at in Chapter 1 were in fact algorithmic codings, and could be simulated on a suitable Turing Machine.

In Chapter 7, we discuss complexity theory issues such as *how efficient* is an algorithm. In those situations, the *type of coding matters*. However, for now we are only seeking to delineate the algorithmic from the non-algorithmic. Thus, at present, we will always regard our objects as (implicitly) effectively coded in some way a Turing Machine could simulate. If the reader has not seen this before they should do some of the exercises below to gain facility at such coding.

The reader should also note that coding allows us to consider a lot of other models such as *multi-tape* Turing Machines. That is we have n tapes instead of one, with n heads. Then the next "move" is determined by a transition function based on the symbol each head is reading and in which state (or states) the machine is in, that determines the action on each of the tapes. In this model, it is convenient to have a single tape for the input and output, at least for unary functions. Again, this could be simulated by a single tape machine using, for example, Gödel numbers, with n primes devoted to each of the tapes, and their powers to the contents.

3.2.3 Exercises

Exercise 3.2.11 (We have already seen this so it should be trivial.) Show how to effectively code all finite subsets of N.

Exercise 3.2.12 Build a Turing machine which simulates $g(x,y) : N \times N \to \{0,1\}$, which is 1 if x is a factor of y, and 0 if either $x = 0$ or x is not a factor of y.

3.2.4 The Church-Turing Thesis

In the next few sections, we will look at other models of computation. They provably give the same class of functions as the partial functions computable by Turing Machines. We are now in a position to state one of the fundamental claims in computability theory.

Church-Turing Thesis *The class of intuitively computable partial functions coincides with exactly the Turing computable partial functions.*

This is a *claim* that cannot be *proven*, although as usual with the Scientific Method, it could be *disproven*. However, this thesis is almost universally accepted. The *evidence*, so far, is that *all models so far found by people coincide with the class of (partial) Turing computable functions, and whenever one analyses the way a computable process might work it seems that a Turing machine can emulate it.*

The context of Turing's 1936 paper [Tur36] where he introduced Turing Machines was that Turing was trying to model an "abstract human". That is, in 1936, a decision procedure would be have been interpreted to mean "a procedure which could be performed by a human with arbitrary amounts of time". Thus, Turing did a thought experiment and argued an *abstract human computor* (*computors* were people employed to do mathematical computations) by limitations of sensory and mental apparatus we have had the following properties:

1. a finite bound for the number symbols,
2. a finite bound for the number of squares,
3. a finite bound on the number of actions at each step,
4. a finite bound on the number of squares each action could be applied to on each step,
5. a finite bound on the movement, and
6. a finite bound on the number of (mental) states.

Gandy, Soare (and others, see [Dow14]) argue that Turing *proves* (in the sense of proofs in the physical sciences) any function calculable by an *abstract human* is computable by a Turing Machine.

There were other models of computation before Turing's model, such as λ-computable functions (not treated in this book,) and *partial computable functions* we cover soon. But the impact of Turing's work can best be described by the following quote from Gandy in [Her95]:

> "What Turing did, by his analysis of the processes and limitations of calculations of human beings, was to clear away, with a single stroke of his broom, this dependency on contemporary experience, and produce a characterization-within clearly perceived limits- which will stand for all time..... What Turing also did was to show that *calculation can be broken down into the iteration (controlled by a "program") of extremely simple concrete operations*[2]; so concrete that they can easily be described in terms of *(physical) mechanisms*. (The operations of λ-calculus are much more abstract.)"

If the reader is interested in this fascinating subject, I urge them to have a look at Turing's original paper. A guide to this and its historical context, accessible to the readers of this text, can be found in Petzgold [Pre08]. We also refer the reader to Herken [Her95] and Downey [Dow14] which are edited collections discussing the development of the ideas, both pre- and post-Turing[3].

3.2.5 Nondeterminism

Here is one application of the Church-Turing Thesis, which otherwise would be a nightmare involving the explicit construction of quadruples. We will show

---

[2] My emphasis.

[3] Turing was a fascinating figure of the 20th Century. I recommend *The Turing Guide*, [CBSW17], which is an edited collection of accessible accounts of aspects of his life and work.

that computation by nondeterministic Turing Machines and Deterministic ones gives the same class of Turing computable partial functions.

For simplicity we will consider nondeterministic *acceptance* meaning that M accepts x if at least one computation path accepts x. Given a nondeterministic Turing machine M that accepts a language L we show how to build a deterministic Turing machine M' that accepts $x \in L$ if and only if M accepts $x \in L$. The problem of course is that at each step of M there may be many possible next steps. To get around this, we will use *breadth first traversal* of the *computation tree* to figure out if there is an accepting path. That is, at the beginning there we look at the possible first step in the computation. (That is, what are the $(q_0, s_j, \cdot, \cdot)$ quadruples). Then for each possible next move begin a tree of possible computations. M will accept iff some of the "paths" of this tree leads to acceptance. The idea is : generate the tree to length s at stage s, then look at all the paths in the tree of length s and see if one accepts. If one path does, accept. If no path of the computation tree accepts, move on to step $s + 1$ and repeat, generating more of the computation tree. The computer science student would say "use breadth first search".

3.2.6 A Universal Turing machine

The following result is central.

Theorem 3.2.13 (Turing-Enumeration Theorem – Universal Turing Machine). *There is a Turing computable way of enumerating all the partial Turing computable functions. That is, there is a list $\Phi_0, \Phi_1, \ldots$ of all such functions such that we have an algorithmic procedure for passing from an index (code) i to a Turing machine computing Φ_i, and vice-versa. Using such a list, we can define a partial Turing computable function $f(x, y)$ of two variables such that $f(x, y) = \Phi_x(y)$ for all x, y. Such a function, and any Turing machine that computes it, we call* universal.

If the reader is willing to believe that Turing machines have the same computational power as modern computers and vice versa, this result is obvious. That is, given a program in some computer language, we can convert it into ASCII code, and treat it as a number. Given such a binary number, we can decode it and decide whether it corresponds to the code of a program, and if so execute this program. Thus a compiler for the given language can be used to produce a universal program.

We will prove a sharper statement of the result in the next section, but here is a sketch of Turing's proof of the result. First, a Turing machine is simply a finite collection of quadruples $Q_1, \ldots Q_k$ over some alphabet of states $q_1, \ldots, q_m$ and symbols $L, R, S_1, \ldots, S_d$. We can assign unique Gödel numbers to each of the symbols and states, and hence a quadruple Q could be coded by $\#(Q) = 2^{\#(q_i)} 3^{\#(S_j)} 5^{\#(A_{i,j})} 7^{\#(q_{i,j})}$. Then we could code the machine by

$2^{\#(Q_1)}3^{\#(Q_2)}\ldots p_k^{\#(Q_k)}$. Then we could consider the two tape Turing machine which reads the input x,y and on tape 1 decodes the input x onto tape 2 as a collection of quadruples and then executes this machine on y.

The reader who, perhaps rightfully, believes that the above is mere hand-waving should be reassured by the next section, where we will formalize things, and culminate with Kleene Normal Form. At the same time we will introduce another more functional model of computation called *partial recursive functions*.

Turing was the deviser of the idea of a universal machine, a *compiler*. The reader should remember that, before this, models of computation were hand crafted for the particular function. For example, things like "slide-rules" were mechanical devices to add, multiply, divide and other basic arithmetical operations. That is if you wanted to compute some function, a machine would be purpose built for it. The Universal Turing machine makes this unnecessary:

Turing said in a lecture of 1947 with his design of ACE (automated computing engine—one of the first designs for a programmable computer)

"The special machine may be called the universal machine; it works in the following quite simple manner. When we have decided what machine we wish to imitate we punch a description of it on the tape of the universal machine.... . The universal machine has only to keep looking at this description in order to find out what it should do at each stage. Thus the complexity of the machine to be imitated is concentrated in the tape and does not appear in the universal machine proper in any way... [D]igital computing machines such as the ACE ... are in fact practical versions of the universal machine."

Living in our digital world, it is hard for us to imagine the intellectual advance that the idea represents. Everything is data for us now. But in the 1940's and 50's, this was not the case: The idea that a computer could be universal was a long time penetrating. For example, Howard Aitken (1956), a US computer expert of the time, said the following in a lecture:

"If it should turn out that the basic logics of a machine designed for numerical solution of differential equations coincide with the logics of a machine intended to make bills for a department store, I would regard this as the most amazing coincidence that I have ever encountered."

3.3 Partial recursive functions

In this section, we will introduce another model of computation which at first glance seems quite ad hoc. Interestingly, this model pre-dated the Turing model. This model is interesting in its own right and is very useful in certain coding situations. We will see this when we look at Hilbert's 10th problem in §4.3.7 which was originally proven using the partial recursive functions as a base for coding. For the student of computer science, this model is the ancestor of all *functional programming languages*, along with yet another model called *Lambda Calculus*.

3.3.1 Primitive recursive functions

We begin with a smaller class of functions which is itself important, though
we do not treat this class in any detail in the present book.

Definition 3.3.1 (Primitive Recursive Functions). The primitive
recursive functions are the smallest class $\mathcal{P}$ of functions (from $\mathbb{N}$ to $\mathbb{N}$)
including 1-4 and closed under the schemes 5-6:

1. *Zero Function* $Z(x) = 0$.
2. *Successor Function* $S(x) = x + 1$.
3. (Dotted) *Predecessor Function* $P(x) = x \dot{-} 1 = \begin{cases} 0 \text{ if } x = 0 \\ x - 1 \text{ if } x \geqslant 1 \end{cases}$.
4. *Projection Function* $P_j^m(x_1, \ldots, x_m) = x_j$.
5. *Substitution* If $g(y_1, \ldots, y_m) \in \mathcal{P}$ and $f_i(x_i, \ldots, x_k) \in \mathcal{P}$ for $i = 1, \ldots, m$ then $g(f_1(x_1, \ldots, x_k), \ldots, f_m(x_1, \ldots, x_k)) \in \mathcal{P}$.
6. *Primitive Recursion* If $g(x_1, \ldots, x_n) \in \mathcal{P}$ and $h(x_1, \ldots, x_{n+2}) \in \mathcal{P}$ then the function f defined below is also in $\mathcal{P}$:

$$f(x_1, \ldots, x_n, 0) = g(x_1, \ldots, x_n)$$
$$f(x_1, \ldots, x_n, n+1) = h(x_1, \ldots, x_n, n, f(x_1, \ldots, x_n, n)).$$

Examples

(i) $f(a, b) = a + b$ is primitive recursive.
Let $h_1(x_1) = S(x)$.
Let $h_2(x_1, x_2) = P_1^2(x_1, x_2) = x_1$.
Let $h_3(x_1, x_2, x_3) = P_3^3(x_1, x_2, x_3) = x_3$.
Let $h_4(x_1, x_2, x_3) = h_1(h_3(x_1, x_2, x_3))$.
Define:

$$f(a, 0) = h_2(a, 0),$$
$$f(a, b+1) = h_4(a, b, f(a, b)).$$

We can check that this definition is correct by induction:
Base case: $f(a, 0) = h_2(a, 0) = a$.
Inductive Hypothesis: Suppose $f(a, b) = a + b$.
Then $f(a, b+1) = h_4(a, b, f(a, b))$
$= h_4(a, b, a + b)$
$= h_1(h_3(a, b, a + b))$
$= h_1(a + b)$
$= a + b + 1$.

(ii) $f(a, b) = ab$ is primitive recursive.
Let $g(x_1, x_2) = x_1 + x_2$.

Let $h_1(x_1) = Z(x_1)$.
Let $h_2(x_1, x_2) = P_1^2(x_1, x_2) = x_2$.
Let $h_3(x_1, x_2, x_3) = P_3^3(x_1, x_2, x_3) = x_3$.
Let $h_4(x_1, x_2, x_3) = P_1^3(x_1, x_2, x_3) = x_1$.
Let $h_5(x_1, x_2, x_3) = g(h_3(x_1, x_2, x_3), h_4(x_1, x_2, x_3))$.
Define:

$$f(a, 0) = h_1(h_2(a)),$$
$$f(a, b+1) = h_5(a, b, f(a, b)).$$

Check by induction:
Base case: $f(a, 0) = h_1(h_2(a, 0)) = Z(a) = 0$.
Inductive hypothesis: $f(a, b) = ab$.
The $f(a, b+1) = h_5(a, b, f(a, b))$
$= h_5(a, b, ab)$
$= g(h_3(a, b, ab), h_4(a, b, ab))$
$= g(ab, a)$
$= ab + a)$
$= a(b+1)$.

One important fact is that *all primitive recursive functions are total*. The reader should prove this in exercise 3.3.2 below.

3.3.2 Exercises

Exercise 3.3.1 Give full derivations from the initial functions and the rules to show that the following functions are primitive recursive:

(i) $f(a, b) = a^b$
(ii) $f(a, b) = a!$

(iii) $f(a, b) = a \div b = \begin{cases} 0 \text{ if } b \geqslant a \\ a - b \text{ if } a > b. \end{cases}$

Exercise 3.3.2 Prove that any primitive recursive function is *total*, that is, it is defined on all arguments. (Hint: Use induction on the way that the functions are defined.)

Exercise 3.3.3 Prove by induction on the complexity of the definition, that all primitive recursive functions are Turing computable.

It is often clear that a function is primitive recursive without writing down the full derivation from the rules and initial functions. The next list of examples illustrates this point: We will use the function called *monus* from Exercise 3.3.1 above.

$$f(a, b) = a \dot{-} b = \begin{cases} 0 \text{ if } b \geqslant a \\ a - b \text{ if } a > b. \end{cases}$$

Examples

(i) $min(a, b) = b \dot{-} (b \dot{-} a)$.

(ii) $max(a, b) = (a + b) \dot{-} min(a, b)$.

(iii) Define the function $sg(a)$ as follows:

$$sg(a) = \begin{cases} 0 & \text{if } a = 0 \\ 1 & \text{if } a > 0 \end{cases}$$

Then $sg(a) = 1 \dot{-} (1 \dot{-} a)$. This follows since if $a > 0$, $1 \dot{-} a = 0$, and hence $1 \dot{-} (1 \dot{-} a) = 1 \dot{-} 0 = 1$, and if $a = 0$, then $1 \dot{-} (1 \dot{-} a) = 1 \dot{-} 1 = 0$.

(iv) The function called *bounded sum* is quite important: $g(x_1, \ldots, x_n, z + 1) = \sum_{y \leqslant z+1} f(x_1, \ldots, x_n, y)$:

$$\sum_{y \leqslant 0} f(x_1, \ldots, x_n, y) = f(x_1, \ldots, x_n, 0),$$

$$\sum_{y \leqslant z+1} f(x_1, \ldots, x_n, y) = \sum_{y \leqslant z} f(x_1, \ldots, x_n, y) + f(x_1, \ldots, x_n, z + 1).$$

(v) As is *bounded product* $g(x_1, \ldots, x_n, z + 1) = \prod_{y \leqslant z+1} f(x_1, \ldots, x_n, y)$:

$$\prod_{y \leqslant 0} f(x_1, \ldots, x_n, y) = f(x_1, \ldots, x_n, y),$$

$$\prod_{y \leqslant z+1} f(x_1, \ldots, x_n, y) = \prod_{y \leqslant z} f(x_1, \ldots, x_n, y) \times f(x_1, \ldots, x_n, z + 1).$$

(vi) $|a - b| = (a \dot{-} b) + (b \dot{-} a)$. You can prove this by considering whether $a > b$ or not. If $a > b$ then $a \dot{-} b = a - b$ and $b \dot{-} a = 0$. If $b \geqslant a$ then the converse holds.

(vii) Let $rm(a, b)$ denote the function which outputs the remainder when b is divided by a. This function is defined by primitive recursion as follows:

$$rm(a, 0) = 0,$$
$$rm(a, b + 1) = (rm(a, b) + 1) \cdot sg|a - (rm(a, b) + 1)|.$$

Remark 3.3.1. * We remark that any (total) function on $\mathbb{N}$ which the reader has ever encountered in "normal" mathematics or computing will be primitive recursive. An explicit function which is total and computable but not primitive recursive is called the *Ackermann Function* $A(n)$ (actually a class of functions) defined as follows: We let $f^{(k)}(n)$ denote the k-th iterate of f, via $f^{(1)}(n) = f(n)$ and $f^{(k+1)}(n) = f(f^{(k)}(n))$. Then define $A_0(n) = n + 1$ and $A_{m+1}(n) = A_m^{(n+1)}(1)$. Then finally $A(n) = A_{n+1}(n)$. It can be shown that $A(n)$ grows too quickly to be primitive recursive. This is hardly the first function that comes to mind. The proof that Ackermann's Function is not primitive recursive is an implicit diagonalization where we use the fact that there is an computable enumeration of all the primitive recursive functions, and this function is engineered to grow faster than the e-th on argument e. (Ackermann [Ack28].) Ackermann's Function, along with even faster growing examples have arisen in modern mathematics as the growth rates of certain combinatorial principles. We refer the reader to, for example, Ketonen and Solovay [KS81] for examples used to show that certain functions associated with combinatorial witnesses grow "too fast" to have proofs in a standard logical system of arithmetic such as Peano Arithmetic. These facts are beyond the scope of this book.

3.3.3 Primitive Recursive Relations

Recall that a *relation* or *predicate* is a set R such that $R(x) = \{x | x$ has property $R\}$. We say that $R(x)$ *holds* if and only if x has property R. We also define the function χ_R to be the following.

$$\chi_R(x_1, \ldots, x_n, y) = \begin{cases} 1 & \text{if } R(x_1, \ldots, x_n, y) \text{ holds} \\ 0 & \text{if } R(x_1, \ldots, x_n, y) \text{ does not hold} \end{cases}$$

Then we can say a relation R is primitive recursive if and only if χ_R is a primitive recursive function. We call χ_R the characteristic function of R, in the same was as the characteristic function of a set.

Examples

(i) Let $D(a, b)$ hold if and only if a divides b. Then

$$\chi_D(a, b) = 1 \dotdiv \text{sg}(\text{rm}(a, b)).$$

(ii) Let $E(a, b)$ hold if and only if $a = b$. Then $\chi_E(a, b) = 1 \dotdiv \text{sg}|a - b|$.

Now we will give some rules for making new primitive recursive functions from old ones. Let R and S be two given primitive recursive relations.

(i) The relation "$T = R$ and S", written $R \wedge S$, is primitive recursive since $\chi_T = \chi_R \cdot \chi_S$.
(ii) The relation $T =$ not R written as $\neg R$ is primitive recursive since the function $\chi_T = 1 \dotdiv \chi_R$ is primitive recursive.
(iii) Hence any boolean combination of primitive recursive functions is primitive recursive.

3.3.4 Bounded quantification

Bounded quantification over primitive recursive relations is also primitive recursive:

$$\forall y \leqslant z R(x_1, \ldots x_n, y) \text{ if and only if } \prod_{y \leqslant z} \chi_R(x_1, \ldots x_n, y) = 1,$$

$$\exists y \leqslant z R(x_1, \ldots x_n, y) \text{ if and only if } \mathrm{sg}(\sum_{y \leqslant z} \chi_R(x_1, \ldots x_n, y)) = 1.$$

The notion of *bounded search*, namely the "least y less than or equal to x such that some relation holds" defined below is also primitive recursive:

$$\mu y \leqslant x R(x_1, \ldots x_n, y) = \begin{cases} \mu y R(x_1, \ldots, x_n, y) & \text{if } (\exists y \leqslant z) R(x_1, \ldots, x_n, y) \\ 0 & \text{otherwise} \end{cases}$$

We invite the reader to prove this in Exercise 3.3.4 below. The primitive recursiveness of bounded quantification and bounded search is useful to quickly observe the primitive recursiveness of other functions and relations.

Examples

(i) Let $\Pr(x)$ hold if and only if x is prime. Then $\Pr(x)$ if and only if

$$x \geqslant 2 \wedge (\forall y \leqslant x)(\mathrm{rm}(y, x) = 0 \implies y = 1 \vee y = x).$$

(ii) Any number x has a unique representation as a product of powers of prime numbers, that is $x = p_0^{k_0} p_1^{k_1} \ldots p_i^{k_i}$. The function $\exp(x, p_j)$ which outputs the exponent p_j in the representation of x is primitive recursive:

$$\exp(x, p_j) = \mu y \leqslant x [\mathrm{rm}(p_j^y, x) = 0 \wedge \neg(\mathrm{rm}(p_j^{y+1}, x) = 0)].$$

3.3.5 Exercises

Exercise 3.3.4 Prove that bounded search is primitive recursive.

Exercise 3.3.5 Let $f_p(x)$ denote the power of p in the prime decomposition of x. Thus, if $x = 6$, then $f_2(x) = 1$ and $f_{11}(x) = 0$, for example. Show that for any prime p, $f_p(x)$ is primitive recursive. Conclude that the relation

$$R_p(x) = \begin{cases} 1 \text{ if the power of } p \text{ in } x \text{ is } 4 \\ 0 \text{ otherwise.} \end{cases}$$

is primitive recursive.

3.3.6 Partial recursive functions

Are the primitive recursive functions enough to characterize all (total) computable functions? Using diagonalization, the answer is clearly no: Since the primitive recursive functions are built up in a (computable) and hierarchical way, we could easily assign a Gödel number to each primitive recursive function. Let $\{f_n : n \in \mathbb{N}\}$ be such an enumeration. Consider the function $g(n) = f_n(n)+1$. This is clearly a (total) and intuitively computable function, but cannot be primitive recursive. If it was, it would need to have $g = f_m$ for some m, but then $g(m) = f_m(m) + 1$, so $0 = 1$! The masochistic reader might be keen to formalize this using Turing Machines. Notice that this is the *diagonalization* technique we met in Chapter 1 to show that the reals were uncountable, but now applied in the context of computation.

Thus, primitive recursive functions are *insufficient* to capture the notion of *intuitively computable functions*. The problem lies in the fact that searches for primitive recursive functions are *bounded*. In fact there is a kind of Church-Turing Thesis for primitive recursive functions:

Church-Turing Thesis for Primitive Recursive Functions. The class of primitive recursive functions are exactly the class of functions which are computable without any "until" loops. This is a search which is looking for a "witness" for some relation from the class to hold. Keep looking until you find this witness. That is, primitive recursive functions are those computable *without* any search which halts only when a witness is found.

As it is not central to our story, we won't dwell on this thesis; but mention it as the primitive recursive functions do occupy a significant place in the theory of computation. Any function the reader has encountered in "normal" mathematics will be primitive recursive.

Thus, to capture the notion of computability we need "until" searches.

Definition 3.3.2 (Partial Recursive Functions). The class of *partial recursive functions* $\mathcal{C}$ is the smallest class of functions containing the primitive recursive functions and closed under the addition rule:

7. Least number (or unbounded search). If $g(x_1, \ldots, x_m, y) \in \mathcal{C}$ then the function f defined below is also in $\mathcal{C}$:

$$f(x_1, \ldots, x_m) = \mu y[g(x_1, \ldots, x_m, y) \downarrow= 0 \ \& \ (\forall z \leqslant y)[g(x_1, \ldots, x_m, z) \downarrow]].$$

We remark that the last clause in the least number rule is necessary to ensure that the class $\mathcal{C}$ is closed under this rule. All of these definitions are due to the logician Stephen Cole Kleene. The (likely apocryphal) legend is that he thought of the least number operator at the dentist whilst having wisdom teeth extracted.

The reader might well ask, why stop here? Can't we play the same trick as before and get something which is intuitively computable but not partial recursive? We can still enumerate the partial recursive functions, similarly as $\{f_n : n \in \mathbb{N}\}$. Then we could define $g(n) = f_n(n) + 1$. Why don't we *diagonalize* out of the class as we did before. The answer is that it is possible that $g(n) = f_n(n) + 1$ because $f_n(n)$ *is undefined!* We are dealing with *partial* functions here. It could be (and is true) that $g(n) = f_n(n) + 1$ is partial recursive. This is because both sides are undefined, and remember $f(x) = g(x)$ if both sides are undefined.

Kleene knew that we could diagonalize out of the class of primitive recursive functions, and realized that he could not diagonalize out of the class of partial recursive functions. The partial recursive functions gave one of the original models of computation. Church and others claimed that the class of partial recursive functions captured all intuitively computable functions; and they made this claim well before the Turing model had been constructed. Imagine you had only seen the definition of partial recursive functions. I believe that you would be rightfully suspicious of this claim that this class *captured all intuitively computable functions.* It was only when Turing proposed his model of Turing machines that people accepted Church's claim. The point was that Turing machines are so *obviously* computable, and genuinely seem to reflect the actions of the human computor of Turing's thought experiment.

Turing [Tur37a] also showed that the class of partial recursive functions and the class of partial Turing computable functions coincide. We will do this in the next subsection. Earlier, Kleene [Kle36] had proven that the models of λ-computable (the even more obscure model of computation developed by Church) and partial recursive functions coincided.

3.3.7 Gödel Numbering Turing Computations

The aim of the next two sections is to prove that the class $\mathcal{C}$ of partial recursive functions is exactly the class of functions $\mathcal{T}$ computable by a Turing machine. The following exercise is one direction. It extends Exercise 3.3.3.

Exercise 3.3.6 Show how to compute a given partial recursive function with a Turing machine program.

We will now work towards the other direction, namely that $\mathcal{T} \subseteq \mathcal{C}$. That is, any partial function computed by some Turing machine program is also a partial recursive function.

To do this we again use the coding technique of *Gödel numbering* to produce a unique code for each Turing machine program.

Recall that a Turing machine program P is a finite set of quadruples, say $Q_0, Q_1, \ldots, Q_n$, of the general form $\langle q_i, x_i, A_{i,j}, q_{i,j} \rangle$. Also recall that the set of states from which q_i and $q_{i,j}$ is finite, as is the alphabet Σ. For our purposes we fix the alphabet Σ to be $\{0, 1, B\}$.

We define a function g to assign a number to the different parameters in a quadruple as follows:

$$
\begin{aligned}
g(0) &= 2 \\
g(1) &= 3 \\
g(B) &= 4 \\
g(L) &= 5 \\
g(R) &= 6 \\
g(q_i) &= 7 + i
\end{aligned}
$$

We can now use this definition to assign numbers to quadruples:

$$g(\langle q_i, x_i, A_{i,j}, q_{i,j} \rangle) = 2^{g(q_i)} \times 3^{g(x_i)} \times 5^{g(A_{i,j})} \times 7^{g(q_{i,j})}.$$

Examples

The following program P when given input x computes $x + 1$.

$Q_0 = \langle q_0, 1, R, q_0 \rangle$
$Q_1 = \langle q_0, B, 1, q_1 \rangle$
$Q_2 = \langle q_1, 1, L, q_1 \rangle$
$Q_3 = \langle q_1, B, R, q_2 \rangle$

where state q_2 is a *halt* state.

The Gödel numbers are:

$g(Q_0) = 2^7 . 3^3 . 5^6 . 7^7$
$g(Q_1) = 2^7 . 3^4 . 5^3 . 7^8$
$g(Q_2) = 2^8 . 3^3 . 5^5 . 7^8$
$g(Q_3) = 2^8 . 3^4 . 5^6 . 7^9$

In a similar way we can also use the definition of g so far to number programs. If P is the Turing machine program consisting of the quadruples $Q_0, Q_1, \ldots, Q_n$ then define:

$$g(P) = 2^{g(Q_0)} \times 3^{g(Q_1)} \times \ldots \times p_{n+1}^{g(Q_n)},$$

where p_{n+1} is the $n + 1$th prime number.

If $g(P) = k$ then we say that k is the *Gödel number* of P.

The Gödel numbers of quadruples can be very large, and hence the Gödel numbers of programs extremely large! Notice that not every number codes a quadruple (or program), but it is true that each quadruple (or program) is coded by a *unique* number. We also code each step of a computation by a Turing machine M.

Definition 3.3.3 (Configuration of a Turing Machine). The *instantaneous condition of M at each step of the computation* is completely determined by:

1. the current state q_i of the machine;
2. the symbol s_0 being scanned;
3. the symbols on the tape to the right of s_0, namely $s_1, s_2, \ldots, s_n$;
4. and the symbols on the tape to the left of s_0, namely $s_{-m}, \ldots, s_{-2}, s_{-1}$.

This is the *configuration* of the machine at that step and is written

$$s_{-m} \cdots s_{-2} s_{-1} q_i s_0 s_1 \cdots s_n.$$

Examples

Let Γ be the program in the previous example. Suppose we give P the input $x = 1$. Then the configurations of the computation are as follows:

$c_0 : q_0 1$

$c_1 : 1 q_0 B$

$c_2 : 1 q_1 1$

$c_3 : q_1 11$

$c_4 : q_1 B 11$

$c_5 : q_2 11$

We can think of a Turing machine computation of program P on input x as a sequence of configurations $c_0, c_1, \ldots, c_t$ such that c_0 represents the machine in state q_0 reading the leftmost symbol of the input x, c_t represents the machine halting in some state q_j, and the transition between configurations c_i and c_{i+1} is given by the program P. Here we are thinking of a computation as a calculation that halts. Since at the beginning the tape contains only finitely many non-blank squares, and this is true at any later stage of the calculation whether it halts or not (since at most one square can be changed from a blank to a non-blank in each configuration), then the integers n and m exist for each configuration.

Now we will use the configuration approach to code a computation with Gödel numbering.

Let r be the sequence $s_1, s_2, \ldots, s_n$. Then define

$$g(r) = 2^{g(s_1)} \times 3^{g(s_2)} \times \ldots \times p_n^{g(s_n)}.$$

If $r = \emptyset$ then define $g(r) = 0$.

Let l be the sequence $s_{-m}, \ldots, s_{-2}, s_{-1}$. Then define

$$g(l) = 2^{g(s-m)} \times \ldots \times p_{m-1}^{g(s-2)} \times p_m^{g(s-1)}.$$

If $l = \emptyset$ then define $g(l) = 0$.

Let the Gödel number of a configuration be $g(c_j)$ where

$$g(c_j) = 2^{g(q_i)} \times 3^{g(s_0)} \times 5^{g(r)} \times 7^{g(l)}.$$

Finally, we can Gödel number the entire computation C_P of program P as

$$g(C_P) = 2^{g(P)} \times 3^{g(c_0)} \times \ldots \times p_{t+2}^{g(c_i)}.$$

Exercise 3.3.7 Calculate the Gödel numbers of the configurations of P with input $x = 1$.

3.3.8 The Kleene Normal Form Theorem

Lemma 3.3.1. *The relation $T(e, x, y)$ asserting "y is the code of a computation for Turing machine program P_e on input x running for y many steps" is primitive recursive.*

Proof. (This is an extended sketch, and the details should be filled in by the reader if they are dubious.) By examining the coding technique for g above, we can see that only primitive recursive functions are used, for example, multiplication. To decode a number y or e we use functions such as *exp* for finding the power of a prime number in the representation of y or e. We have already seen that *exp* is primitive recursive. Hence the decoding process is primitive recursive. □

The predicate in Lemma 3.3.1 is traditionally called *the Kleene T-predicate*, and is almost always denoted by "$T(e, x, y)$" because that is the notation Kleene used. This notational tradition of using Kleene's notation also extends to the "*s-m-n*-Theorem, an important theorem we meet in Chapter 5. We can enumerate the Turing machines by examining integers and seeing if they correspond to Gödel numbers of Turing machines. Then we can record the first one, the second one etc.

Definition 3.3.4. *We let φ_e denote the function computed by the e-th machine on the list.*

Theorem 3.3.8 (Kleene Normal Form Theorem). *There exists a primitive recursive relation $T(e, x, y)$ and a primitive recursive function $U(y)$ such that:*

$$\varphi_e(x) = U(\mu y T(e, x, y)).$$

Proof. Let $T(e, x, y)$ be as in Lemma 3.3.1. To see whether $T(e, x, y)$ holds we first decode the index e and determine the set of quadruples that it represents. Then we recover the list of configurations $c_0, \ldots, c_t$ from the code y if in fact y does correspond to such a computation. Now we check that $c_0, \ldots, c_t$ is a computation according to P_e with x as input in c_0. If this is the case, the $U(y)$ outputs the number of 1's in the final configuration c_n. We have already seen that $T(e, x, y)$ is primitive recursive and it is not too difficult to see how to define U to be primitive recursive via bounded search and addition. □

It follows from the Normal Form Theorem that every Turing computable partial function is partial recursive. In fact, the proof shows that any partial recursive function can be obtained from two primitive recursive functions by *one* application of the μ-operator. The following is the promised proof of the existence of a universal Turing machine.

Theorem 3.3.9 (Enumeration Theorem). *There is a partial recursive function of two variables $\varphi_z^{(2)}(e, x)$ such that $\varphi_z^{(2)}(e, x) = \varphi_e(x)$. (Indeed, the Enumeration Theorem holds for partial recursive functions of n variables.)*

Proof. Let $\varphi_z^{(2)}(e, x) = U(\mu y T(e, x, y))$. □

We call $\varphi_z^{(2)}(e, x)$ a universal Turing machine because it can simulate the computation of any other Turing machine. An informal proof of the Enumeration Theorem is: a program P_z given inputs e and x effectively recovers the program P_e and runs it on input x until, if ever, an output is obtained.

It is fascinating that the model of partial recursive functions were found *first*, and it was only after Turing's model that people became convinced by Church's now Church-Turing Thesis[4].

What we will do in the next couple of section is look at one further other model. There are two good reasons for this.

- First, remember our goal is to look at computability of problems in "normal" mathematics. To do this we would like to code problems from models of computation into the mathematical objects being discussed. Some models are easier than others. For instance, later we will look at Hilbert's 10th problem and this was originally done using partial recursive functions, and later (as we see) using Register Machines. To my knowledge, there is no proof using Turing machines, and it has been speculated that the the replacement of formal models such as partial recursive functions by informal models appealing to the Church-Turing Thesis, was an impediment to the discovery of the proof.
- Second, we see other models of computation, which, like varying programming languages, can seem more appropriate for considering computation in different settings. More importantly, we prove these other models compute

---

[4] The reader might like to have a look at the articles in [Dow14] and [Her95] or discussion of the evolution of these ideas.

the same class of partial functions as the partial computable functions, and
hence support the Church-Turing Thesis.

3.4 Minsky and Register Machines

Here we will introduce yet another model of computation. This model is very
useful in modelling complexity (questions of size and running times) as it
is more faithful to the idea of a RAM (Random Access Machine, the kind
that modern computers are). Generally if we wish to show that, for example,
multiplication takes a certain time for n-bit numbers, we would choose a
model like a register machine rather than a Turing machine. The point is that,
while Turing machines and register machines have the same computational
power, there is a certain overhead of simulating one by another. Thus what
might take a linear number of steps on register machine might take, say,
cubic time on a Turing machine. This is because adding one to a register
takes one step on a Register Machine, whereas a Turing Machine would need
to set aside a part of the tape to record the contents of the register, move to
that part and add one to the block and then move back. This involves lots
of criss-crossing the tape and hence *many* steps. We remark in passing that
the overhead in this simulation is polynomial at worst, and can be done so
that one register machine step corresponds to about a quadratic number of
Turing Machine steps. We will discuss this point more later in Chapter 7.

Definition 3.4.1. A Minsky Machine or (**register machine**) is a
finite set of registers $\langle R_1, \ldots, R_n \rangle$ with a finite list of numbered instructions, each being one of the following.

- R_i^+, (i.e. $R_i \leftarrow R_i + 1$) go to $\langle$line number$\rangle$
- R_i^-, Leave R_i as it is (i.e. $R_i \leftarrow R_i$) if the contents of $R_i = 0$ go to
  $\langle$line number$\rangle$

  $R_i \leftarrow R_i - 1$ if the contents of R_i is > 0 go to $\langle$line number$\rangle$
- Halt

The *interpretation* here is that, for example, "R_i^+, go to j" means "add
one to the contents of register i and then go to line j, and that the two "R_i^-"
instructions first see if the contents of the register are bigger than 0, if so
reduce the register's content by 1, and go to line j_1, else leave the register
alone and go to j_2.

Convention 3.4.1 We will use the following useful convention: We will not
allow "go to $\langle 0 \rangle$ as part of any instruction. Thus, for example, "0. R_1^+ go

to 0" would be invalid as a first line. This makes certain proofs simpler in Chapter 4. (See also Lemma 3.4.1.)

A register machine simulates a partial function $f : \mathbb{N} \to \mathbb{N}$ if when started on line 0 with $\langle x, 0, 0, \ldots, 0 \rangle$ it produces $\langle f(x), 0, 0, \ldots, 0 \rangle$ if and when it reaches the Halt line.

Example 3.4.1. $f(x) = 2x$

0. R_1^- if $= 0$ go to 6
    if > 0 go to 1
1. R_2^+ go to 2
2. R_2^+ go to 3
3. R_1^- if $= 0$ go to 4
    if > 0 go to 1
4. R_2^- if $= 0$ go to 6
    if > 0 go to 5
5. R_1^+ go to 4
6. Halt

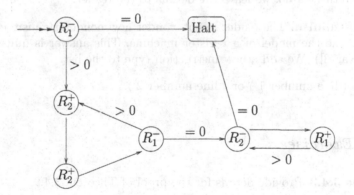

The convention implicit in the following result will be useful in Chapter 4

Lemma 3.4.1. *A partial function f can be simulated by a Minsky Machine M iff f can be simulated by a Minsky Machine N, such that, for $n \in \ \text{dom}(f)$, $\langle f(n), 0, \ldots, 0 \rangle$ is the first occurrence of the registers of N being the form $\langle p, 0, \ldots, 0 \rangle$, when N is run on input $\langle n, 0, \ldots, 0 \rangle$ for at least one step.*

Proof. (Sketch) Suppose that M is given. We need to construct N. Add an extra register R_p addition to those of M. The first action after we execute line 0 of M is to move to a new line $\langle b \rangle$ added to the list, which requests that we add one to R_p, and then return to executing M. If we were to hit the Halt line in M, then we will instead first move to line $\langle b+1 \rangle$ which first asks us to subtract one from register R_p and the go to the Halting line of M.   □

As with partial recursive functions, again we arrive at the same class of functions as we did with Turing Machines.

Theorem 3.4.2. *f is partial Turing computable iff f can be simulated by a register machine.*

Proof. We sketch the proof, and leave the details to the reader. The easiest model to simulate using register machines is the partial recursive functions. It is trivial to simulate zero, predecessor, successor. For projection, we can either use Gödel numbers to represent k-ary functions or define register machines to devote their first k registers for inputs $\langle x_1, \ldots, x_k \rangle$. In either case, projection is also trivial, or at least easy. Composition is straightforward, since we can start the second machine on the halt line of the first machine. This only leaves recursion and least number. For recursion we would devote a register for the parameter being recursed (i.e. used in applications of the recursion scheme) upon, and least number similarly, increasing the parameter when the machine returns a no. we leave the details to the reader.   □

Nondeterminism The reader might wonder how nondeterminism could be added to the the model of a register machine. This answer is quite simple (Karp [Kar73]). We add a new instruction type to the list.

- Go to $\langle$ line number 1 $\rangle$ or $\langle$ line number 2 $\rangle$.

3.4.1 Exercises

Exercise 3.4.3 Provide details for the proof of Theorem 3.4.2.

Chapter 4
Undecidable Problems

Abstract We prove a number of natural problems are undecidable. We do this by coding the halting problem into them. These problems include Conway's generalization of the Collatz function, word problems in formal languages, the Entscheidungsproblem, word problems in semigroups and groups, and we finish with a proof of the undecidability of Hilbert's 10th Problem for exponential Diophantine equations.

Key words: Conway functions, halting problem, undecidability, Hilbert's 10th Problem, word problems

4.1 Introduction

In Chapter 3, we introduced several models of computation and showed them to generate the same class of partial functions we regard as computable. We will regard the Church-Turing Thesis as true, at least for the results of this chapter. The goal of this chapter is to use the mathematical definition of computability to show that a number of natural problems are algorithmically undecidable.

Definition 4.1.1. Let $A \subseteq \mathbb{N}$ code a mathematical problem. That is, for a decision problem P, $n \in A$ iff n codes a *yes* instance of P; and $n \notin A$ if $P(n)$ is a *no* instance.

1. Then we say that P is *(algorithmically) decidable* iff there is a Turing Machine such that the characteristic function of A is computable (and total).
2. If the characteristic function of A is computable, we say that A is a computable set, or computable language, depending on context.

© The Author(s), under exclusive license to Springer Nature Switzerland AG 2024
R. Downey, *Computability and Complexity*, Undergraduate Topics
in Computer Science, https://doi.org/10.1007/978-3-031-53744-8_4

3. If P is not algorithmically decidable, we say that P is *undecidable* (sometimes called *incomputable* in the literature).

Convention 4.1.1 Henceforth we will not distinguish between a set (language) and its characteristic function.

Notice that every problem with only a finite number of *yes* instances, or a finite number of *no* instances is decidable. For example, if the instances were a finite number of yes ones at $\{k_1, \ldots, k_m\}$, then the algorithm for the problem P coded by $\{k_1, \ldots, k_m\}$ would be

$$M_{\langle k_1, \ldots, k_m \rangle}(x) = \begin{cases} 1 \text{ if } x \in \{k_1, \ldots, k_m\} \\ 0 \text{ otherwise.} \end{cases}$$

The reader should be careful to differentiate between

- There *exists* an algorithm, and
- We *know* the algorithm.

If we only know the set is finite, there *exists* an algorithm, but we only know the algorithm if we *know* the elements of the set. So we are saying something "can be decided" although we don't actually know the algorithm. There is a branch of logic where this kind of reasoning is not allowed, called *constructive* mathematics. In that area only the latter kind of algorithm would be accepted. If we used only constructive reasoning we would argue, for instance, that something should only be a theorem if we can prove it to be true, and it is not enough to have a proof that it is not false. Whilst it might seem strange to hamper developing mathematics by denying the Law of Excluded Middle, constructive mathematics is important in software engineering, for example, where we want to extract programs from proofs. We refer the reader to, for example, Beeson [Bee85], for a logical treatment of constructivism, or to Poernomo, Crossley and Wirsing [PCW05] (amongst many books) for recent work on "proof extraction", which is the extraction of programmes from proofs using something called the "Curry-Howard Correspondence".

We invite the reader to prove several algorithmic problems relating to regular languages are decidable with Exercise 4.1.3 below.

4.1.1 Exercises

Exercise 4.1.2 Suppose that A is a set whose characteristic function is primitive recursive. Show that A is computable.

Exercise 4.1.3 Suppose that L, L_1, and L_2 are finite state. Prove that the following are decidable if we are given finite state automata M, M_1, and M_2 with $L = L(M), L_1 = L(M_1)$, and $L_2 = L(M_2)$.

(a) Is $L = \emptyset$?
(b) Is $L = \Sigma^*$?
(c) Is $L_1 \subseteq L_2$?
(d) Is $L_1 = L_2$?

4.1.2 The Halting Problem

The method we use to prove natural algorithmic problems P are undecidable is called the *method of reduction*. What we do is start with some undecidable problem Q. We show that for each instance x of Q we can produce an instance $f(x)$ of P, in such a way that $f(x)$ is a yes instance of P iff x is a yes instance of Q. Therefore, if we could algorithmically decide P, we would be able to algorithmically decide Q, which we can't because Q is undecidable. Therefore P is also undecidable.

The simplest incarnation of such reductions are called m-reductions.

Definition 4.1.2 (m-reduction).
Let $A, B \subset \mathbb{N}$. We say that $A \leqslant_m B$, "A is m-reducible to B" iff $A \in \{\emptyset, \mathbb{N}\}$, or there is a computable (total) function f such that for all $x \in \mathbb{N}$,

$$x \in A \text{ iff } f(x) \in B.$$

Again we will drop the trivial cases of $A \in \{\emptyset, \mathbb{N}\}$, as we did in Chapter 1.
In Definition 4.1.2, we think of A as coding the problem Q and B as coding P in the explanation above. The following Lemma is immediate.

Lemma 4.1.1. *If B is computable and $A \leqslant_m B$, then A is computable.*

Proof. Given x to decide if $x \in A$ compute $f(x)$ and see if $f(x) \in B$. If yes, $x \in A$, if no, $x \notin A$. $\square$

You have certainly met the idea of a reduction in mathematics. Given an $n \times n$ matrix M, you know that to determine if it has an inverse, you convert this problem into calculating the determinant of M, $\det(M)$, and seeing it it is non-zero. The name "m-reduction" may seem strange, but it stands for "many-one reduction". As with the matrix case, several matrices might give the same determinant, so many go to one. If there is a 1-1 computable function f such that $x \in A$ iff $f(x) \in B$, then we can say that $A \leqslant_1 B$, but this refined reducibility plays no explicit part in the present book.

Now to complete our preliminaries we need some core problem to reduce *from*.

Definition 4.1.3 (Halting Problem).
The following problem is algorithmically undecidable. Recall from the existence of a Universal Turing Machine, Theorem 3.3.8, that we have an enumeration

$$\{\varphi_e : e \in \mathbb{N}\}$$

of all partial computable functions.

HALTING PROBLEM
Input: $\langle x, y \rangle \in \mathbb{N} \times \mathbb{N}$.
Question: Does $\varphi_x(y) \downarrow$?

Theorem 4.1.4 (Gödel (in some sense), Turing [Tur36]). *The halting problem is undecidable.*

Proof. The proof resembles the proof that the primitive recursive functions do not exhaust the total computable functions. It is a diagonalization. Towards a contradiction, suppose that the Halting Problem was decidable. Then we would have an algorithm showing that the set A was computable: where $\langle x, y \rangle \in A$ iff $\varphi_x(y) \downarrow$. Now we define a partial computable function g as follows:

$$g(x) = \begin{cases} 1 \text{ if } \varphi_x(x) \uparrow \\ \varphi_x(x) + 1 \text{ if } \varphi_x(x) \downarrow . \end{cases}$$

Then g would be a total and computable function. To compute $g(x)$ see if $\langle x, x \rangle \in A$. We have an algorithm for this by assumption. If the answer is no, then $g(x) = 1$. If the answer is yes, we know that $\varphi_x(x) \downarrow$, and hence we can use the Universal Turing Machine, to run $\varphi_x(x)[s]$ over stages s until it halts. When we find the answer, we add one.

But now we arrive at a contradiction. Since g is computable, for some z, $g = \varphi_z$. But then $g(z) \downarrow$ as g is total, and hence $\varphi_z(z) \downarrow = g(z) = \varphi_z(z) + 1 = g(z) + 1$, so that $0 = 1$ a contradiction!

Therefore the algorithm deciding the halting problem cannot exist. □

Actually the proof above yields the following sharper result. Not only can't we decide whether $\varphi_x(y) \downarrow$ for *arbitrary* $\langle x, y \rangle$ but we can't even decide this for $x = y$. In Theorem 5.1.5 we will prove that both problems are more or less the same, as they have the same m-degree.

Corollary 4.1.1. *There is no algorithm to decide if $\varphi_x(x) \downarrow$.*

Since all the models of computation are the same, there are similar results for the partial recursive functions and the register machines. That is, for instance, there is no algorithm to decide if a given register machine on input $\langle n, 0, \ldots, 0 \rangle$ halts.

Remark 4.1.1. The reader will note that we have been a wee bit naughty in referring to *the* halting problem, when which pairs $\langle x, y \rangle$ have $\varphi_x(y) \downarrow$ *thoroughly depends on the choice of universal Turing Machine.* The point is that it is irrelevant *which enumeration we choose,* since *any* will yield an undecidable problem. In fact it is possible to show that all halting problems are m-reducible to each other[1], so more-or-less the same problem, and we'll see how to do this in Chapter 5.

4.1.3 Exercises

Exercise 4.1.5 Let $f, g : \mathbb{N} \to \mathbb{N}$, be partial functions. We say that g *extends* f if for all $x \in \mathbb{N}$, if $x \in \text{dom}(f)$, then $x \in \text{dom}(g)$. (That is, if $f(x) \downarrow$ then $g(x) \downarrow = f(x)$ if they are both partial computable. It may be that $g(y) \downarrow$ for some y with $f(y) \uparrow$.)

Show that there is a partial computable function which cannot be extended to a (total) computable function. (Hint: Think about the proof of the algorithmic undecidability of the halting problem.)

Exercise 4.1.6 Use the method of the proof of the undecidability of the halting problem to show that there is no computable function f such that

$$\{\varphi_{f(e)} : e \in \mathbb{N}\}$$

lists all *and only* the total computable functions.

Exercise 4.1.7 Suppose that $A \leqslant_m B$ and $B \leqslant_m C$. Show that $A \leqslant_m C$.

Exercise 4.1.8 Define $A \oplus B = \{z : (z = 2x \text{ and } x \in A) \text{ or } (z = 2y + 1 \text{ and } y \in B)\}$.

Show that if $C \leqslant_m A_1 \oplus A_2$ then there exist disjoint C_1, C_2 with $C = C_1 \sqcup C_2$ and $C_1 \leqslant_m A_1$ and $C_2 \leqslant_m C_2$.

Exercise 4.1.9 Suppose that $A \leqslant_m B_1 \times B_2$. Show that if B_1 is finite, $A \leqslant_m B_2$.

---

[1] Relative to what are called *acceptable* enumerations, which has a technical definition we will suppress.

4.2 Minsky Machines and Generalized Collatz Functions

The goal of this chapter is to look at a fascinating undecidability result from
the 1970's. The reduction will be based on the HALTING PROBLEM but coded
using Register Machines. Register Machines are often quite good models for
proving undecidability as they have few "moving parts". They are also good
for analysing the actual complexity (number of steps to compute) things
like multiplication, as we see in Chapter 7, as a Register Machine's action
resembles a random access machine.

In this section we look at a generalization of a classical number theoretical
question.

4.2.1 Collatz functions

Definition 4.2.1. The *Collatz function* is defined as follows:

$$g(n) = \begin{cases} n/2 & \text{if } n \text{ even} \\ 3n+1 & \text{if } n \text{ odd} \end{cases}$$

We consider iterates of g applied to itself. That is, consider

$$g(n), g(g(n)), g(g(g(n))), \ldots, g^{(k)}(n), \ldots.$$

Lothar Collatz conjectures the following:

Conjecture 4.2.1 (Collatz). $\forall n \, \exists k \; g^{(k)}(n) = 1$. Equivalently, that $g^{(t)}(n)$ is a
power of 2, for some t.

The famous number theorist Paul Erdös said of the Collatz problem:
"Mathematics may not be ready for such problems.[2]" An apparently simpler question is the following. This question clearly has a positive solution if
the Collatz conjecture has a positive solution, since $C = \mathbb{N}^+$ in that case.

Question 4.2.1. Is the set $C = \{n | \exists k \; g^{(k)}(n) = 1\}$ computable?

To my taste, the Collatz conjecture is somewhat *ad hoc*, although apparently very difficult. However, there is a remarkable generalization of this
problem due to John Conway.

---

[2] Although Terry Tao [Tao22] has recently made substantial progress on the problem using
probabilistic methods.

Definition 4.2.2 (Conway's generalization of Collatz Functions-Conway [Con72]).
Conway's generalisation is to give $d \in \mathbb{N}$, $\mathbf{a} = a_0 \ldots a_{d-1}$, $\mathbf{b} = b_0 \ldots b_{d-1}$
with $a_i, b_i \in \mathbb{Q}^+$ *and define*

$$g_{\langle \mathbf{a},\mathbf{b} \rangle}(n) = \widehat{g}(n) = a_i n + b_i \text{ if } n \equiv i (\mathrm{mod}\ d).$$

Here we choose d and the rationals $\{a_i, b_i : i = 0, \ldots, d-1\}$ so that g is always a positive integer. For example, for the original Collatz function, $d = 2$, $a_0 = \frac{1}{2}, b_0 = 0$ and $a_1 = 3, b_1 = 1$. Conway asked the following strange question:

Question 4.2.2. What, if anything, can be said of problems with input $\langle a_0, \ldots, a_{d-1}, b_0, \ldots b_{d-1} \rangle$. For example, is the set of n such that $\exists k \ \widehat{g}^{(k)}(n)$ is a power of 2 a computable set?

Even more, consider $\widehat{g} = a_i n$ if $n \equiv i (\mathrm{mod}\ d)$ so all the $b_i = 0$. What can we say about such systems? Remarkably, Conway proved:

This is another universal model of computation!

Theorem 4.2.1 (Conway). *If* φ *is partial computable, then there is a* $\widehat{g}$, *as defined above, such that*

(i). $\forall n \in dom\,\varphi$
        $2^{\varphi(n)}$ *is the first power of 2 in* $\widehat{g}(2^n), \widehat{g}^{(2)}(2^n), \ldots$
(ii). $\forall n \notin dom\,\varphi$
        there is no power of 2 in $\widehat{g}(2^n), \widehat{g}^{(2)}(2^n), \ldots$

Corollary 4.2.1. *Hence, there is a choice of rationals* $\{a_0, \ldots, a_{d-1}\}$ *such that there is no algorithm to decide the following problem:*

Input: n.
Question: *Does there exist a* k *such that* $\widehat{g}^{(k)}(2^n)$ *is a power of 2?*

Conway's theorem is proven using a series of m-reductions. We define several new algorithmic games, and show Minsky machines reduce to VECTOR GAMES which reduce to RATIONAL GAMES and these to GENERALISED COLLATZ. Hence, in particular, the halting problem for Minsky Machines reduces to the power of two question for GENERALIZED COLLATZ FUNCTIONS; meaning that this question is undecidable.

In fact, we show that these reductions are so faithful that they are *equivalences* in the same way that Turing computable is equivalent to partial recursive for partial functions. That is, we show that each game can be used as a universal programming language[3].

4.2.2 Vector games

Definition 4.2.3. A VECTOR GAME is a finite ordered list of vectors $L = \{v_1, \ldots, v_n\} \subseteq \mathbb{Z}^d$.
   To play the game:
Take as input some vector $v \in \mathbb{N}^d$ (i.e. all entries non-negative).
Define $g_L(v) = v + v_i$, where i is least so that $v + v_i$ is non-negative, that is $v + v_i \in \mathbb{N}^d$.
Compute $g_L(v), g_L^{(2)}(v), \ldots$, etc.

Convention 4.2.2 The game will stop if we can't find any such vector. This condition will be understood in all the games below.

We have defined the action of the game, and, like Turing Machines and Register Machines, now we need to define what we mean by a game simulating a function.

Definition 4.2.4. A game g_L *simulates* f if on input $\langle k, 0, \ldots, 0 \rangle$

(i). $\forall k \in \text{dom } f$
        $\langle f(k), 0, \ldots, 0 \rangle$ is the first vector of form $\langle k', 0, \ldots, 0 \rangle$ in the
    sequence $g_L(\langle k, 0, \ldots, 0 \rangle), g_L^{(2)}(\langle k, 0, \ldots, 0 \rangle), \ldots$.
(ii). $\forall k \notin \text{dom } f$
        there never appears a vector of form $\langle k', 0, \ldots, 0 \rangle$ in the sequence
    $g_L(\langle k, 0, \ldots, 0 \rangle), g_L^{(2)}(\langle k, 0, \ldots, 0 \rangle), \ldots$.

Convention 4.2.3 We will be using Register Machines of the form of Lemma 3.4.1, where they only simulate a function when the first occurrence of $\langle m, 0, \ldots, 0 \rangle$ it its execution on input $\langle n, 0, \ldots, 0 \rangle$ is $\langle f(n), 0, \ldots, 0 \rangle$.

For reasons that soon become apparent, we will distinguish the last two entries by a vertical bar |. With the simulation of Theorem 4.2.4, they are there to distinguish the parts of the vectors corresponding to line numbers in the register machines.

---

[3] Which Conway called FRACTRAN.

Example 4.2.1.

$$L : \langle 0, 0, 0 | 1, -1 \rangle = v_1$$
$$\langle -1, 0, 0 | -3, 1 \rangle = v_2$$
$$\langle 0, 0, 0 | -3, 0 \rangle = v_3$$
$$\langle 0, 0, 1 | -2, 3 \rangle = v_4$$
$$\langle 0, 1, 0 | -1, 2 \rangle = v_5$$
$$\langle -1, 0, 0 | 0, 1 \rangle = v_6$$

So, on input $x = \langle 3, 0, 0 | 0, 0 \rangle$, we get the sequence

$$g(x) \quad = \langle 2, 0, 0 | 0, 1 \rangle \qquad (v_6)$$
$$g^{(2)}(x) = \langle 2, 0, 0 | 1, 0 \rangle \qquad (v_1)$$
$$g^{(3)}(x) = \langle 2, 1, 0 | 0, 2 \rangle \qquad (v_5)$$
$$g^{(4)}(x) = \langle 2, 1, 0 | 1, 1 \rangle \qquad (v_1)$$
$$g^{(5)}(x) = \langle 2, 1, 1 | 2, 0 \rangle \qquad (v_1)$$
$$g^{(6)}(x) = \langle 2, 1, 1 | 0, 3 \rangle \qquad (v_4)$$

... etc.

Theorem 4.2.4 (Conway [Con72]).

- *For any partial computable φ there is a list of vectors L yielding a vector game g_L such that the partial function simulated by the game g_L is exactly φ.*
- *Hence* VECTOR GAMES *are a universal programming language.*

Proof. Given any partial computable φ, there is a Minsky Machine to compute it. We can then transform this machine into a vector game as follows:

Suppose we have a Minsky Machine with n registers, then we construct vectors of arity $n + 2$.

For $1 \leqslant i \leqslant n$, the ith position in each vector corresponds to addition/subtraction on register R_i.

The $n + 1^{\text{st}}$ position corresponds to the current Minsky Machine instruction, and the $n + 2^{\text{nd}}$ position corresponds to the 'next' Minsky Machine instruction.

To construct L, first re-order the line numbers of the Minsky Machine to begin with line 0 (this makes the simulation work), put $v_0 = \langle 0, \ldots, 0 | 1, -1 \rangle$, and then derive the other v_i from the Minsky Machine, ordering them in *decreasing order of k.*

Let $L = \{v_0, \ldots, v_n\}$. If instruction k of the Register machine is R_i^+ and go to line t, then this is represented by

$$v_j = \langle 0, 0, \ldots, \underset{\text{position } i}{1}, \ldots, 0 | -k, t \rangle.$$

(Here, the vertical line here has no meaning beyond distinguishing the last two positions in the vector.)

If the instruction at line k is R_i^- ; if > 0 go to line t_1, and if $= 0$ go to line t_2. Then this instruction is represented by two vectors:

$$v_j = \langle 0, 0, \ldots, \underset{\text{position } i}{-1}, \ldots, 0| - k, t_1 \rangle,$$

placed on the list above

$$v_{j+1} = \langle 0, 0, \ldots, 0, \ldots, 0| - k, t_2 \rangle.$$

The halt instruction which occurs at line h is represented by $\langle 0, \ldots, 0| - h, 0 \rangle$.

Example 4.2.2. Before we give a formal proof that this works, we pause for an example; our old friend $f(x) = 2x$. Recall that this had the Minsky program

0. R_1^- if $= 0$ go to 6
       if > 0 go to 1
1. R_2^+ go to 2
2. R_2^+ go to 3
3. R_1^- if $= 0$ go to 4
       if > 0 go to 1
4. R_2^- if $= 0$ go to 6
       if > 0 go to 5
5. R_1^+ go to 4
6. Halt

The corresponding Minsky Machine is the following:

$$v_0 = \langle 0, 0|1, -1 \rangle$$
$$v_1 = \langle 0, 0| - 6, 0 \rangle$$
$$v_2 = \langle 1, 0| - 5, 4 \rangle$$
$$v_3 = \langle 0, -1| - 4, 5 \rangle$$
$$v_4 = \langle 0, 0| - 4, 6 \rangle$$
$$v_5 = \langle -1, 0| - 3, 1 \rangle$$
$$v_6 = \langle 0, 0| - 3, 4 \rangle$$
$$v_7 = \langle 0, 1| - 2, 3 \rangle$$
$$v_8 = \langle 0, 1| - 1, 2 \rangle$$
$$v_9 = \langle -1, 0|0, 1 \rangle$$
$$v_{10} = \langle 0, 0|0, 6 \rangle$$

Returning to the proof, we prove by induction that each Minsky Machine is emulated by by a vector game constructed in this way.

A register machine emulates a function f in a finite number of steps, from $\langle x, 0, 0, ..., 0 \rangle$ to $\langle f(x), 0, 0, ..., 0 \rangle$. An equivalent vector game will convert $\langle x, 0, 0, ..., 0 \,|\, 0, 0 \rangle$ to $\langle f(x), 0, 0, ..., 0 \,|\, 0, 0 \rangle$. Suppose our Minsky machine is M and we construct a vector game G from it.

Let $\mathbf{v}$ be the "state" of our vector game. Immediately after a vector $\mathbf{u}$ in G corresponding to an instruction of M is added to $\mathbf{v}$, so that the new $\mathbf{v} = \langle x_1, x_2, ..., x_k \,|\, 0, a \rangle$, a is the number of the instruction which the register machine would have executed after executing the instruction corresponding to $\mathbf{u}$.

Now we can move to the details. Suppose that $S \subseteq \mathbb{N}$ and $S = \{n \in \mathbb{N} : R(n)\}$, where R is the relation which evaluates to "true" if after n steps in the vector game corresponding to the first n steps of the register machine, the states of the register machine and the vector game are equivalent.

Base case: $n = 0$. At 0 steps of the register machine, we have the state: line 0 is next, and registers $\langle x, 0, ..., 0 \rangle$; the vector game starts with state vector $\mathbf{v} = \langle x, 0, ..., 0 \,|\, 0, 0 \rangle$. So the states are equivalent, and thus we see that $0 \in S$.

Hypothesis: suppose that $k \in S$, so after k steps in the vector game, corresponding to the first k steps of execution of M, the states of the two are equivalent. Suppose the state of the register machine is $\langle x_1, x_2, ..., x_r \rangle$ with next instruction c, and the state of G is $\mathbf{v} = \langle x_1, x_2, ..., x_r \,|\, 0, c \rangle$.

Since M was used to construct G, what happens next in G depends on what would happen next in M. Observe that the first vector in G will be added in $\mathbf{v}$ c times to get a "new" $\mathbf{v} = \langle x_1, x_2, ..., x_3 \,|\, c, 0 \rangle$. Then there are 3 cases:

- Instruction c of M is "R_i+, go to line a". By the construction of G, after the first vector
$$\mathbf{v_0} = \langle 0, 0, ..., 0 \,|\, 1, -1 \rangle$$
we get vectors of form $\langle ... \,|\, -p, q \rangle$, ranked in decreasing order of p. The vector corresponding to instruction c is $\mathbf{u} = \langle 0, 0, ..., 0, 1, 0, ..., 0 \,|\, -c, a \rangle$, with a 1 in position i. The vector game finds the first vector after $\mathbf{v_0}$ which can be added to $\mathbf{v}$ to keep all its components non-negative. All of the vectors (aside from $\mathbf{v_0}$) before $\mathbf{u}$ have $p > c$, so they can't be selected; the first applicable vector is therefore $\mathbf{u}$.
$\mathbf{v} + \mathbf{u}$ gives $\langle x_1, ..., x_{i-1}, x_i + 1, x_{i+1}, ..., x_r \,|\, 0, a \rangle$, and this corresponds to the registers of M containing $\langle x_1, ..., x_{i-1}, x_i + 1, x_{i+1}, ..., x_r \rangle$ and next instruction a, which is consistent with executing instruction c of M.

- Suppose instruction c of M is "R_j-, and if > 0 go to a, else go to b"; this gives vectors of G: $\mathbf{u} = \langle 0, ..., 0, -1, 0, ..., 0 \,|\, -c, a \rangle$ (the -1 is in position j), and $\mathbf{u'} = \langle 0, 0, ..., 0 \,|\, -c, b \rangle$. The vectors appear in G in this order. Suppose $x_j > 0$ in $\mathbf{v}$. Then, as in the previous case, searching for the first applicable vector in G (after transforming $\mathbf{v}$ to $\langle x_1, ..., x_r \,|\, c, 0 \rangle$) is

u. $\mathbf{v} + \mathbf{u}$ gives $\langle x_1, ..., x_{j-1}, x_j - 1, x_{j+1}, ..., x_r \mid 0, a \rangle$ — this is consistent with the state of M after executing instruction c: next instruction a, state $\langle x_1, ..., x_j - 1, ..., x_r \rangle$.

If $x_j = 0$, **u** can't be applied, since then the value in position j of the vector would be -1. By the construction of G, **u**$'$ is next in line, which can be applied, giving a new vector $\langle x_1, ..., x_r \mid 0, b \rangle$. In this case, the state of M after executing instruction c will be: registers $\langle x_1, ..., x_r \rangle$, next instruction b, which is equivalent.

- Instruction h of M is "Halt". This gets translated to $\mathbf{u} = \langle 0, 0, ..., 0 \mid -h, 0 \rangle$. By the same reasoning as above, **v** becomes $\mathbf{u} + \mathbf{v} = \langle x_1, ..., x_r \mid 0, 0 \rangle$. M emulates a function f, so when it executes instruction h, it must be that $x_2 = x_3 = ... = x_r = 0$, and $x_1 = f(x)$. Then by the definition of vector games, since the vector $\mathbf{v} + \mathbf{u}$ must have form $\langle f(x), 0, ..., 0 \mid 0, 0 \rangle$, the vector game finishes, with answer $f(x) = x_1$. So both G and M terminate with the same answer.

So therefore, the vector game emulates the Minsky machine at step $k+1$ of M, and $k \in S \Rightarrow k+1 \in S$. Thus, by the principle of mathematical induction, $S = \mathbb{N}$ and so the vector game G emulates the Minsky machine M. □

4.2.3 Rational Games

The next step of the equivalences concerns a simple translation into another game. Remember that, multiplying the same number to different powers is the same as adding exponents, and so, for example $5^n 5^m = 5^{n+m}$ and more generally, $(2^x 3^y) \cdot (2^z 3^m) = 2^{x+z} 3^{y+m}$. That is, when we multiply numbers which are the same prime to two exponents, the multiplication translates as addition of exponents. This observation is the motivation of the next game.

Definition 4.2.5. A RATIONAL GAME is a finite ordered list of nonzero rationals $L = \{r_0, \ldots, r_k\}$.
    To play the game:
Take as input some $n \in \mathbb{N}$.
Define $g_L(n) = r_i n$, where i is least so that $r_i n \in \mathbb{N}$.
Compute $g_L(n), g_L^{(2)}(n), \ldots$ etc.

Definition 4.2.6. A game g_L *simulates* a partial function φ if on input 2^n

(i). $\forall n \in \operatorname{dom} \varphi$

$2^{\varphi(n)}$ is the first power of 2 in $g_L(2^n), g_L^{(2)}(2^n), \dots$

(ii). $\forall n \notin \operatorname{dom} \varphi$

there is no power of 2 in $g_L(2^n), g_L^{(2)}(2^n), \dots$

We can emulate vector games by rational games, where $v_i = \langle a, b, c, \dots \rangle$ is replaced by $r_i = 2^a 3^b 5^c \dots$

Example 4.2.3. $\langle 0, 0, 0 | 1, -1 \rangle$ becomes $2^0 3^0 5^0 7^1 11^{-1} = \frac{7}{11}$.

Theorem 4.2.5 (Conway [Con72]). *Every function simulated by a vector game can be simulated by an equivalent rational game.*

Proof. First we show that every Vector Game can be emulated by a Rational Game. For a vector game of arity k, let $2, 3, 5, \dots, p_k$ be the first k prime numbers Then, for all vector $v_i = \langle v_{i1}, v_{i2}, \dots, v_{ik} \rangle$ in the game, encode as $2^{v_{i1}} 3^{v_{i2}} \cdots p_k^{v_{ik}} = r_i$, hence preserving order of the vectors in the order of the rationals. So the input vector $v = \langle x, 0, \dots, 0 \rangle$ translates to 2^x, as expected by the rational game. Moreover, the output vector $v = \langle f(x), 0, \dots, 0 \rangle$ translates to $2^{f(x)}$ as expected also, for all $x \in \operatorname{dom} f$. Choosing the first vector v_i in order to preserve $\langle v_1, \dots, v_k \rangle \in \mathbb{Z}^+$ corresponds to choosing the first rational to preserve $2^{v_1} 3^{v_2} \cdots p_k^{v_k} \in \mathbb{N}^+$.

Every Rational Game can be simulated by a Vector Game: Find the highest prime needed for representation of a rational in the rational game, say p_k. Then make a vector game of arity k by constructing a vector v_i from each r_i as follows:

Let $r_i = 2^{r_{1i}} 3^{r_{2i}} \cdots p_k^{r_{ki}}$, then $v_i = \langle r_{1i}, r_{2i}, \dots, r_{ki} \rangle$.

Then the input rational $r = 2^x$ corresponds to the vector $\langle x, 0, \dots, 0 \rangle$ as expected by the vector game. And the output rational $r = 2^{f(x)}$ corresponds to the vector $\langle f(x), 0, \dots, 0 \rangle$ as expected also, for all $x \in \operatorname{dom} f$. Choosing the first rational to preserve $2^{r_{1i}} 3^{r_{2i}} \cdots p_k^{r_{ki}} \in \mathbb{N}^+$ corresponds to choosing the first vector to preserve $\langle r_{1i}, r_{2i}, \dots, r_{ki} \rangle \in \mathbb{Z}^k$. $\square$

We remark that it is unusual to see Gödel numbers used with *negative* exponents.

4.2.4 Generalized Collatz functions

Thus our final task is to demonstrate that rational games can be translated into a system of congruences for a suitably chosen p.

Lemma 4.2.1. *Given* $r_0, \ldots, r_n \in \mathbb{Q}^+$, $x \in \mathbb{N}^+$ *and a rational game* $f(x) = r_i x$ *(where* i *least such that* $r_i x \in \mathbb{N}^+$*), then there is a number* p *and there are pairwise disjoint sets* $D_1, D_2, \ldots, D_n$ *with*

$$\bigcup_{i=1}^n D_i = \{0, 1, \ldots, p-1\}$$

such that

$$f(x) = r_i x$$

where i *is uniquely determined by* $x \equiv y \pmod{p}$ *for some* $y \in D_i$.

That is, every Rational Game is step by step equivalent to a Generalised Collatz function.

Proof. From a rational game we can construct a generalised Collatz function as follows.

Let $r_i = \frac{a_i}{b_i}$, $a_i, b_i \in \mathbb{N}^+$, and $p = b_0 b_1 b_2 \cdots b_n$.

Note that $r_i x \in \mathbb{N}$ iff $b_i | a_i x$ iff $p | (p/b_i) a_i x$. Thus if $r_i x \in \mathbb{N}$ and $z \equiv x \pmod{p}$, then this happens iff $r_i z \in \mathbb{N}$.

So we can construct sets D_i by

$$D_0 = \{x < p : p | (\frac{p}{b_0}) a_0 x\}$$

$$D_{j+1} = \left\{ x < p : x \notin \bigcup_{i=1}^j D_i \text{ and } p \left| \frac{p}{b_{j+1}} a_{j+1} x \right. \right\}.$$

By construction, this system of congruences will step-by-step emulate the rational game. $\square$

Remark 4.2.1. So if we took the Minsky Machine we constructed earlier for $f(x) = 2x$, the vectors were all of the form $\langle i, j | k, m \rangle$ and the first vector was $\langle 0, 0 | 1, -1 \rangle$ giving the rational $\frac{5}{7}$. Calculating the value of p we get $p = 7 \cdot 5^6 \cdot 5^5 \cdot (3 \cdot 5^4) \cdot 5^4 \cdot (2 \cdot 5^3) \cdot 5^3 \cdot 5^2 \cdot 5 \cdot 2$ as the product of the derived denominators. Then, for example, $r_0 z \in \mathbb{N}$ iff $\frac{5}{7} z \in \mathbb{N}$ iff 7 is a factor of z iff p is a factor of $\frac{p}{7} 5z$ (as $a_0 = 5$ in this case). This means that x must be congruent to some $x < p$ in D_0 modulo p. D_1 then looks at eligible numbers not in D_0, similarly. A fully worked example can be found in the solutions to Exercise 4.2.6 below.

4.2.5 Exercises

Exercise 4.2.6 Construct a Minsky Machine for the function $f(x) = x^2 + 1$, turn it into a Vector Game, then into a Rational Game, and then into a generalized Collatz function.

Exercise 4.2.7 Construct a Minsky Machine for the function

$$f(x) = \begin{cases} 1 \text{ if } x \text{ is a power of 3} \\ \uparrow \text{ otherwise.} \end{cases}$$

turn it into a Vector Game, then into a Rational Game, and then into a generalized Collatz function.

Exercise 4.2.8 What games would the original Collatz function generate?

4.3 Unsolvable Word problems

In the last section, we looked at coding Register Machines into number-theoretical problems. In this section we will look at a class of problems called *word problems*, which have a general set up as follows: We have an algebraic system of some kind with some method of generating equivalences within that system. The elements of the system are called words. The algorithmic questions we look at are of the kind :

Can we transform one word into another?

These problems have quite a long history in sofar as undecidability results in "general mathematics" are concerned. Indeed the word problem for groups, which we meet in subsection 4.3.5, goes back to fundamental work of Dehn [Deh11] who studied what are called finitely presented groups[4] and asked whether there is an algorithm to decide if a given product of generators is equivalent to the identity in the group. Again, this question was in 1911, and pre-dated the development of computability theory.

4.3.1 Semi-Thue Processes

We begin by studying systems close to Turing Machines.

From Chapter 2.2, recall that given an alphabet $A = \{a_1, \ldots, a_k\}$, of *symbols* a finite sequence of elements of A is called *string* or a *word* on A. For this section, we will tend to refer to strings as words because of history. We will study what are called *semi-Thue* systems which are algebraic systems whose main action is substitution.

---

[4] Groups expressed as the quotient of a free group on a finite number of generators by a (free) finitely generated normal subgroup. More on this in §4.3.5.

Definition 4.3.1. A *semi-Thue production* on A is an ordered pair $\langle g, \bar{g} \rangle$ of words on A, written $g \to \bar{g}$. A *semi-Thue process* on A is a finite non-empty set Π of semi-Thue productions on A.

If p is the semi-Thue production $g \to \bar{g}$ then we write $u \Rightarrow_p v$ if for any words a, b we have that $u = agb$ and $v = a\bar{g}b$. This simply expresses the fact that we can substitute $\bar{g}$ for g in the word.

If Π is a semi-Thue process we write $u \Rightarrow_\Pi v$, if $u \Rightarrow_p v$ for some $p \in \Pi$. We write $u \Rightarrow_\Pi^* v$ if there is a finite sequence $u = u_1 \Rightarrow_\Pi u_2 \Rightarrow_\Pi \cdots \Rightarrow_\Pi u_n = v$, for $n \geqslant 1$. That is $\Rightarrow_\Pi^*$ is the transitive closure of $\Rightarrow_\Pi$. Note in particular that $u \Rightarrow_\Pi^* u$ for all words u.

Definition 4.3.2 (Word Problem for Semi-Thue Processes).
The *word problem* for a given semi-Thue process Π is the decision problem:

Input : Words u, v
Question : Does $u \Rightarrow^* v$?

Theorem 4.3.1 (Post [Pos47]). *There are semi-Thue systems whose word problems are unsolvable.*

Proof. To prove this, we show how to obtain from any Turing machine T a corresponding semi-Thue system $\Pi(T)$ such that an algorithm for solving the word problem for $\Pi(T)$ can be used to solve the halting problem for T. If T is a Turing Machine with states Here B is the blank symbol. $q_0, q_1, \cdots q_n$ and alphabet $\Sigma = \{0, 1\}$, then $\Pi(T)$ will be a semi-Thue process on the alphabet $A = \{B, 0, 1, q_0, q_1, \cdots q_n, q, q', h\}$. We refer to the symbols $q_0, q_1, \cdots q_n, q, q'$ as *q-symbols*, and B is the blank symbol.

The following echoes the definition of configurations we used in the proof that Turing Machines could be emulated by partial recursive functions in Definition 3.3.3.

Definition 4.3.3. A *Post word* is a word on A of form huq_ivh, where

$$u, v \text{ are words on } \Sigma$$
$$v \neq \lambda, \text{ and}$$
$$0 \leqslant i \leqslant n.$$

Thus, the successive configurations of T during a computation will be represented by Post words where q_i is the current instruction, the tape contents are uv, and the initial symbol of v is being scanned. The effect of T on successive configurations will be simulated by productions of $\Pi(T)$, which will have corresponding effects on the Post words.

The productions of $\Pi(T)$ are:

(i). For each quadruple $\langle q_i, S_i, \beta, q_{i,j} \rangle$ in T, (i.e., in state q_i, reading S_i, do $\beta \in \{S_{i,j}, L, R\}$, change to state q_j) add to $\Pi(T)$: Let the symbols of the Turing Machine be $\{S_1, \ldots, S_n\}$. The for each pair of symbols S_i, S_k, and quadruple $\langle q_i, S_j, S_k, q_{i,j} \rangle$, we would have:

$$q_i\, S_i \longrightarrow q_{i,j}\, S_k$$

(ii). For each quadruple $\langle q_i, S_j, R, q_{i,j} \rangle$ we would have, for each symbol S_d of T,

$$q_i\, S_j\, S_d \longrightarrow S_j\, q_{i,j}\, S_d$$
$$q_i\, S_j\, h \longrightarrow S_j\, q_{i,j}\, B\, h$$

Here B denotes the blank symbol.

(iii). For each quadruple $\langle q_i, S_j, L, q_{i,j} \rangle$ in T, and S_d, a symbol of T, add to $\Pi(T)$:

$$S_d\, q_i\, S_j \longrightarrow q_{i,j}\, S_d\, S_j$$
$$h\, q_i\, S_j \longrightarrow h\, q_{i,j}\, B\, S_j.$$

(iv). For each pair $\langle i, j \rangle$ where there is no quadruple in T beginning q_i, S_j, add to $\Pi(T)$:

$$q_i\, S_j \longrightarrow q\, S_j.$$

(This uses the convention that if I get to a place the machine has no corresponding quadruple for $q = q_{halt}$, it halts.)

(v). For each symbol S_d, always add to $\Pi(T)$:

$$q\, S_d \longrightarrow q$$
$$q\, h \longrightarrow q'\, h$$
$$S_d\, q' \longrightarrow q'$$

Each production added by (i), (ii) and (iii) causes changes in a Post word exactly corresponding to the effect on a machine configuration of applying the quadruple from the machine specified. Suppose $h\ u\ q_i\ S_j\ v\ h$ is a Post word corresponding to a given configuration of T, and that T contains, for example, the quadruple $\langle q_i, S_j, S_{i,j}, q_{i,j} \rangle$. Then we have

$$h\ u\ q_i\ S_j\ v\ h \Rightarrow_{\Pi(T)} h\ u\ q_{i,j}\ S_{i,j}\ v\ h$$

where $h\ u\ q_{i,j}\ S_{i,j}\ v\ h$ is in fact the Post word corresponding to the next configuration of T. The other kinds of quadruples translate similarly, remembering the h denotes the end markers for the portion of the tape addressed so far. The symbol q' is a kind of "eating" symbol which converts a halt state to a single symbol q'. This follows since each production added for (iv) operates on a Post word corresponding to a configuration of T when T has just been forced to halt. The production replaces instruction symbol q_i by q.

That is, the productions from (i), (ii), and (iii), the semi-Thue process emulates the actions of the Turing Machine until the Turing Machine halts. If the Turing Machine eventually halts, the productions from (iv) and (v) delete the representation of the tape contents until only $h\ q'\ h$ is left, giving us a single specific word to ask the word problem about.

Now, if T begins at q_0 scanning the leftmost symbol of x, the corresponding Post word is $h\ q_0\ x\ h$. Suppose that T eventually halts. Then

$$h\ q_0\ x\ h\ \Rightarrow^*_{\Pi(T)}\ h\ u\ q\ v\ h\ \Rightarrow^*_{\Pi(T)}\ h\ q'\ h.$$

Conversely, if we suppose $h\ q_0\ x\ h\ \Rightarrow^*_{\Pi(T)}\ h\ q'\ h$, then in the same sequence:

$$h\ q_0\ x\ h = u_1 \Rightarrow u_2 \Rightarrow \cdots \Rightarrow u_n = h\ q'\ h.$$

Note each u_i must contain exactly one q-symbol, since the initial word contained exactly one and each production $g \to \bar{g}$ has exactly one q-symbol in each of g, $\bar{g}$. Now, (i) and (iii) replace each q_i with q_j and (v) never replaces a q_i. However, to get to q' some q_i must be replaced by q, and q by q'. Hence, a production from (iv) must have been used (exactly) once. This implies that T halts. $\quad\square$

Hence we have proven the following.

Theorem 4.3.2. *The Turing Machine T, beginning with instruction q_0 on input x, eventually halts if and only if $h\ q_0\ x\ h \Rightarrow_{\Pi(T)} h\ q'\ h$.*

Proof. Above. $\quad\square$

Corollary 4.3.1. *There is no algorithm to determine, for any $x \in \mathbb{N}$, whether*

$$h\ q_0\ x\ h \Rightarrow^*_{\Pi(T)} h\ q'\ h.$$

Corollary 4.3.2. *The word problem for $\Pi(T)$ is unsolvable.*

4.3.2 Thue Processes and Word Problems in Semigroups

Definition 4.3.4. The *inverse* of a semi-Thue production $g \to \bar{g}$ is the production $\bar{g} \to g$.

A semi-Thue process Π is called a *Thue process* if for each $p \in \Pi$, the inverse of p is also in Π.

Let Π be any semi-Thue process. Then $u \Rightarrow_{\Pi}^{*} v$ is clearly reflexive and transitive.

If Π is a Thue process then it is also symmetric, since

$$\text{whenever } u = u_1 \Rightarrow_{\Pi} u_2 \Rightarrow_{\Pi} \cdots \Rightarrow_{\Pi} u_n = v$$
$$\text{also } v = u_n \Rightarrow_{\Pi} u_{n-1} \Rightarrow_{\Pi} \cdots \Rightarrow_{\Pi} u_1 = u.$$

So $u \Rightarrow_{\Pi}^{*} v$ is an equivalence relation, written $u \sim_{\Pi} v$. Then $[u]$ is the equivalence class containing u.

The equivalence classes yield a structure called a *semigroup*.

- A semigroup $G = (G, \circ)$ is structure where $\circ$ is an associative binary relation on G, and G is closed under $\circ$.
- A semigroup G is a group if it has an identity and every element has a $\circ$-inverse. Namely, there is an element $\mathbf{1}$ such that for all $g \in G$, there is an element $z \in G$, such that $g \circ z = z \circ g = \mathbf{1}$, and such that $z \circ \mathbf{1} = \mathbf{1} \circ z = z$.

All groups are semigroups. For example $(\mathbb{N}, +)$ with $\circ$ interpreted as $+$, is a semigroup and $(\mathbb{Z}, \cdot)$ is a semigroup with $\circ$ interpreted to mean $\cdot$ (multiplication). Neither are groups as, for example, 2 has no additive inverse in $\mathbb{N}$ and, for example, 3 has no multiplicative inverse in $(\mathbb{Z}, \cdot)$. However, $(\mathbb{Z}, +)$ is a semigroup as it is also a group, where $\circ$ is interpreted as $+$. In the case of $(\mathbb{Z}, +)$ the reader might note that, confusingly, $\mathbf{1} = 0$, since $x^{-1} = (-x)$ in that group.

Semigroups do not need an identity element. Not every semigroup is a group.

The reason that the equivalence classes of the Thue system form a semigroup is as follows. First we interpret $\circ$ as concatenation. Then we observe the following:

Whenever $u \Rightarrow_{\Pi}^{*} v$, then for all words w :

$$wu \Rightarrow_{\Pi}^{*} wv$$

and $uw \Rightarrow_{\Pi}^{*} vw$.

So if $u \sim_{\Pi} v$ and $u' \sim_{\Pi} v'$, $uu' \sim_{\Pi} uv' \sim_{\Pi} vu' \sim_{\Pi} vv'$, so the operation $[u] \circ [u'] = [uu']$ is well-defined as it does not depend on the element of the equivalence classes chosen. Since concatenation is associative, the resulting structure is a semigroup. Typically, if G is a semigroup, we would write $x = y$ using the symbol "$=$" to mean an equivalence relation satisfying the axioms of $(G, \circ)$ being an associative binary operation on G. So we'd write $u = v$ to mean $[u] = [v]$.

Definition 4.3.5.

- If S is the semigroup defined from a Thue Process as above, we call Π a (finite) *presentation* of S. Each pair $\langle g, \bar{g} \rangle$ such that $g \to \bar{g}$ and $\bar{g} \to g$ are productions of Π is called a *relation* of the presentation, written $g \sim \bar{g}$. The symbols of A are called *generators* of the presentation. Thus a presentation will be of the form $\langle a_1, \ldots, a_m : \Lambda_j = \Gamma_j$ for $j \in I \rangle$ where Λ_j, Γ_j are words in the alphabet $\Sigma = \{a_1, \ldots, a_m\}$.
- The WORD PROBLEM for finitely presented semigroups asks, given a finitely presented semigroup G, is there a an algorithm to decide if, given $u, v \in G$, does $u = v$?

If Π is a semi-Thue process, write $\bar{\Pi}$ for the Thue process obtained from Π by adjoining to it the inverse of each production of Π. Let $\Pi(T)$ be the semi-Thue process obtained from the Turing Machine T as in 4.3.1. The following lemma yields a further undecidability result. It says that the fact that the Turing Machine computations are faithfully emulated by productions in the semi-Thue process are *also* emulated in the Thue process.

Lemma 4.3.1. *For any Turing machine T and $x \in \mathbb{N}$,*

$$h \, q_0 \, x \, h \Rightarrow_{\Pi(T)}^{*} h \, q' \, h \; \textit{iff} \; h \, q_0 \, x \, h \sim_{\bar{\Pi}(T)} h \, q' \, h.$$

Proof. The lemma is clearly true in direction $\to$, since $\Pi(T) \subseteq \bar{\Pi}(T)$, and hence all the productions from $\Pi(T)$ are present in those of $\bar{\Pi}(T)$.

Conversely, suppose

$$h \, q_0 \, x \, h = u_1 \Rightarrow_{\bar{\Pi}(T)} u_2 \Rightarrow_{\bar{\Pi}(T)} \cdots \Rightarrow_{\bar{\Pi}(T)} u_n = h \, q' \, h.$$

Denote the production giving $u_i \Rightarrow u_{i+1}$ by P_i, and consider i such that $P_{i+1}, \dots, P_{n-1} \in \Pi(T)$ but $P_i \notin \Pi(T)$. If there is no such i we are done, as then all productions are in $\Pi(T)$.

Note for each j, either P_j or its inverse must be in $\Pi(T)$. By definition of $\Pi(T)$, no production of $\Pi(T)$ applies to $h \, q' \, h$, so the inverse of P_{n-1} cannot be in $\Pi(T)$. Hence P_{n-1} itself is, and $i < n-1$. Let Q be the inverse of P_i, so $Q \in \Pi(T)$ and $u_{i+1} \Rightarrow_Q u_i$. Now $u_{i+1} \Rightarrow_{P_{i+1}} u_{i+2}$ and $Q, P_{i+1} \in \Pi(T)$. At most one production of $\Pi(T)$ applies to a given Post word, so Q and P_{i+1} must be the same, so we have $u_i = u_{i+2}$. We may therefore excise u_{i+1} and u_{i+2} from the sequence, leaving

$$u_1 \Rightarrow_{p_1} u_2 \Rightarrow_{p_2} \cdots \Rightarrow_{p_{i-1}} u_i \Rightarrow_{p_{i+2}} u_{i+3} \cdots \Rightarrow_{p_{n-1}} u_n.$$

By repeating the above steps we can remove all productions not in $\Pi(T)$, and so end up with $h \, q_0 \, x \, h \Rightarrow^*_{\Pi(t)} h \, q' \, h$ as required. $\square$

From the corollaries in §4.3.1, we get:

Corollary 4.3.3 (Post [Pos47]).

- *There is no algorithm to determine, for any $x \in \mathbb{N}$, whether $h \, q_0 \, x \, h \sim_{\overline{\Pi}(T)} h \, q' \, h$.*
- *There is a finitely presented semigroup with an unsolvable word problem.*

4.3.3 Exercises

Exercise 4.3.3 Let $A = \{a, b, c, d, e\}$ be an alphabet, and let the semi-Thue process Π be $\{aaa \to abab, ab \to ba, c \to dd, bd \to e, e \to b\}$.

1. Show that $cabcde \Rightarrow^*_\Pi cbadde$.
2. Give the Thue process $\overline{\Pi}$ obtained from Π.
3. What is the semigroup obtained from Π?

Exercise 4.3.4 A TILING SYSTEM is a finite collection of unit square tiles with coloured edges and a set of rules saying which colours can be next to which. The TILING PROBLEM (for the plane) asks whether a given finite partial tiling of the plane can be extended to a full tiling. Describe a reduction from the HALTING PROBLEM (actually the "non-halting" problem) which shows that the TILING PROBLEM is undecidable. (Hint: Think about the squares of the tiles as corresponding to squares on the tape. The next configuration of a Turing Machine would correspond to the row of tiles that can be legally placed above the present row. The "colours" on the edges of the tiles can be the symbols of the alphabet with the exception of the three critical tiles, one for the symbol being read-colour "$q_i S_j$"-and one coding the symbols on the tape each side of the symbol being read. The rules for compatibility will be determined by the quadruples of the Turing Machine. If h is the "end marker" of the configuration, then you would also add tiles for h, and ones saying that it is compatible to add tiles with colours h_l and h_r which can be added to the left and right respectively of a tile coloured h and extended arbitrarily left and right[5] DNA models another universal programming language.)

4.3.4 Post's correspondence problem

Definition 4.3.6. A **Post correspondence system** consists of an alphabet A and a finite set of ordered pairs $\langle h_i, k_i \rangle$, $1 \leqslant i \leqslant m$, of words on A. A word u on A is called a **solution** of the system if for some sequence $i \leqslant i_1, i_2, \ldots, i_n \leqslant m$ (the i_j need not be distinct) we have $u = h_{i_1} h_{i_2} \cdots h_{i_n} = k_{i_1} k_{i_2} \cdots k_{i_n}$.

That is, given two lists of m words, $\{h_1, \ldots, h_m\}$ and $\{k_1, \ldots, k_m\}$, we want to determine whether any concatenation of words from the h list is equal to the concatenation of the *corresponding* words from the k list. A solution is such a concatenation.

Given a semi-Thue process Π and words u, v, we can construct a Post correspondence system that has a solution iff $u \Rightarrow_\Pi^\star v$. Then we can conclude

Theorem 4.3.5. *There is no algorithm for determining of a given arbitrary Post correspondence system whether or not it has a solution.*

Proof. Let Π be a semi-Thue process on alphabet $A = \{a_1, \ldots, a_n\}$, and let u, v be words on A. We construct a Post correspondence system P on the alphabet
$$B = \{a_1, \ldots, a_n, a_1', \ldots, a_n', [,], \star, \star'\}$$
of $2n + 4$ symbols. For any word w on A, write w' for the word obtained from w by placing $'$ after each symbol of w.

Let the productions of Π be $g_i \to \bar{g}_i$, $i \leqslant i \leqslant k$, and assume these include the n identity productions $a_i \to a_i$, $1 \leqslant i \leqslant n$. Note this is without loss of generality as the identity productions do not change the set of pairs u, v such that $u \Rightarrow_\Pi^\star v$. However, we may now assert that $u \Rightarrow_\Pi^\star v$ iff we can write $u = u_1 \Rightarrow_\Pi u_2 \Rightarrow_\Pi \cdots \Rightarrow_\Pi u_m = v$, where m is *odd*.

---

[5] Tiling systems provide a fascinating area to represent many, often very complicated, undecidable problems. For example, you might change the shapes, ask for aperiodicity, etc. An old account, written for the lay audience can be found in Wang [Wan65]. Remarkably, these ideas found further realizations in modelling DNA *self assembly*, the basis of life, beginning with Winfree's remarkable PhD Thesis [Win98]. For a more recent analysis, see Doty et. al. [DLP$^+$12].

P is to consist of the pairs:

$$\langle [u\star, [\rangle, \langle \star, \star' \rangle, \langle \star', \star \rangle,$$

$$\left.\begin{array}{l} \langle \bar{g}_j, g'_j \rangle \\ \langle \bar{g}'_j, g_j \rangle \end{array}\right\} \text{ for } 1 \leqslant j \leqslant k, \text{ and}$$

$$\langle], \star' v] \rangle$$

Let $u = u_1 \Rightarrow_\Pi u_2 \Rightarrow_\Pi \cdots \Rightarrow_\Pi u_m = v$, where m is odd. Then the word

$$w = [u_1 \star u'_2 \star' u_3 \star \cdots \star u'_{m-1} \star' u_m]$$

is a solution of P, as is obvious from the two decompositions

$$w = [u_1 \star |u'_2 \star' |u_3 \star| \cdots |]$$
$$= [|u_1 \star |u'_2 \star' | \cdots | \star' u_m]$$

Note that $u'_2 \star'$ corresponds to $u_1 \star$ as we can write $u_1 = rg_j s$, $u_2 = r\bar{g}_j s$ for some $1 \leqslant j \leqslant k$. Then $u'_2 = r'\bar{g}'_j s'$ and the correspondence is clear.

Conversely, if $\bar{w}$ is a solution, then to avoid mismatches on the ends, $\bar{w}$ must begin with [and end with]. Hence, letting w be the portion of $\bar{w}$ up to the first],

$$w = [u \star \cdots \star' v]$$

where to begin with we have the correspondences

$$w = [u \star | \cdots \star' v|]$$
$$= [|u \star \cdots | \star' v]$$

Hence we must have $u\star$ corresponding to some $r'\star'$ and $\star'v$ to some $\star s'$, where $u \Rightarrow_\Pi^* r$ and $s \Rightarrow_\Pi^* v$. Continuing this procedure, we see that $u \Rightarrow_\Pi^* v$.

This completes the construction and hence the proof of the theorem. □

4.3.5 Groups*

In this section, we will look at decision problems for "finitely presented" groups. We will assume that the reader is familiar with the rudiments of group theory, so that the discussion below makes sense. Do not be concerned if some of the concepts are unfamiliar, as this section can be skipped, and only gives a top-level view of the main ideas, anyway.

A *much* more difficult result is to show that there is a finitely presented *group* (i.e. rather than a semigroup) with an undecidable word problem. Classically, a presentation of a group is an expression of the form

$$G = \langle a_1, \ldots, a_n : g_1 = g_2, \ldots, g_k = g_{k+1} \rangle.$$

Here the g_i are words in the generators $\{a_1, \ldots, a_n\}$. In the same way as
the semigroup above, we are declaring that we can substitute g_{i+1} whenever
we see g_i as a subword of v and conversely. Groups will automatically have
additional relations since they must have an identity element $\mathbf{1}$, and for each
$v \in G$, there must also be an inverse v^{-1}; hence we will also add lots of new
elements. As we have seen, we can use Post words to construct semigroups
which are very close to Turing Machines.

However, because of the additional structure guaranteed by the presence of
inverses many, many new "productions" will be present if we try to somehow
emulate the proof for semigroups to give an undecidability proof for groups
as we discuss below.

The definition of a finitely presented group above correlates with another
classical equivalent formulation of group presentations. For that formulation
we would have a *free* group F based on the symbols $\{a_1, \ldots, a_n\}$. One definition for a free group on such symbols is that F is the group where the
only relations holding between the symbols are the ones guaranteed by the
definition of being a group. So if $x, y \in F$ then $xy \in F$, and this "product" is
concatenation. Every $x \in F$ must have an inverse x^{-1} with $x^{-1}x = xx^{-1} = \mathbf{1}$;
and this is induced for the elements of F by giving each of $a_1, \ldots, a_n$ their
own inverse. So if $x = a_2 a_1 a_3^{-1}$, say, then x^{-1} would be $a_3 a_1^{-1} a_2^{-1}$.

Then we can add the relations $g_k = g_{k+1}$ by thinking of them as saying
$g_k g_{k+1}^{-1} = \mathbf{1}$. Using group theory, the way that we can force this to happen in
F is to consider the *quotient group*, $G = F/N$, where N is a normal subgroup
generated by $\{g_i g_{i+1}^{-1} : i = 1, \ldots, k\}$.

In this equivalent classical form, in Max Dehn [Deh11] asked three fundamental questions (expressed in more modern language), in some sense found-
ing the area of *combinatorial group theory*.

- THE WORD PROBLEM[6]. Is there an algorithm to decide if given $g \in G$
  whether $g = \mathbf{1}$ in G. Because of the existence of inverses, this is equivalent
  to asking whether we can decide, given u, v whether $u = v$ since $u = v$ iff
  $uv^{-1} = \mathbf{1}$.
- THE CONJUGACY PROBLEM. Given u and v in G, can we decide if there
  exists a z such that
  $$u = zvz^{-1}?$$
- THE ISOMORPHISM PROBLEM. Is there an algorithm, which given two
  finitely presented groups, G_1 and G_2, decides whether $G_1 \cong G_2$?

---

[6] On the first page of this paper, Dehn states all three problems. For instance, the word
problem is stated as "Itrgend ein Element der Gruppe ist dutch seine Zusammensezung aus
den Erzeugenden gegeben. Man soll eine Methode angeben, um mit einer endlichen Anzahl
yon Schritten zu entscheiden, ob dies Element der Identität gleich ist oder nicht." This
translates (more or less) as "Some element of the group is given by its composition of the
generators. One should specify a method to decide in a finite number of steps whether this
element of the identity is the same or not." As with Hilbert and his Entscheidungsproblem,
Dehn implicitly believes that there will be such an algorithm.

Clearly, the word problem is m-reducible to the conjugacy problem, since, $u = 1$ in G iff there exist z with $u = z1z^{-1}$.

In [Deh11] Dehn used geometric methods to give decision algorithms for several special cases of the questions above. Some of Dehn's ideas still penetrate today. However, all of Dehn's problems are famously undecidable, but their solution led to a huge amount of new group theory. One starting point for proving the undecidability of the word problem for groups is to start with the semigroup we derived using Post words in §4.3.2. We take A containing an even number of symbols $a_1, \ldots, a_k, b_1, \ldots, b_k$, and a Thue process Π that contains all relations $a_i b_i \sim \lambda$ for $1 \leqslant i \leqslant k$ (That is, λ acts the part of the identity and a_i, b_i as pairs of inverses). Then what we do is add a lot of new relations $g_i = g_{i+1}$ to the structure and "turn it into a group" with an undecidable word problem. Here is one such group.

It starts with a presentation of the Post semigroup of §4.3.2:

$$\langle s_0, \ldots, s_m, q_0, \ldots, q_k, h, h' : \Lambda_j = \Gamma_j \text{ for } j \in I \rangle.$$

We add additional generators k, t, x, y, ℓ_i, r_i for $i \in I$. Then for each $a \in \{0, \ldots, k\}$ and $b \in \{0, \ldots, m\}$, we have relations $s_b y = y y s_b$, $x s_b = s_b x x$, $s_b \ell_i = y \ell_i y s_b$, $r_i s_b = s_b x r_i x$, $\Lambda_i = \ell_i \Gamma_i r_i$, $t \ell_i = \ell_i t$, $t y = y t$, $r_i k = k r_i$, $r k = k x$, and $(q_0^{-1} t q_0) k = k(q_0^{-1} t q_0)$. The reader can see that this is not the first group they would think of!

The difficulty is proving that within the group the Turing Machine computations are faithfully emulated. Such emulation is kind of obvious in the case of a semi-Thue system as such systems based on Post words are basically Turing Machine configuration transitions. Such faithful emulation is reasonably easy in the case of Thue systems and semigroups, but each additional structure makes things harder. For example, when we turned the semi-Thue system into a Thue one, we had to check that adding inverses did not add new methods of generating $hq'h$. In the case of groups, we are *forced* to add 1 and inverses for every member of G. This means that *many new computations* might be possible. It might be possible for $u = 1$ without this reflecting *any computations in some underlying Turing Machine T*. In fact, this is almost certainly true, but it is also irrelevant, since the undecidability proof of the word problem for groups entails showing that *for certain special words u from a set U*, if $u = 1$ in G then it must be that for some input $x = x(u)$ of T, x leads to a halting computation in T; and conversely. In this way, the use of special words show that every computation from T on a y is faithfully represented by a special word $u(y)$ in U. Roughly speaking, what we do is put "buffer" symbols around x which protect the Turing computations from the underlying group structure. This is the point of all the new generators and their relations. All undecidability proofs of the word problem for groups are far from trivial and seem hard even to this day. All proofs rely on some quite nontrivial group theory. In fact, it is easily argued that the special group theory: HNN extensions, free products with amalgamation, Britton's

Lemma, and the like, were by-products of the quest for undecidability. These
methods are now mainstays of combinatorial group theory. Rotman [Rot65],
Chapter 12 gives a very good account of the standard proof. The hidden
message here is that more involved undecidability proofs will often rely on
some highly nontrivial structure theory.

Theorem 4.3.6 (Novikov-Boone [Nov55, Boo59]). *There is a finitely
presented group with an unsolvable word problem.*

In the case that we allow an *infinite* set of relations[7] $g_i = g_{i+1}$ then the
following group has an unsolvable word problem

$$G_W = \langle x, y, u, t : u^i x u^{-i} = t^i y t^{-i} \text{ if } \varphi_i(i) \downarrow \rangle$$

The reason that this is true is that $u^i x u^{-i} = t^i y t^{-i}$ holds iff $\varphi_i(i) \downarrow$, so if we
could decide whether $u^i x u^{-i} \cdot (t^i y t^{-i})^{-1} = 1$, then we could solve the halting
problem. The proof of this fact is purely group-theoretic, and is beyond the
scope of this book.

We will briefly discuss a few more results in group theory. We can deduce
the undecidability of the word problem from the example above and the
following famous result:

Theorem 4.3.7 (Higman Embedding Theorem [Hig61]). *A group has
a presentation with a finite number of generators and a computably enumerable set of relations (that is, they are given as codes as the range of a
computable function) iff it is isomorphic to a subgroup of a finitely presented
group.*

In some sense this is a bit of a cheat, since the *proof* of the Higman
Embedding Theorem relies on the coding method in the proof of the Unde-
cidability of the Word Problem plus a certain technique called the "Higman
Rope Trick."

While remaining restricted to groups, we can also expand the problems
shown to be algorithmically unsolvable. Let $\mathcal{G}$ be a collection of groups. We
say $\mathcal{G}$ is

non-trivial if it and its complement are both non-empty (as usual),

invariant if it is closed under isomorphism, and

hereditary if it is closed under taking subgroups.

Theorem 4.3.8 (Adian-Rabin Theorem [Adi55, Rab58]). *If $\mathcal{G}$ is a collection of groups which is invariant, non-trivial, and hereditary, then it is
unsolvable whether a given group presentation is of a group belonging to $\mathcal{G}$.*

---

[7] As we later see this is a *computably enumerable* set of relations, and can be generated
by a computable function.

For example, $\mathcal{G}$ may consist of only the trivial group $\{\mathbf{1}\}$. It might consist of exactly the groups which are cyclic, finite, free, or abelian. Thus there is no algorithm allowing us to tell from a presentation whether the associated group has any of those properties. The Adian-Rabin Theorem says that most "reasonable" properties of finitely presented groups are algorithmically unsolvable. Later in the book we will look at Rice's Theorem, Theorem 5.2.1, we will see that the same is true for deciding properties by their machine description alone. The proof of the Adian-Rabin Theorem is not that difficult, but would take us a bit far. We refer the reader to, for example, Lyndon and Schupp [LS01] for more on this fascinating subject.

4.3.6 The Entscheidungsproblem*

Recall that one of Hilbert's problem was to give a decision procedure for first order logic. We are in a position to sketch the proof that there is no such procedure. That is, we give a detailed sketch of the proof of the undecidability of predicate logic.

For this section, we will assume that the reader is familiar with the rudiments of predicate logic. Here we start with variables $x, y, z, \ldots$, predicates $P\bar{x}$ for $\bar{x} = x_1 \ldots x_k$ (for $k \in \mathbb{N}$) and the usual rules of boolean algebra (for formulas A, B, C) such as $A \wedge (B \vee C) = (A \wedge B) \vee (A \wedge C)$, $\neg(A \vee B) = \neg A \vee \neg B$, etc, and then enrich with quantifiers $\forall x$ and $\exists y$. The Entscheidungsproblem is the following

Is predicate logic decidable? That is, is there an algorithm which, when given (the Gödel number of) a formula ψ of predicate logic, will decide if ψ is valid?

The proof is to take the proof for semi-Thue systems and "interpret it" as something happening in predicate logic. From the proof of the undecidability of semi-Thue systems, we have the following: We can use a unary relation R, and think of $R(q)$ as q'. That is, we can have a set of states $q, q', q'', \ldots$ representing $q_0, q_1, q_2, \ldots$. Similarly symbols will be represented by $B, 1, 1', 1'', \ldots$ in place of $B, S_0, S_1, S_2 \ldots$. We'd need the "axiom" that $x(yz) = (xy)z$, which say that brackets can be forgotten. We also interpret $u \Rightarrow_{\Pi}^* v$ as a binary relation $\Rightarrow^*$. We then need to write out axioms for the behaviour of $\Rightarrow^*$. Namely, we would write for each $u \Rightarrow_{\Pi}^* v$ generated by T,

$$\forall x \forall y [(xu \Rightarrow^* xv) \wedge (uy \Rightarrow^* vy) \wedge (xuy \Rightarrow^* xvy)],$$

plus add the fact that $\Rightarrow^*$ is transitive:

$$\forall u, v, w([(u \Rightarrow^* v) \wedge (v \Rightarrow^* w)] \to u \Rightarrow^* w).$$

For each $u \Rightarrow_\Pi v$, generated by T, we would add the above sentence $\psi_{u,v}$. The particular formula which would not be decidable would be that for a suitably chosen T, that $[\forall x, y, z(x(yz) = (xy)z) \wedge [\wedge_{u \Rightarrow_\Pi v} \psi_{u,v}] \wedge \forall u, v, w([(u \Rightarrow^* v) \wedge (v \Rightarrow^* w)] \rightarrow u \Rightarrow^* w)] \rightarrow (hxh \Rightarrow^* hq'h)$, giving a sentence ρ_x. This is valid iff T halts on x.

4.3.7 Diophantine equations*

Another of Hilbert's problems which turns out to be undecidable is called *Hilbert's 10th Problem*. This is because it is the 10th of his famous list of 20 problems for mathematics for the "new" (20th) Century (Hilbert [Hil12]). This problem asked for an algorithm to determine whether an arbitrary polynomial equation $P = 0$, where the coefficients of P are all integers, has a solution in integers. Again, in 1900, the idea there may not be any such algorithm would not have been considered. Specifically, Hilbert asked

> "Eine diophantische Gleichung mit irgendwelchen Unbekannten und mit ganzen rationalen Zahlkoeffizienten sei vorgelegt; man soll ein Verfahren angeben, nach welchem sich mittels einer endlichen Anzahl von Operationen entscheiden lisst, ob die Gleichung in ganzen rationalen Zahlen lksbar ist[8]."

In 1970, building on extensive work of Davis, Putnam and Robinson, Matijacevič [Mat70] proved the problem to be algorithmically unsolvable.

Definition 4.3.7. A k-ary relation R on $\mathbb{N}$ is *Diophantine* if for some polynomial $P(x_1, \ldots, x_k, y_1, \ldots, y_n)$ with integer coefficients,

$$R = \{\langle x_1, \ldots, x_k \rangle : \exists y_1, \ldots, y_n \in \mathbb{N} \text{ s.t. } P(x_1, \ldots, x_k, y_1, \ldots, y_n) = 0\}.$$

The actual setting for the proof is in the natural numbers, not the integers. This does not result in loss of generality because we are proving unsolvability. If the integer version is solvable, then we may solve the natural number version for $P(x_1, \ldots, x_n) = 0$ by solving the integer version of $P(u_1^2 + v_1^2 + x_1^2 + y_1^2, \ldots, u_n^2 + v_n^2 + x_n^2 + y_n^2) = 0$, because every non-negative integer is the sum of four squares. Therefore if the natural number version is unsolvable, the integer version must also be.

For example, sticking to the naturals, "is a factor of" is Diophantine, since "a is a factor of b" iff $\exists x(a \cdot x = b)$; similarly: "a is composite" iff $(\exists x_1)(\exists x_1)[(x_1 + 2)(x_2 + 2) - a = 0]$; "$a < b$" iff $(\exists x)[a + x + 1 - b = 0]$.

---

[8] A Diophantine equation with any unknowns and with whole rational number coefficients are presented; one should have a procedure indicate after which, using a finite number of operations, decide whether the equation is in whole rational numbers is solvable.

We will consider a generalization of the halting problem as a k-ary relation where $\langle x_1, \ldots, x_k \rangle \in R$ iff $\varphi_{\langle x_1, \ldots, x_k \rangle}(\langle x_1, \ldots, x_k \rangle) \downarrow$, and call this set the k-ary representation of the halting problem[9]

Theorem 4.3.9 (Matijacevič [Mat70]).

1. *There is a k such that the k-ary representation of the halting problem is m-reducible to a Diophantine relation.*
2. *In fact, with the definition of computable enumerable from Definition 5.1.2, a relation on $\mathbb{N}$ is Diophantine if and only if it is computably enumerable[10].*

The proof of this result involves very clever results about the behaviour of polynomials, as well as ingredients like Fibonacci numbers, and other miscellaneous number-theoretic facts. Thus we will not cover this intricate proof in detail.

However, because we have covered Register Machines, there is a relatively accessible proof due to Jones and Matijacevič [JM79] for one of the preliminary results of the proof. This proof is not widely known, and it seems worth studying. Matijacevič's Theorem is the end of a long series of papers, and is regarded as a major theorem in the history of mathematics. The result we prove here is one of the major steps in the proof and we clarify its relationship with the final proof later. Thus we will take the opportunity to look at the Jones-Matijacevič material in this section, and then use hand waving for the last step of the proof.

We will prove the weaker result that the halting problem (any c.e. relation) is *exponential* Diophantine. This is defined precisely as in Definition 4.3.7, but also allowing terms like a^x (exponentials, but only *single* exponentials, so 2^{2^x}, for example, is not allowed), in the definition. Thus, for example, $2^y - 7 = x^2$, $x^x y^y = z^z$ and $2^x + 11^z = 7^y + 1$ are examples of exponential Diophantine equations. Matijacevič's key result was to show that *exponential Diophantine is the same as Diophantine*. The earlier undecidability result for exponential Diophantine relations is a celebrated result due to Davis, Putnam, and Robinson [DPR61].

Theorem 4.3.10 (Davis, Putnam and Robinson [DPR61]). *Theorem 4.3.9 holds if we replace "Diophantine" by "exponential Diophantine".*

We code the action of Register Machines and the goal will be to ensure the following: We want to write equations which have a solution iff the machine

---

[9] Actually this works for any computably enumerable k-ary relation as defined in Chapter 5.

[10] We state this as we will make some remarks about the consequences of the full result at the end. Diophantine to c.e. is the easy direction. Let R be as in Definition 4.3.7. We may computably list all n-tuples $\langle y_1, \ldots, y_n \rangle$. The procedure $\varphi(\langle x_1, \ldots, x_k \rangle)$ evaluates $P(x_1, \ldots, x_k, y_1, \ldots, y_n)$ on successive n-tuples until 0 is obtained, if ever (and we argue by the Church-Turing thesis that φ is partial computable).

accepts x (i.e. halts on x). That is, exponential Diophantine conditions so that x is accepted iff $(\exists s)(\exists Q)(\exists I)(\exists R_1 \ldots R_r)(\exists L_0, \ldots L_\ell)$
[The exponential Diophantine conditions hold].

To follow the paper [JM79], we will slightly modify the definition of register machines we use. The new lines are[11]:

Li GOTO Lk; meaning line i says go to line k.

Li If $Rj < Rm$, GOTO Lk; meaning if the contents of register i is less than that of register j, go to line k.

Li If $Rj \leqslant Rm$, GOTO Lk; meaning if the contents of register i is less than or equal to that of register j, go to line k.
Li $Ri \leftarrow Ri + 1$; meaning add one to the contents of register i.

Li $Ri \leftarrow Ri - 1$; meaning take one to the contents of register i.
We still need the conditional, and we use two commands:

$$Li \text{ If } Rj = 0, \text{ GOTO } Lk.$$

$$Li + 1 \text{ ELSE } Rj \leftarrow Rj - 1.$$

Finally,

$$Li \text{ HALT.}$$

It is not hard to prove that the machines created with these commands are equivalent to those we considered when we looked at Minsky Machines, and Collatz Functions.

We now introduce some new definitions and notations where all definitions have variables in $\mathbb{N}$.

Definition 4.3.8. q pow 2 means that "q is a power of 2";
and as in primitive recursion, " $\text{rem}(x, y)$" is the remainder after x is divided by y.

Lemma 4.3.2. *The following are exponential Diophantine (with reasons):*

(i) $a < b \leftrightarrow \exists x[a + x + 1 = b]$
(ii) If A and B are Diophantine the both $A \wedge B$ and $A \vee B$ are, since

$$[(A = 0) \vee (B = 0)] \leftrightarrow AB = 0 \text{ and}$$

$$[(A = 0) \wedge (B = 0)] \leftrightarrow (A^2 + B^2 = 0).$$

---

[11] For the purposes of this section, following [JM79], we will use upper case letters for register machine instructions, to differentiate them from text in the proof. We will use Li to denote line i, in this context.

(iii) $a \equiv b(\bmod c) \leftrightarrow \exists x[(a = b + cx) \vee (b = a + cx)]$.
(iv) For $0 < c$, $a = \mathrm{rem}(b, c) \leftrightarrow [b \equiv a(\bmod c) \wedge a < c)]$

The next step is to show that $\binom{n}{k}$ is Diophantine. This is a result of Julia Robinson [Rob52]. It is also where where things become more complicated. The proof uses the fact that $\binom{n}{k}$ is defined by the binomial theorem that $(u + 1)^n = \sum_{i=0}^{n} \binom{n}{k}u^i$. Therefore the binomial coefficients are simply the "digits" in the base u expansion of $(u + 1)^n$, when u is sufficiently large: $u > 2^n \geqslant \binom{n}{k}$. These digits are unique. Therefore, $m = \binom{n}{k}$ iff there exist unique u, w, v such that

$$u = 2^n + 1 \wedge (u + 1)^n = wu^{k+1} + mu^k + v \wedge v < u^k \wedge m < u.$$

This expression is exponential Diophantine. The two place exponentials in this definition can be replaced by one-place ones using

$$\mathrm{rem}(2^{xy^2}, 2^{xy} - x) = x^y.$$

This is true since $2^{xy} \equiv x(\bmod 2^{xy} - x)$.

The key idea we will use to allow definitions of the action of register machine is the relation $\preccurlyeq$ defined for this section as follows. Suppose that s and r are written in base 2 notation.

$$r = \sum_{i=0}^{n} r_i 2^i (0 \leqslant r_i \leqslant 1); s = \sum_{i=0}^{n} s_i 2^i (0 \leqslant s_i \leqslant 1).$$

Definition 4.3.9. $r \preccurlyeq s$ iff $r_i \leqslant s_i$ for $0 \leqslant r_i \leqslant n$.

The following are some useful properties of $\preccurlyeq$.

- $a \preccurlyeq b \leftrightarrow a \wedge b = a$ (Thinking of the binary numbers as being lines of a truth table.)
- $a \wedge b = c \leftrightarrow c \preccurlyeq b$ and $b \preccurlyeq a + b - c$.
- pow2 $\leftrightarrow a \preccurlyeq 2a - 1$.
- If Q pow2 and $a < Q$, then

$$[a \preccurlyeq b \text{ and } c \leqslant d] \leftrightarrow [a + cQ \preccurlyeq b + dQ].$$

Lemma 4.3.3 (Jones and Matijacevič [JM79]). $\preccurlyeq$ *is single exponential, because*

$$r \preccurlyeq s \text{ iff } \binom{s}{r} \equiv 1(\bmod 2).$$

The lemma's proof relies on a classical result from number theory called Lucas' Theorem [Luc78], which says that

$$\binom{s}{r} = \prod_{i=0}^{n} \binom{s_i}{r_i} \pmod{2},$$

where the r_i and s_i are the binary digits of r and s, respectively (taken to have common length n).

We turn to coding register machines. We will define the version of the halting problem for M with registers $R1, \ldots, Rr$ to have lines labelled $L1, \ldots, L\ell$. $M(x)$ will halt when started on x in R_1 finished in the halt line with zero in all other registers. This makes no essential difference to the definition of halting. We will pick a suitably large number Q, and use digit of numbers, written in base Q, to emulate the action of the machine. Let s be the number of steps used by the machine when it stops, let $r_{i,t}$ denote the content or register r_i at step t, and $\ell_{i,t}$ be 1 if we execute line Li at time t, and 0 if we do not. Now we define

$$R_i = \sum_{t=0}^{s} r_{j,t} Q^t (0 \leqslant r_{j,t} \leqslant \frac{Q}{2}), \text{ and,}$$

$$L_i = \sum_{t=0}^{s} \ell_{i,t} Q^t (0 \leqslant \ell_{j,t} \leqslant 1).$$

The base Q we use will be a sufficiently large power of 2. At step t the contents of any register cannot be larger than $x + t$, and hence $r_{j,t} \leqslant x + s$. So we can choose Q so that

$$x + s < \frac{Q}{2} \wedge \ell + 1 < Q \wedge Q \text{ pow2}.$$

Thus take $Q = 2^{x+s+\ell}$.

Example 4.3.1. The following example is taken from [JM79] (so that I don't get it wrong!)

```
L0   R2← R2+1
L1   R2← R2+1
L2   IF R3=0 GOTO L5
L3   R3← R3-1
L4   GOTO L2
L5   R3← R3+1, R4← R4+1, R2← R2-1
L6   IF R2> 0 GOTO L5
L7   R2← R2+1, R4← R4-1
L8   IF R4> 0 GOTO L7
L9   IF R3< R1 GOTO L5
L10  IF R1< R3 GOTO L1
L11  IF R2< R1 GOTO L10
L12  R1← R1-1, R2← R2-1, R3← R3-1
L3   IF R1> 0 GOTO L12
L14  HALT
```

In [JM79], they combine commands on single lines for compactness of presentation, so long as the same register is not used twice in the same line. This makes no difference.

Using this machine, and Q calculated above, we get the following table representing the halting computation. Time is proceeding leftwards, with the numbers $R_1, \ldots, R_4, L_0, \ldots, L_{14}$ presented in base Q notation.

```
0 0 1 1 2 2 2 2 2 2 2 2 2 2 2 2 2 2 R1
0 0 1 1 2 2 2 2 1 1 0 0 1 1 2 2 1 0 R2
0 0 1 1 2 2 2 2 2 2 2 2 1 1 0 0 0 0 R3
0 0 0 0 0 0 0 0 1 1 2 2 1 1 0 0 0 0 R4
0 0 0 0 0 0 0 0 0 0 0 0 0 0 0 0 0 0 L0
0 0 0 0 0 0 0 0 0 0 0 0 0 0 0 0 1 1 L1
0 0 0 0 0 0 0 0 0 0 0 0 0 0 0 1 0 0 L2
0 0 0 0 0 0 0 0 0 0 0 0 0 0 0 0 0 0 L3
0 0 0 0 0 0 0 0 0 0 0 0 0 0 0 0 0 0 L4
0 0 0 0 0 0 0 0 0 0 0 0 0 1 0 1 0 0 0 L5
0 0 0 0 0 0 0 0 0 0 0 0 1 0 1 0 0 0 0 L6
0 0 0 0 0 0 0 0 0 1 0 1 0 0 0 0 0 0 L7
0 0 0 0 0 0 0 0 1 0 1 0 0 0 0 0 0 0 L8
0 0 0 0 0 0 0 1 0 0 0 0 0 0 0 0 0 0 L9
0 0 0 0 0 0 1 0 0 0 0 0 0 0 0 0 0 0 L10
0 0 0 0 0 1 0 0 0 0 0 0 0 0 0 0 0 0 L11
0 0 1 0 1 0 0 0 0 0 0 0 0 0 0 0 0 0 L12
0 1 0 1 0 0 0 0 0 0 0 0 0 0 0 0 0 0 L13
1 0 0 0 0 0 0 0 0 0 0 0 0 0 0 0 0 0 L14
```

For the proof we need a number I which is all 1's when written base Q. If $1 + (Q-1)I = Q^{s+1}$, then

$$I = \sum_{t=0}^{s} Q^t \text{ will work.}$$

We remark that $\sum_{t=0}^{s} a_t Q^t \preccurlyeq \sum_{t=0}^{s} b_t Q^t$ for any $a_t, b_t < Q$. It remains to show how how these parameters $x, s, Q, I, R_1, \ldots, R_r, L_0 \ldots, L_\ell$ can be used to write conditions that have a solution iff

$$M(x) \downarrow .$$

That is, there is some s with $M(x) \downarrow [s]$. The numbers r, ℓ are constants, x is the exponential Diophantine parameter and $s, Q, I, R_1, \ldots, R_r, L_0 \ldots, L_\ell$ are unknowns.

The first set of conditions is

$$x + s < \frac{Q}{2} \wedge \ell + 1 < Q \wedge Q \text{ pow2.}$$

The next are $1 + (Q - 1)I = Q^{s+1}$ and $I = \sum_{t=0}^{s} Q^t$. To force an arbitrary number R_j to have the form $R_i = \sum_{t=0}^{s} r_{j,t} Q^t (0 \leqslant r_{j,t} \leqslant \frac{Q}{2})$, we use $\preccurlyeq$, as Q is a power of 2. This condition and $r_{j,t} \leqslant \frac{Q}{2}$ is implied by

$$R_j \preccurlyeq (\frac{Q}{2} - 1)I.$$

The next action is to force one digit to be 1 in each column of the columns. Since $\ell < Q$, we use

$$I = \sum_{i=0}^{t} L_i, \text{ and,}$$

$$L_i \preccurlyeq I (i = 0, \ldots, \ell).$$

The starting condition is stipulated by $1 \preccurlyeq L_0$. Because of GOTO commands, we can assume there is only one halting line, and this happens at step s, so

$$L_\ell = Q^s.$$

If the line is Li GOTO Lj, the condition to be included is

$$QL_i \preccurlyeq L_k.$$

This forces the machine to go from location L_i to L_k whenever L_i is executed. Now we need to deal with conditional line transfer. If the line is

$$Li : \text{ If } Rj = 0 \text{ GOTO } Lk,$$

we use

$$QL_i \preccurlyeq L_k + L_{i+1} \text{ and } QL_i \preccurlyeq L_{i+1} + QI - 2R_j.$$

This works for $k \neq i + 1$, and if $k = i + 1$ use $QL_i \preccurlyeq L_k$.
If $k \neq i + 1$, and the command is

$$Li \text{ If } Rj < Rm \text{ GOTO } Lk,$$

we simulate with

$$QL_i \preccurlyeq L_k + L_{i+1} \text{ and } QL_i \preccurlyeq L_k + QI + 2R_j - 2R_m.$$

If the command is

$$Li \text{ If } Rj \leqslant Rm \text{ GOTO } Lk,$$

simulate with

$$QL_i \preccurlyeq L_k + L_{i+1} \text{ and } QL_i \preccurlyeq L_{i+1} + QI + 2R_j + 2R_m - 2R_j.$$

For each of these 3 types of commands, we need to include a condition directing the machine to take the next instruction, to execute $Li + 1$ after L_i. That is, we include

$$QL_i \preccurlyeq L_{i+1}.$$

The construction is finished by including an equation saying that the contents of register R_t is correct at time t, that is the t-th Q-ary digit of R_j represents the contents of Rj. These register equations are:

$$R_j = QR_j + \sum_k QL_k - \sum_i QL_i + x \text{ for } j = 1, \text{ and,}$$

$$R_j = QR_j + \sum_k QL_k - \sum_i QL_i \text{ for } 2 \leqslant j \leqslant r.$$

The k-sum is taken over all k for which the programme has an instruction of the type $Lk \ldots Rj \leftarrow Rj + 1 \ldots$. The i sum is taken over all i where the programme has an instruction of the form $Li \ldots Rj \leftarrow Rj - 1 \ldots$, or one of the form $Li + 1 \ldots \text{ ELSE } Rj \leftarrow Rj - 1$. This simulation proves the theorem. More details can be found in Jones and Matijacevič [JM79]. In that paper thay also discuss applications in computational complexity theory and NP-completeness, via miniaturizations, and placing bounds on Register Machines.

What about Diophantine relations, rather than exponential Diophantine relations? The final step in the full proof of Hilbert's 10th Problem being undecidable is to extend the undecidability result for exponential Diophantine relations to (normal) Diophantine relations. What is needed is a proof that *exponential Diophantine relations are Diophantine*. This proof is nontrivial mathematics but involves no new theory of computation. Rather, Matijacevič's proof involves very clever results about the behaviour of polynomials, as well as ingredients like Fibonacci numbers, and other miscellaneous *number-theoretic* facts. The reader should be excited enough to chase the accessible proof in the book Matijacevič [Mat93].

We remark that the uniformity of the proof shows that many sets are Diophantine, i.e. solutions to polynomial equations, which one would not expect to be. For example, there is a Diophantine relation whose solutions are exactly the prime numbers. This polynomial has many variables and high degree, and there would be no obvious way to find such a polynomial without the solution to Hilbert's 10th Problem.

4.3.8 Exercises

Exercise 4.3.11 Quantifying over $\mathbb{N}$, show that "$\sqrt{a}$ is rational" is Diophantine.

Exercise 4.3.12 (Putnam) Prove that every Diophantine set is the positive part of the range of a polynomial over $\mathbb{Z}$.

Exercise 4.3.13 Prove by induction that if $r \leqslant s$ then $r \preccurlyeq s$ iff $\binom{s}{r}$ is odd. (Hint: Use induction on s.)

Exercise 4.3.14 Apply the method of this section to obtain a table corresponding to the following register machine for $x = 2$.

> L0 IF R1=0 GOTO L6
> L1 R1← R1-1
> L2 IF R1=0 GOTO L5
> L3 R1← R1-1
> L4 GOTO L0
> L5 R1← R1+1
> L6 HALT

4.3.9 Coda

There are many fascinating decision problems in mathematics, some of which are decidable and some undecidable. We have seen several in this chapter, but will now move on to other topics. The interested reader is referred to articles such as Davis [Dav77] or Poonen [Poo14] for a guide to decision problems in many other areas; especially in knot theory, algebra and topology Borger, Grädel and Gurevich [BCC07] gives an account of undecidability and decidability results in logic; heirs to the Entscheidungsproblem.

Chapter 5
Deeper Computability

Abstract In this Chapter, we will develop a number of more advanced tools we can use to tackle issues in computability theory. In particular, we will be able to deal with problems more complex than the halting problem, as delve more deeply into the fine structure of reducibilities and noncomputable sets. We introduce Turing reducibility. We prove the s-m-n theorem and recursion theorem. We will look at computable structure theory via computable linear orderings. We introduce the arithmetical hierarchy and show how definability aligns with computation. Finally we will look at constructions in the Turing degrees including the finite extension and finite injury methods, showing how Post's Problem was solved.

Key words: Turing degree, degrees of unsolvability, arithmetical hierarchy, priority method, finite extension method, Kleene-Post, jump operator, limit lemma, computable linear orderings, recursion theorem

5.1 Introduction

In this Chapter, we will develop a number of more advanced tools we can use to tackle issues in computability theory. In particular, we will be able to deal with problems more complex than the halting problem, as delve more deeply into the fine structure of reducibilities and noncomputable sets.

5.1.1 Computably Enumerable Sets

One of our themes is a wish to regard all problems as coded by subsets of $\mathbb{N}$. For example, the halting problem can be coded by

© The Author(s), under exclusive license to Springer Nature Switzerland AG 2024   109
R. Downey, *Computability and Complexity*, Undergraduate Topics
in Computer Science, https://doi.org/10.1007/978-3-031-53744-8_5

$$K = \{x : \Phi_x(x){\downarrow}\}$$

Alternatively, we could code using the two-variable formulation, by

$$K_0 = \{\langle x, y \rangle : \Phi_x(y){\downarrow}\}).$$

In Theorem 5.1.5, we will see that they are the "same" problem as $K \equiv_m K_0$.

We can use simular codings for other problems:

NONEMPTINESS

Input : e

Question : Is $\mathrm{dom}(\varphi_e) \neq \emptyset$.

This could be coded by

$$N = \{e : \exists y \exists s (\varphi_{e,s}(y) {\downarrow}\}.$$

The hardness of the halting problem stems from the fact that we are seeking a witness and perhaps one never occurs. Such problems have a very special character. Although they are not decidable (proven below for N), they are "semi-decidable". This means that a "yes" instance will be observed should one ever happen, as time goes by. For example, if we began a process:

Observing the Halting Computations

- **Stage 0** Compute $\varphi_0^0(0)$. This means "run partial computable function φ_0 on input 0 for 0 many steps.
- **Stage s** For $e, j \leqslant s$, compute $\varphi_e^s(j)$. That is, run the first s many machines for s steps for inputs $j \leqslant s$

Definition 5.1.1 (Dovetail Enumeration). The process above is called a *dovetail enumeration* of all Turing Machine computations.

Because of the existence of a universal Turing machine, the Kleene T-predicate, this process is stage by stage computable (in fact primitive recursive), and we could see which machines had halted by stage s on inputs $\leqslant s$ where the machines are only run for s steps. The key point is that if $\varphi_e(y)$ halts, we will, at some stage, observe this happening. That is, at some stage s, we would see $\varphi_e(y) \downarrow [s]$. Thus something is "semi-decidable" if there is an algorithm which halts on *YES* instances, and won't halt on *NO* instances. This idea is formalized in the following definition:

Definition 5.1.2. A set $A \subseteq \mathbb{N}$ is called

(i) *computably enumerable (c.e.)* if $A = \mathrm{dom}(\Phi_e)$ for some e, and
(ii) *computable* if A and $\overline{A} = \mathbb{N} \setminus A$ are both computably enumerable.
(iii) If a decision problem P has the property that the (codes of the) "yes" instances form a c.e. set with an algorithm which does not halt on the "no" instances, we say that P is *semi-decidable*.

The following definition makes sense once we consider the dovetail enumeration above.

Definition 5.1.3. We let W_e denote the e-th computably enumerable set, that is, $\mathrm{dom}(\varphi_e)$, and let $W_e[s] = \{x \leqslant s : \varphi_e(x)[s] \downarrow\}$, where $\varphi_e(x)[s]$ is the result of running the Turing machine corresponding to φ_e for s many steps on input x. We think of $W_{e,s} = W_e[s]$ as the result of performing s steps in the enumeration of W_e.

Remark 5.1.1. We remark that we will use *both* the notations $W_{e,s}$ and $W_e[s]$, as well as similar ones like $\varphi_{e,s}$ and $\varphi_e[s]$ deliberately, as *both* are commonly used in the literature. The notation with $[s]$ is due to Lachlan and is particularly convenient as we take it to mean "approximate everything for s many steps."

If A is c.e., then it clearly has a *computable approximation*, that is, a uniformly computable family $\{A_s\}_{s \in \mathbb{N}}$ of sets such that $A(n) = \lim_s A_s(n)$ for all n. (For example, $\{W_e[s]\}_{s \in \omega}$ is a computable approximation of W_e.) $W_e[s]$ is our old friend $T(e, x, s)$, where T is the Kleene T-predicate of Lemma 3.3.1. The halting problem, or problems, occupy a central place in the theory of c.e. sets.

Definition 5.1.4 (Standard m-Complete Sets). The sets below are computably enumerable:

$$K = \{e : \varphi_e(e) \downarrow\}.$$

$$K_0 = \{\langle e, y \rangle : \varphi_x(y) \downarrow\}.$$

Note that "K" in this definition is the standard notation. The etymology is that "K" is the first letter of *komplett*, "complete" in German. K and K_0

are called *m-complete* since they are c.e. and if W is a c.e. set the $W \leqslant_m C$ where $C \in \{K, K_0\}$.

This is obviously true for K_0 since

$$y \in W_x \text{ iff } \langle x, y \rangle \in K_0.$$

The m-reduction is given by $y \mapsto \langle x, y \rangle$ and for a fixed x, this is primitive recursive. We will prove the m-completeness of K below, once we have the *s-m-n* Theorem as a tool.

A set B is *co-c.e.* if its complement, $\mathbb{N} \setminus B$, is c.e. Thus a set is computable iff it is both c.e. and co-c.e. Of course, it also makes sense to say that A is computable if its characteristic function χ_A is computable, particularly since, as mentioned in Chapter 1, we identify sets with their characteristic functions. It is straightforward to check that A is computable in the sense of Definition 5.1.2 if and only if χ_A is computable.

Proposition 5.1.1. *A is computable iff χ_A is computable; That is definition 5.1.2 holds.*

Proof. If A is c.e. and co-c.e. then we have partial computable functions f and g such that $A = dom(f)$ and $\overline{A} = dom(g)$. To compute $\chi_A(x) = A(x)$, we run both $f(x)$ and $g(x)$ simultaneously, using dovetailing: Compute $f(x)[0], g(x)[0]$ and at stage s, compute $f(x)[s]$ and $g(x)[s]$. Whichever halts first (and exactly one must, as $A \sqcup \overline{A} = \mathbb{N}$), will tell us whether $x \in$ (if $f(x)][s] \downarrow$ first), or $x \in \overline{A}$ (if $g(x)[s] \downarrow$ first). The other direction is trivial. □

The name *computably enumerable* comes from a notion of "effectively countable", via the following characterization, whose proof is straightforward[1].

Proposition 5.1.2. *A set A is computably enumerable iff either $A = \emptyset$ or there is a total computable function f from $\mathbb{N}$ onto A. (If A is infinite then f can be chosen to be injective.)*

Proof. Again, we suppose that $A \neq \emptyset$. Suppose that A is c.e. with $A = \operatorname{dom} \varphi$. We define a computable function f with $f(\mathbb{N}) = A$. Dovetail the enumeration of $\varphi(x)$ so that at stage s, we have computed $A_s =_{\text{def}} \{z : z \leqslant s \wedge \varphi(z) \downarrow [s]\}$. To define $f(0)$ wait for the first stage s where some $x_0 \in A_s$. Then for the least such x_0, we define $f(0) = x_0$. For simplicity, we can suppose that at most one element enters $A_{s+1} \setminus A_s$, so that x_0 is uniquely determined as the first element of A to occur. Then for numbers $n \geqslant 1$ we keep defining $f(n) = x_0$, until we see some $x_1 \neq x_0$ with $x_1 \in A_1 \in A_t \setminus A_s$ at some $t > s$. The we would define $f(t) = x_1$. We could continue to define $f(n) = x_1$ for $n > t$ until some $x_2 \in A_v \setminus A_t \ x_2 \notin \{x_0, x_1\}$ is found, and then define

---

[1] This was the formulation we foreshadowed and used in, for example, the Higman Embedding Theorem in §4.3.5.

$f(v) = x_2$, etc. Then since $x \in A$ iff $x \in A_u$ for some u, we see that the range of f will be A, as required.

Conversely, if $f(\mathbb{N} = A$, is computable, we build a partial computable function φ, by declaring $\varphi(x) \downarrow [s]$ if $f(s) = x$. Then $A = \operatorname{dom} \varphi$. $\square$

The reader might note that this proof is basically an "effectivization" (i.e. making computable step by step) of arguments from Chapter 1, §1.1.2.

Some important refinements of the results above, are the following two results:

Proposition 5.1.3. *An infinite A is c.e. iff there is a computable injective function g such that $g(\mathbb{N}) = A$.*

Proposition 5.1.4.

1. *An infinite A is computable iff there is a computable increasing function g such that $g(\mathbb{N}) = A$.*
2. *Hence show that if A is an infinite c.e. set, there is a computable infinite B with $B \subseteq A$.*

The reader should establish these results as exercises (see below) to make sure that they have followed the material above. Solutions can be found in the back of the book if they are stuck.

5.1.2 Exercises

Exercise 5.1.1 Show that if A and B are c.e. then so are $A \cap B$ and $A \cup B$.

Exercise 5.1.2 Show that if f is a computable function then $\cup_e W_{f(e)}$ is computably enumerable.

Exercise 5.1.3 Modify the proof of Proposition 5.1.2 to prove Proposition 5.1.3.

Exercise 5.1.4 Modify the proof of Proposition 5.1.2 to prove Proposition 5.1.4.

5.1.3 The s-m-n Theorem

We begin with a simple result which looks complicated. It is essentially an analog of the fact used in calculus that a two variable function can be considered as a family of one variable functions, each function being obtained by fixing the first argument. We will use this result relentlessly.

Lemma 5.1.1 (The s-m-n Theorem).

1. *(This is the case $m = n = 1$.) Let $g(x,y)$ be a partial computable function of two variables. Then there is a primitive recursive function[2] $s(x)$ such that, for all x, y,*

$$\varphi_{s(x)}(y) = g(x,y).$$

2. *Let $g(x_1,\ldots,x_m,y_1,\ldots,y_n)$ be a partial computable function of $m + n$ many variables. Then there is a primitive recursive function $s(x)$ such that, for all $x_1,\ldots,x_m,y_1,\ldots,y_n$,*

$$\varphi_{s(x_1,\ldots,x_m))}(y_1,\ldots,y_n)) = g(x_1,\ldots,x_m,y_1,\ldots,y_n).$$

Proof (Sketch). The proof of Lemma 5.1.1 runs as follows: For 1, we will be given (the quadruples and hence the index of) a Turing machine M computing g and a number x, we can build a Turing machine N that on input y simulates the action of writing the pair (x,y) on M's input tape and running M. We can, in a primitive recursive way, calculate an index $s(x)$ for the function computed by N from the index of that of M. $\square$

We give two typical proofs using the s-m-n Theorem. The first is the promised proof of the m-completeness of K

Proposition 5.1.5. $K \equiv_m K_0$.

Proof. Since K is c.e., $K \leqslant_m K_0$. To show that $K_0 \leqslant_m K$, we work as follows. We define a partial computable function $f(\cdot,\cdot)$ via

$$f(\langle x,y\rangle, z) = \begin{cases} 1 \text{ if } \varphi_x(y) \downarrow. \\ \uparrow \text{ otherwise.} \end{cases}$$

Note that $f(a,z)$ does not depend on z. Also note that $f(\langle x,y\rangle, z) \downarrow$ for all z iff $f(\langle x,y\rangle, z) \downarrow$ for *some* z iff $\varphi_x(y) \downarrow$.

By the s-m-n-Theorem, there is a computable function s such that for each a, $f(a,\cdot) = \varphi_{s(a)}$. It follows that $\varphi_x(y) \downarrow$ iff $\varphi_{s(\langle x,y\rangle)}(s(\langle x,y\rangle)) \downarrow$. That is $\langle x,y\rangle \in K_0$ iff $s(\langle x,y\rangle) \in K$. $\square$

Proposition 5.1.6. *Let E be the problem defined as follows:*

Input : $x \in \mathbb{N}$
Question: *Is* $\operatorname{dom}\varphi_x$ *empty?*

Then there is is no algorithm to decide yes instances of E.

Proof. We will show that the halting problem is m-reducible to this emptiness problem. To do this, we define a partial computable function of two variables via

---

[2] Actually, this function is linear for any reasonable programming system.

$$g(x, y) = \begin{cases} 1 & \text{if } \varphi_x(x)\downarrow \\ \uparrow & \text{if } \varphi_x(x)\uparrow. \end{cases}$$

Notice that g ignores its second input.

Via the s-m-n theorem, we can consider $g(x, y)$ as a computable collection of partial computable functions. That is, there is a computable (primitive recursive) $s(x)$ such that, for all x, y, Now

$$\text{dom}(\varphi_{s(x)}) = \begin{cases} \mathbb{N} & \text{if } \varphi_x(x)\downarrow \\ \emptyset & \text{if } \varphi_x(x)\uparrow, \end{cases}$$

so if we could decide for a given x whether $\varphi_{s(x)}$ has empty domain, then we could solve the halting problem. That is $K \leqslant_m N$. □

5.2 Index Sets and Rice's Theorem

We have seen in Exercise 4.1.3 that some algorithmic problems are decidable. For example, if α is a regular expression, "Is $L(\alpha)$ nonempty?" is algorithmically decidable. However, in this section we will show that for a nontrivial problem to be decidable, it *cannot* be a problem given purely by a Turing machine description. We want to capture the idea that we are representing a class of (partial computable) functions and not *particular ways* of computing the functions. This idea leads to the notion of an index set defined below.

Definition 5.2.1 (Index Set).
An *index set* is a set A such that if $x \in A$ and $\varphi_x = \varphi_y$ then $y \in A$.

For example, $\{x : \text{dom}(\varphi_x) = \emptyset\}$ is an index set. On the other hand the set of indices of Turing machine with 12 states is not an index set. Keep in mind: an index set can be thought of as coding a problem about *computable functions* (like the emptiness of domain problem) whose answer does not depend on the *particular algorithm* used to compute a function. Generalizing Proposition 5.1.6, we have the following result, which shows that nontrivial index sets are never computable. Its proof is very similar to that of Proposition 5.1.6.

Theorem 5.2.1 (Rice's Theorem [Ric53]). *An index set A is computable (and so the problem it codes is decidable) iff $A = \mathbb{N}$ or $A = \emptyset$.*

Proof. Let $A \notin \{\emptyset, \mathbb{N}\}$ be an index set. Let e be such that $\text{dom}(\varphi_e) = \emptyset$. We can assume without loss of generality that $e \in \overline{A}$ (the case $e \in A$ being symmetric). Fix $i \in A$. By the s-m-n theorem, there is a computable $s(x)$ such that, for all $y \in \mathbb{N}$,

$$\varphi_{s(x)}(y) = \begin{cases} \varphi_i(y) & \text{if } \varphi_x(x)\downarrow \\ \uparrow & \text{if } \varphi_x(x)\uparrow . \end{cases}$$

If $\varphi_x(x)\downarrow$ then $\varphi_{s(x)} = \varphi_i$ and so $s(x) \in A$, while if $\varphi_x(x)\uparrow$ then $\varphi_{s(x)} = \varphi_e$ and so $s(x) \notin A$. Thus, if A were computable, $\emptyset'$ would also be computable.
$\square$

How should we reconcile Rice's Theorem with the fact that, for example, for a regular expression α we can decide if $L(\alpha)$ is finite? The point is that for that result, we supposed that L was given as a $L(\alpha)$. Suppose that some regular L is only given by M a Turing machine with $L(M) = L$. Because of Rice's Theorem, from *only a Turing Machine description of L*, rather than the particular representation by a regular expression (or by an automaton), we can deduce nothing about a regular language.

Notice that in the proof of Rice's Theorem we are establishing an m-reduction from the halting problem K. Thus, as corollary we have.

Corollary 5.2.1. *Suppose that I is a nontrivial index set. Then either $K \leqslant_m I$ or $K \leqslant_m \bar{I}$. In particular, if I is a c.e. nontrivial index set such as $\{e : \mathrm{dom}(\varphi_e) \neq \emptyset\}$, then $K \equiv_m I$.*

5.3 The Recursion Theorem

A very important and subtle theorem which is a consequence of the s-m-n-Theorem is *The Recursion Theorem*. Kleene's Recursion Theorem (also known as the Fixed Point Theorem) is a fundamental result in classical computability theory. It allows us to *use an index for a computable function or c.e. set that we are building in a construction as part of that very construction*. Thus it forms the theoretical underpinning of the common programming practice of having a routine make recursive calls to itself.

Theorem 5.3.1 (Recursion Theorem, Kleene [Kle38]). *Let f be a total computable function. Then there is a number n, called a* fixed point *of f, such that*

$$\varphi_n = \varphi_{f(n)},$$

and hence

$$W_n = W_{f(n)}.$$

Furthermore, such an n can be computed (in a primitive recursive way) from an index for f.

Proof. First define a total computable (primitive recursive) function d via the s-m-n Theorem so that

$$\varphi_{d(e)}(k) = \begin{cases} \varphi_{\varphi_e(e)}(k) & \text{if } \varphi_e(e)\downarrow \\ \uparrow & \text{if } \varphi_e(e)\uparrow . \end{cases}$$

Let i be such that

$$\varphi_i = f \circ d$$

and let $n = d(i)$. Notice that φ_i is total and its index can be computed in a primitive recursive way from that for f. The following calculation shows that n is a fixed point of f.

$$\varphi_n = \varphi_{d(i)} = \varphi_{\varphi_i(i)} = \varphi_{f \circ d(i)} = \varphi_{f(n)}.$$

The explicit definition of n given above can clearly be carried out computably given an index for f. $\square$

The reader, like the author, might find proof above to be quite mysterious. *It is.* A longer but more perspicuous proof of the recursion theorem was given by Owings [Owi73]; see also Soare [Soa87, pp. 36–37]. I will give another proof in a starred section §5.3.1.

There are many variations on this theme. For example, if $f(x, y)$ is computable, then there is a computable total function $n(y)$ such that, for all y,

$$\varphi_{n(y)} = \varphi_{f(n(y), y)}.$$

This result is called the *Recursion Theorem with Parameters*, and is also due to Kleene [Kle38]. (See Exercise 5.3.7.)

Here is a very simple application of the Recursion Theorem.

Proposition 5.3.1. K *is not an index set.*

Proof. Let f be a computable function such that $\varphi_{f(n)}(n)\downarrow$ and $\varphi_{f(n)}(m)\uparrow$ for all $m \neq n$. We can obtain such an f by a standard application of the s-m-n-Theorem: That is, we can define

$$g(n, y) = \begin{cases} 1 \text{ if } y = n \\ \uparrow \text{ otherwise} \end{cases}$$

Now take f with $\varphi_{f(n)}(\cdot) = g(n, \cdot)$, using the s-m-n-Theorem.

Let n be a fixed point for f, so that $\varphi_n = \varphi_{f(n)}$. Let $m \neq n$ be another index for φ_n. Then $\varphi_n(n)\downarrow$ and hence $n \in K$, but $\varphi_m(m)\uparrow$ and hence $m \notin K$. So $\emptyset'$ is not an index set. $\square$

Note that this example also shows that there is a Turing machine that halts only on its own index.

5.3.1 An Alternative Proof of the Recursion Theorem*

Let's do another proof of the Recursion Theorem. This proof was communicated to Noam Greenberg and is by Iskander Kalimullin from Kazan.

We will work up to it by first looking at an apparently weaker form.

Our goal is to prove that if f is total computable then there is an e with

$$W_e = W_{f(e)},$$

where $W_e = \operatorname{dom} \varphi_e$ is the e-th c.e. set.

Here's an "easy" but FALSE proof

Recall that $\langle , \rangle : \mathbb{N} \times \mathbb{N} \to \mathbb{N}$ the computable bijection. We can denote the n-th *slice* of a set A, $A^{[n]} = \{\langle n, y \rangle \mid \langle n, y \rangle \in A\}$. Clearly, if A is c.e., so is $A^{[n]}$, and moreover, from the numbers e, n we can compute a number $g(e, n)$ where $W_e^{[n]} = W_{g(e,n)}$. (In fact g is primitive recursive.) That is, there is a uniformly computable procedure taking an index e and a slice number n, and computing an index $g(e, n)$ for the n-th slice of W_e. This can be seen using the s-m-n theorem.

Okay, so here's the false proof: Let f be total computable. Let's try to construct a c.e. set A such that, on the assumption that $W_e \neq W_{f(e)}$ for all e, A would be different from all c.e. sets (plainly impossible). To organize thing we would make A look like $K_0 = \{\langle x, y \rangle \mid x \in W_y\}$.

With this idea in mind, we will define a c.e. set

$$A = \{\langle e, y \rangle \mid y \in W_{f(e)}^{[e]}\}.$$

Then A is clearly c.e.. Thus there in some index n where $A = W_n$. Consider the n-th slice of A, and notice since $A = W_n$, $A^{[n]} = W_n^{[n]}$. Then

$$W_n^{[n]} = A^{[n]} = \{\langle n, y \rangle \mid y \in W_{f(n)}^{[n]}\} = W_{f(n)}^{[n]}.$$

Then the desired fixed point is $W_n^{[n]}$, since this is certainly a c.e. set and $W_n^{[n]} = W_{f(n)}^{[n]}$.

The Problem What's the problem? The problem is that the index of $W_n^{[n]}$ is *not* n, but $g(n, n)$.

The Fix, and correct proof

Okay let's modify the proof to make it work: We do this by including g in the definition. We define a c.e. set

$$B = \{\langle e, y \rangle \mid \langle e, y \rangle \in W_{f(g(e,e))}\}.$$

Now B is c.e. thus $B = W_m$ some m and we can compute m from the given data. We have

$$W_{g(m,m)} = W_m^{[m]} = B^{[m]} = \{\langle m, y \rangle \mid \langle m, y \rangle \in W_{f(g(m,m))}\}.$$

hence $g(m, m)$ is a fixed point.

Now, to prove the classical Recursion Theorem:

Theorem 5.3.2 (Recursion, Kleene). *For every computable function f we can compute a fixed point.*

We can use the same proof, but instead of having a c.e. set B, define a partial computable

$$h(\langle e, y \rangle) = \varphi_{f(g(e,c))}(\langle e, y \rangle),$$

and the proof goes as above. (Strictly speaking, g here would now be the index of the partial computable function $h(x, y) = \varphi_e(y)$ for $x = e$ and undefined otherwise.)

I don't know if this is easier, but it gives some insight. The false proof does it best. We try to move every slice by f, but since the underlying set has a number n, it's n-th slice won't move. This can be seen as a diagonalization argument that fails.

5.3.2 Exercises

Exercise 5.3.3 Let g be a computable function. Prove that there is an n such that $W_n = \{n, \ldots, n + g(n)\}$.

Exercise 5.3.4 Prove that there is an index e such that $\varphi_e(e) = e^2$.

Exercise 5.3.5 Let f be a computable function. Show that there is an e such that W_e is computable, and the least index y with $W_y = \overline{W_e}$ has $y > f(e)$.

Exercise 5.3.6 Prove that the collection $Fin = \{n : \mathrm{dom}(\varphi_n) \text{ is finite}\}$ is not computably enumerable. (There are several ways to do this. One is to suppose that $Fin = \mathrm{ra}(g)$ for 1-1 computable g, and then use the Recursion Theorem to build $\varphi_{s(n)}$ which has empty domain until we see $s(n) \in \mathrm{ra}(g)[s]$, at which point make $\varphi_{s(n)}$ total. Then take a fixed point. This method also allows us to show that $Fin \not\leq_T \emptyset'$ using the Limit Lemma.)

Exercise 5.3.7 Prove that if $f(x, y)$ is computable, then there is a computable function $n(y)$ such that, for all y,

$$\varphi_{n(y)} = \varphi_{f(n(y),y)}.$$

Exercise 5.3.8 (Double Recursion Theorem-Smullyan) Let $f(\cdot, \cdot), g(\cdot, \cdot)$ be two computable functions. Prove that there exist a, b such that $\varphi_a = \varphi_{f(a,b)}$ and $\varphi_b = \varphi_{g(a,b)}$.

Exercise 5.3.9 Prove that there is no nontrivial index set I with $I \leq_m \overline{I}$.

Exercise 5.3.10 Use the Recursion Theorem to show that there is an infinite computable set C with $K \cap C = \emptyset$.

5.4 Wait and See Arguments

In this section, we will look a class of often quite simple arguments commonly used in computability theory, called *wait and see arguments*, or sometimes *bait and snatch arguments*. The idea is that we want to build some objects, typically computably enumerable sets, and want to perform some diagonalization. We think of the diagonalization as meeting *requirements*.

As a prototype for such proofs, think of Cantor's [Can79] proof that the collection of all infinite binary sequences is uncountable (essentially Theorem 1.3.1). We can conceive of this proof as follows.

Suppose we could list the infinite binary sequences as $\mathcal{S} = \{S_0, S_1, \ldots\}$, with $S_e = s_{e,0} s_{e,1} \ldots$. It is our goal to construct a binary sequence $U = u_0 u_1 \ldots$ that is not on the list $\mathcal{S}$. We think of the construction as a game against our opponent who must supply us with $\mathcal{S}$. We construct u in stages, at stage t specifying only $u_0 \ldots u_t$, the initial segment of U of length $t + 1$. Our list of *requirements* is the decomposition of the overall goal into subgoals of the form

$$\mathcal{R}_e : U \neq S_e.$$

There is one such requirement for each $e \in \mathbb{N}$. Of course, we know how to satisfy these requirements. At stage e, we simply ensure that $u_e \neq s_{e,e}$ by setting $u_e = 1 - s_{e,e}$. This action ensures that $U \neq S_e$ for all e; in other words, all the requirements are met. This fact contradicts the assumption that $\mathcal{S}$ lists all infinite binary sequences, as U is itself an infinite binary sequence.

5.4.1 Some Examples

In the construction of a c.e. set with certain properties, we will lay some kind of trap to meet requirements R_e of some kind. We observe the dovetail enumeration of the universe and make some decision.

The simplest example is a form of the halting problem. The following is an immediate consequence of the existence of $\emptyset'$, but let's pretend we did not know that result.

Proposition 5.4.1 (The Halting Problem-Revisited). *There is a c.e. noncomputable set A.*

Proof. We think about building A using requirements

$$R_e : \overline{W_e} \neq A.$$

If we build A as a c.e. set and for each e, we satisfy R_e, then A is non-computable, since it will have no c.e. complement. To make A c.e. we will construct $A = \cup_s A_s$ in a computable construction in stages.

The bait for satisfying R_e is the number e.

Construction.

At stage 0 we declare that $e \notin A_0$. We will update the definition of A to define A_{s+1} from A_s. Nothing enters A between stages s and $s+1$ unless *we* declare that it should so enter.

At stage s, R_e is (definitely) satisfied if $A_s \cap W_{e,s} \neq \emptyset$. At stage $s+1$ of the construction, for $e \leqslant s$, if R_e is not yet definitely satisfied, and we see that $e \in W_{e,s}$, then put $e \in A_{s+1} \setminus A_s$. Note that R_e will be satisfied at stage $s+1$, since $A_{s+1} \cap W_{e,s} \neq \emptyset$. At stage s, we will say that R_e *receives attention*.

End of Construction

It is clear that all R_e will be satisfied. *Either $e \notin A \cup W_e$ since R_e never receives attention; or R_e receives attention and $e \in A \cap W_e$.* □

Of course there is nothing new here, but the proof constitutes a new way of viewing the construction of K.

Here is an another example. In this construction, we will construct an example of a variety of c.e. sets we have not yet met in this book.

Definition 5.4.1 (Post [Pos47]). We say that a c.e. set $A \subseteq \mathbb{N}$ is *simple* if $\overline{A}$ is infinite and for all infinite c.e. sets W, $A \cap W \neq \emptyset$.

Theorem 5.4.1 (Post [Pos47]). *Simple sets exist.*

Proof. We will build $A = \cup_s A_s$ in stages. At each stage s we will decide whether to put certain elements into $A_{s+1} \setminus A_s$. Since the construction will be computable, this makes $A = \cup_s A_s$ computably enumerable.

We have two tasks. We must keep $|\overline{A}| = \infty$ and second we need to meet the requirements

$$R_e : |W_e| = \infty \text{ implies } W_e \cap A \neq \emptyset.$$

We think of $\mathbb{N}$ as being divided into *boxes*. The size of the e-th box is $e + 2$, so that $B_0 = \{0, 1\}$, $B_1 = \{2, 3, 4\}$, ...

Construction At stage s, for $e \leqslant s$, if R_e is not yet satisfied, meaning that $W_{e,s} \cap A_s = \emptyset$, and we see some $y \in W_{e,s}$ with $y \notin \cup_{j \leqslant e} B_j$, put $y \in A_{s+1} \setminus A_s$. This action makes R_e satisfied at stage $s+1$.

End of Construction

Notice that since we need to act at most once for each R_e, and the condition $y \notin \cup_{j \leqslant e} B_j$ means that if our action for R_e takes something from box B_j and puts it into A, then $j > e$. Therefore at most $e - 1$ elements can be taken from B_e and put into A. Thus at least one remains. Therefore $\overline{A}$ is infinite as witnessed by the elements of the boxes not put into A. R_e is met since, if W_e is infinite, it must have some element $y \notin \cup_{j \leqslant e} B_j$. This will be observed at some stage $s > y$ and we will put it into A_{s+1}, if necessary.   □

Simple sets were constructed by Post to try to show that there are undecidable semi-decidable problems which are *not* either the halting problem or its complement. Simple sets do this at least for m-reducibility.

Theorem 5.4.2 (Post [Pos47]). *If a c.e. set A is simple, then $\emptyset <_m A <_m K$.*

Proof. Suppose that $K \leqslant_m A$ via computable f. Using the Recursion Theorem, we build an infinite computable $C \subseteq \overline{K}$ as follows (cf. Exercise 5.3.10). C will be $\{e_i : i \in \mathbb{N}\}$, and we will monitor $f(e_i)$. At each stage s, we keep $\varphi_{e_i}(e_i)[s] \uparrow$ making $e_i \notin K_s$. Should ever we observe $f(e_i) = f(e_j)$ for $i \neq j$, at some stage t, then we declare $\varphi_{e_i}(e_i) \downarrow$ (for one i), and this will give a contradiction as $e_i \in K$ and $e_j \notin K$, yet $f(e_i) = f(e_j)$. On the other hand, if we never do this then for all $i \neq j$, $f(e_i) \neq f(e_j)$, and $C \subseteq \overline{K}$ is therefore a computable, implying $f(C)$ is an infinite computable subset of $\overline{A}$.   □

5.4.2 Computable Structure Theory: Computable Linear Orderings

This subsection is devoted to using wait and see; but aimed at a new arena of computability theory: *computable structure theory*. Computable structure theory is a broad area where we seek to understand the *effective (computable) content* of well-known algebraic structures. We have already seen this in Chapter 4, where we looked at word problems in structures such as groups and semigroups. Those structure had *finite* presentations, but there is no reason we not to look at *computable* presentations. Historically, one of the classical examples was that of Frölich and Shepherdson [FS56] who studied computable procedures in field theory[3], such as whether there is an algorithm to classify the algebraic closure. This paper clearly shows the historical context of the subject, the clear intuition of van der Waerden (which apparently came from Emmy Noether's lecture notes) and the fact that isomorphic computable structures (here fields) can have distinct algorithmic properties, and hence *cannot* be *computably* isomorphic. Here we quote from the abstract.

"Van der Waerden (1930a, pp. 128–131) has discussed the problem of carrying out certain field theoretical procedures effectively, i.e. in a finite number of steps. He defined an 'explicitly given' field as one whose elements are uniquely represented by distinguishable symbols with which one can perform the operations of addition,

---

[3] The reader should not concern themselves if they have not had a course in abstract algebra, since this is only a historical example. Suffice to say that fields are central algebraic objects studied by mathematicians, and understanding the effective (algorithmic) content of their theories seems important.

multiplication, subtraction and division in a finite number of steps. He pointed out that if a field K is explicitly given then any finite extension K' of K can be explicitly given, and that if there is a splitting algorithm for K i.e. an effective procedure for splitting polynomials with coefficients in K into their irreducible factors in $K[x]$, then (1) there is a splitting algorithm for K'. He observed in (1930b), however, that there was no general splitting algorithm applicable to all explicitly given fields K, [...] We sharpen van der Waerden's result on the non-existence of a general splitting algorithm by constructing (§7) a particular explicitly given field which has no splitting algorithm. We show (§7) that the result on the existence of a splitting algorithm for a finite extension field does not hold for inseparable extensions, i.e. we construct a particular explicitly given field K and an explicitly given inseparable algebraic extension $K(x)$ such that K has a splitting algorithm but $K(x)$ has not."

In modern terms, Frölich and Shepherdson [FS56] showed that the halting problem is many-one reducible to the problem of having a splitting algorithm.

Since this book is a first course in the theory of computation, we will concentrate on the effective content of a class of structures where we don't need to concern ourselves with complicated operations. That is, we will study the class of linear orderings (linearly ordered sets). Recall that a linear ordering a set A and an ordering $<_A$ which is transitive, antisymmetric and antireflexive. We will only be concerned with *infinite* linear orderings.

Definition 5.4.2. An (infinite) computable linear ordering L is a a linear ordering $L = (\mathbb{N}, <_L)$ where the binary relation $x <_L y$ is a computable relation, and forms a linear ordering. Strictly speaking this is called a *computable presentation* of L.

For example, ω, which is the standard ordering on $\mathbb{N}$ is a computable presentation. Similarly $(\mathbb{Z}, <)$ has a computable presentation where we have the domain $\mathbb{N}$ with $0 \mapsto 0$, $n \mapsto 2n$ and $-n \mapsto 2n + 1$. The ordering L would have all then odd numbers $<_L 0 <_L 2m$ for all m, $2n <_L 2m$ and $2(m+1) <_L 2(n+1)$ for all $m > n$. The ordering would appear as

$$\cdots 5 <_L 3 <_L 1 <_L 0 <_L 2 <_L 4 <_L 6 <_L \cdots.$$

The rationals with the standard ordering has a computable presentation obtained from some Gödel numbering of $\mathbb{Q}$. The rationals have a special role in the theory of countable orderings. All such orderings embed into $(\mathbb{Q}, <_\mathbb{Q})$; indeed all countable orderings embed into the $(\mathbb{Q}, <_\mathbb{Q})$. This result is also computable in the following sense:

Proposition 5.4.2. *If* $L = (L, <_L)$ *is a computable linear ordering then there is a computable subset A of $\mathbb{Q}$ such that L is isomorphic to $(A, \leqslant_\mathbb{Q})$.*

Proof. Although the domain of L is $\mathbb{N}$, it is convenient to denote it by $\{\ell_0, \ell_1, \dots\}$. Consider the presentation $L = \cup_s L_s$. At each stage s we will have

L_s with domain $\{\ell_0, \ell_1, \ldots \ell_s\}$. Suppose that we have mapped $\ell_i \mapsto a_i \in \mathbb{Q}$, for $0 \leqslant i \leqslant s$. At stage $s + 1$, ℓ_{s+1} enters $L_{s+1} \setminus L_s$. One of the following will hold:

1. $\ell_{s+1} <_L \ell_i$ for all $i \in L_s$. In this case, there are infinitely many elements of Q, $q <_{\mathbb{Q}} a_i$ for all $0 \leqslant i \leqslant s$, as $\mathbb{Q}$ has no least element. Choose such a b with Gödel number $> s$, and set $\ell_{s+1} \mapsto b$, so that $a_{s+1} = b$.
2. $\ell_{s+1} >_L \ell_i$ for all $i \in L_s$. In this case, there are infinitely many elements of Q, $q >_{\mathbb{Q}} a_i$ for all $0 \leqslant i \leqslant s$, as $\mathbb{Q}$ has no least element. Choose such a b with Gödel number $> s$, and set $\ell_{s+1} \mapsto b$, so that $a_{s+1} = b$.
3. For some pair ℓ_i, ℓ_j with no $\ell_i <_L \ell_k <_L \ell_j$ in L_s, it is the case that $\ell_i <_L \ell_{s+1} <_L \ell_j$ in L_{s+1}. In that case, as $(\mathbb{Q}, <_{\mathbb{Q}})$ is a dense linear ordering, there are infinitely many elements b with $a_i <_{\mathbb{Q}} b <_{\mathbb{Q}} a_j$. Choose some b with Gödel number $> s$, and set $a_{s+1} = b$, mapping $\ell_{s+1} \mapsto b$.

In either case we have extended the map. The reason we always choose b to have Gödel number bigger than s, is that this ensures that $A = \{a_i : i \in \mathbb{N}\}$ is a computable set. To decide if $b \in A$, first compute the Gödel number of b. If $s = \#(b)$, then go to stage s in the construction above. $b \in A$ iff $b \in A_s = \{a_0, \ldots, a_s\}$, since after stage s only elements with Gödel numbers $> s$ (and hence bigger than that of b) will be placed in the range of the map. $\square$

Definition 5.4.3. The *adjacency relation* $S(L)$ is the collection of pairs in L such that $x <_L y$ and $\forall z \neg (x <_L z <_l y)$.

For example, in the standard ordering of the natural numbers, $(n, n + 1)$ forms an adjacency, for each $n \in \mathbb{N}$. The reader might wonder why $S(L)$ is chosen as the notation. It is because the adjacency relation is sometimes called the *successivity relation*.

Here is a wait and see argument in the theory of computable linear orderings. Often, when we are thinking about the natural numbers as an ordering we will write ω in place of $\mathbb{N}$.

Theorem 5.4.3. *There is a computable presentation $(L, <_L)$ of ω such that the adjacency relation on L is not computable.*

Proof. We construct $L = \cup_s L_s$ in stages. We meet the requirement

$$R_e : \varphi_e(\langle \cdot, \cdot \rangle) \text{ does not compute the adjacency relation of } L.$$

The construction of L works as follows. The bait (witness) for R_e is the pair $(2e, 2(e + 1))$

Construction. At each stage s, we will add at least one new (odd) element to $L_{s+1} \setminus L_s$. It is convenient to think of $L_0 = 0 <_L 2 <_L 4 <_L 4 <_L \ldots$. We could deal with a finite number of points at stage s, but this will always be the skeleton of the orderings, so we might as well as regard this as the initial ordering. For $e \leqslant s$, we will say that R_e requires attention at stage s if it is not yet satisfied, and $\varphi_{e,s}(\langle 2e, 2(e+1) \rangle) \downarrow [s] = 1$, so that it is telling us that $(2e, 2(e+1))$ is an adjacency, then at step s for the least such e, we will add the least odd number $2k + 1$ into the adjacency: that is, declare

$$2e <_L (2k+1) <_L 2(e+1) \text{ in } L.$$

The order type of L is ω since we only have at most $e + 1$ new elements added below $2(e+1)$, one for each R_e. Again it is clear that R_e is satisfied, since *either* $\varphi_e(\langle 2e, 2(e+1) \rangle)$ does not halt to say that $(2e, 2(e+1))$ is an adjacency, and nothing will be added between them, so φ_e is wrong there, *or* $\varphi_e(\langle 2e, 2(e+1) \rangle)$ does halt and say $(2e, 2(e+1))$ is an adjacency, and we switch, because we will make it a non-adjacency once it gets its turn. □

Three of the basic operations performed on linear orderings are reversal, addition and multiplication.

  Thus if L_1 and L_2 are linear orderings, L_1^* denotes the reversal of L_1, and $L_1 + L_2$ is the ordering where we have L_1 preceding L_2. Thus if ω^* is an ordering of type $3 > 2 > 1 > 0$, so that the integers has order type $\omega^* + \omega$. Let **n** be the finite ordering consisting of n points. Then $\mathbb{N} + 2$ would be a copy of the natural numbers followed by an adjacency. However, $2 + \mathbb{N}$ would be an adjacency followed by $\mathbb{N}$. Clearly $2 + \mathbb{N}$ is another copy of $\mathbb{N}$, showing that addition is not commutative!

$L_1 \cdot L_2$ if the ordering obtained by replacing every point of L_2 by a copy of L_1. So the ordering $(\mathbb{Q} + 2 + \mathbb{Q}) \cdot L$ is the ordering obtained by replacing every point in L by an adjacency with a copy of the rationals each side. Note that the adjacencies of $(\mathbb{Q} + 2 + \mathbb{Q}) \cdot L$ would be isomorphic to L.

5.4.3 Exercises

Exercise 5.4.4 We say that a collection of canonical finite sets (finite sets given by Gödel numbers)

$$\mathcal{F} = \{F_x : x \in \mathbb{N}\}$$

is a *standard array* if $F_0 = \{0\}$, $F_1 = \{1, 2\}$, $F_2 = \{3, 4, 5\}$, etc, with F_n having $n + 1$ elements. We say that A is $\mathcal{F}$ array noncomputable if for all e, there is an x such that $W_e \cap F_x = A \cap F_x$. Use a wait and see argument to construct a c.e. $\mathcal{F}$ array noncomputable set A.

Exercise 5.4.5 A disjoint pair A_1, A_2 of c.e. sets is called *computably inseparable* if there is no computable set C with $A_1 \subseteq C$ and $C \cap \overline{A_2} = \emptyset$. Use a wait and see argument to construct a computably inseparable pair of c.e.

sets. (Hint: Build $A_i = \cup_S A_{i,s}$ in stages. Note that W_e is computable iff there is some j with $W_e \sqcup W_j = \mathbb{N}$. Thus, it suffices to meet the requirements

$$R_e : (e = \langle i, j \rangle \wedge W_i \sqcup W_j = \mathbb{N}) \text{ implies } [(A_1 \not\subseteq W_i) \vee (W_i \cap A_2 \neq \emptyset)].$$

Think about setting aside a witness for this requirement, and see if it occurs in W_i or W_j, or neither.)

Exercise 5.4.6 A real number α is called a *computable real* (Turing [Tur36] (1937 correction [Tur37b])) if there is a computable sequence $\{q_i : i \in \mathbb{N}\}$ of rationals such that, for all n,

$$|\alpha - q_n| < 2^{-n}.$$

Equivalently, there is a procedure φ which, on input n computes a rational q_n, with $|\varphi(n) - \alpha| < 2^{-n}$. Familiar reals like π and e and $\sqrt{2}$ are all computable reals, as they have such fast approximating sequences. For example, for e, let $q_n = \sum_{i=0}^{n} \frac{1}{i!}$.

A real number β is called a *left-c.e.* *real* if there is a computable sequence of rationals $\{r_i : i \in \mathbb{N}\}$, with $r_{i+1} \geqslant r_i$ for all i, and

$$\lim_{n \to \infty} r_n = \beta.$$

Construct a left-c.e. real β which is not a computable real. (Hint: This is a wait and see argument. Build $\beta = \lim_n r_n$, in stages $n \in \mathbb{N}^+$. Meet the requirements $R_e : \exists n(\varphi_e(n)$ is wrong on $\beta)$. You can do this by adding a little, whilst still keeping the sequence convergent.)

Exercise 5.4.7 Show that if L is a computable linear ordering, then we can construct a computable presentation of $(\mathbb{Q} + 2 + \mathbb{Q}) \cdot L$. (In Exercise 5.6.5, we will extend this result showing that L can be highly noncomputable and yet $(\mathbb{Q} + 2 + \mathbb{Q}) \cdot L$ can have a computable presentation.)

Exercise 5.4.8 *(This exercise is not easy.) Let $f(\mathbb{N}) = A$ be a 1-1 computable function so that A is a c.e. set. Suppose that A is noncomputable. We can construct a *dump set* D of A as follows. At each stage s, we let $\{d_{i,s} : i \in \mathbb{N}\}$ list in order $\overline{D_s}$. Then we define $D_{s+1} = D_s \cup \{d_{f(s),s}, \ldots, d_{f(s)+s,s}\}$. Thus $d_{i,s+1} = d_{i,s}$ for $i < f(s)$, and $d_{i,s+1} = d_{i+f(s)+s+1,s}$ for $i \geqslant f(s)$.

Show that D is a c.e. simple set. (Hint: You will need to show that $\lim_s d_{i,s} = d_i$ exists. To prove that if W is an infinite c.e. set then $D \cap W \neq \emptyset$, show by contradiction that if not, then A would be computable.)

Exercise 5.4.9 Construct a computable copy of $(\mathbb{N}, <)$, i.e. a computable ordering of order type ω, where the adjacency relation

$$\text{Adj}_L(x, y) = \begin{cases} 1 \text{ if } (x, y) \text{ forms an adjacency,} \\ 0 \text{ otherwise} \end{cases}$$

computes $\emptyset'$. (Hint: The bait should be a dedicated pair for the question "?Is $e \in \emptyset'$?". Make these non-adjacent iff e enters $\emptyset'[s]$. This shows that $\emptyset'$ is m-reducible to the nondajacency relation.)

Exercise 5.4.10 We say that a computable linear ordering L has a *computable distance function* of there is a computable $f(\langle \cdot, \cdot \rangle)$ such that for all x, y, $f(\langle x, y \rangle) = n$ ($n \in \mathbb{N}^+ \cup \{\infty\}$), means that x and y have exactly n elements $<_L$ between them. Construct a computable linear ordering of order type ω with a computable adjacency relation, but for which the distance function is not computable.

5.5 Turing Reducibility

We have been using relentlessly the concept of an m-reduction. If $A \leqslant_m B$ via f, then to compute $A(x)$ (i.e. $\chi_A(x)$) we only need to compute $B(f(x))$. That is, compute $f(x)$ and then see if $f(x) \in B$. But also note that we should be able to compute $A(x)$ by consulting $\overline{B}$; namely $x \in A$ if $f(x) \notin \overline{B}$. We used this approach in Exercise 5.4.9. It is clear that there is no reason that, when computing $A(x)$ from B we should have to limit ourselves to single questions of B. There should be a more general notion of *reduction*, that is, the idea that "if we can decide membership of B then this ability also allows us to decide membership of A".

In other words, *questions about problem A are* reducible *to ones about problem B.*

In the same was that we wanted to clarify the notion of computation, we would like to clarify the most general notion of
    "$A \leqslant B$ if the ability to solve B allows us also to solve A."

There is such a reducibility and it is called *Turing reducibility*. We will introduce this in the next section.

Fixing notation, for any reducibility $\leqslant_R$, we write $A \equiv_R B$ if $A \leqslant_R B$ and $B \leqslant_R A$. We write $A <_R B$ if $A \leqslant_R B$ and $B \nleqslant_R A$. Finally, we write $A \mid_R B$ if $A \nleqslant_R B$ and $B \nleqslant_R A$.

5.5.1 Oracle machines and Turing reducibility

Now the question is: *What principles should we use for a general reduction?* We do a mind experiment akin to that done by Turing when he introduced Turing Machines:

1. We should be able to computably perform all of the actions we did for partial computable functions.
2. We should have access to B as some kind of "read only memory". In the original formulation, this was imagined that there will be a an infinite additional tape where we wrote χ_B in unary, separated by 0's. This was called an *oracle tape*.
3. In any computation deciding $A(x)$, we can access only a finite portion of the memory. This accords with our intuition for being *computable with oracle access*.

The first formulation of this idea is the following. This formulation imagines a Turing process Φ with oracle B as a *functional*, $\Phi(B) = A$. A functional is a function mapping functions to functions. Here we are thinking of A and B as their characteristic functions and therefore also as infinite binary sequences.

The definition below looks a bit technical, but the intuition is the following. We know what it means to be a computable function, and hence we can generalize this to a computable function from strings to strings, resorting to Gödel numbering if necessary. The idea is that $A \leqslant_T B$ should mean that

"We should be able to compute the first x bits of A, $A \upharpoonright x = \sigma$, say using a computable procedure, a partial computable function, evaluated on some finite portion $\tau \prec B$, that is $B \upharpoonright m$ for some m".

In the case of an m-reduction $A \leqslant_m B$ via f, τ is simply $B \upharpoonright f(x)$, although we only use $\tau(f(x)) = B(f(x))$. But the spirit of a general Turing procedure is that we should be able to use the information encoded into $B \upharpoonright m$, not just a single bit.

Definition 5.5.1 (Turing reducibility; functional definition).
Let $\Sigma = \{0, 1\}$.

1. A (partial) Turing procedure Φ is induced by a partial computable function $\varphi : \Sigma^* \to \Sigma^*$, with the following properties
   a. φ is continuous on its domain. That is, $\varphi(\tau_0) \downarrow = \sigma_0$, and $\varphi(\tau_1) \downarrow = \sigma_1$ and $\tau_0 \prec \tau_1$ implies that $\sigma_0 \preccurlyeq \tau_1$.
   b. If $\varphi(\tau) \downarrow = \sigma$, then for each $\widehat{\sigma}$ with $|\widehat{\sigma}| < |\sigma|$, there is a $\widehat{\tau} \prec \tau$, and $\varphi(\widehat{\tau}) = \widehat{\sigma}$. (This condition says that we can ask that before we define $\Phi^\tau(x)$, we will ask that we have already defined $\Phi^\tau(y)$ for $y < x$.)
2. We say that $\Phi^B = A$, induced by φ, if for all $\sigma \prec A$, there is a $\tau \prec B$ with $\varphi(\tau) = \sigma$.
3. We would write $\Phi^B(x) = A(x)$ when $|\sigma| > x$ in this definition.

If there is a Φ with $\Phi^B = A$, then we write $A \leqslant_T B$.

An alternative, and perhaps more traditional, formulation is given by the concept of an *oracle Turing Machine*.

Definition 5.5.2 (Turing [Tur39]). An *oracle machine* is a Turing machine with an extra infinite read-only *oracle tape*, which it can access one bit at a time while performing its computation. That is, we are allowed to ask questions such as "is n on the oracle tape" in the course of the computation. If there is an oracle machine M that computes the set A when its oracle tape codes the set B, then we say that A is *Turing reducible* to B, or *B-computable*, or *computable in B*, and write $A \leqslant_T B$.

The reader might wonder how to formalize this definition using quadruples, etc. One model would work as follows. We imagine a 3 tape Turing Machine, with one tape to work on as with the normal Turing Machine, one tape for the oracle set to be written upon as, for example, χ_B as a sequence of 1's and 0's with $B(x) = 1$ iff $x \in B$. Then one question tape to write queries. Each tape has its own head, and quadruples as processed can use the current state, the symbols currently being read on each tape by the head devoted to the tape to decide the action on the tapes. The query tape has blocks of a special character, which also have a symbol $+, -$, which indicate whether it is a positive or negative query. They also have some symbol such as $*$ which indicates that the query is ready to be asked. There will be extra tuples and states devoted to writing on each tape; work states, question states and query states. During the computation on the work tape we can be generating a question on the question tape, and will place a $*$ when the question has the right length. Then the query states will act to see if the position of 1 in the question tape has (for $+$), or does not have ($-$), the same sign as the appropriate block on the oracle tape. This determines the next state, symbol and moves of the rest of the machine, and wiping the contents of the question tape before the next move of the work tape, so that it is ready for a new query. The oracle tape is read-only; meaning that all we can do is decide if entries on the question tape match the contents of the oracle tape.

This all looks extremely messy; and that's because it is. The details of equivalent formulations have been worked out by many authors and are not especially illuminating; so we will omit them. Again, note that the action is fundamentally local, (in fact primitive recursive) the machine itself is defined without recourse to the contents of the oracle tape, but what the machine does depends on the contents of the oracle tape.

We strongly advise the reader to think of the above definition(s) informally, as there is a version of the Church-Turing Thesis for oracle computations, as we see below[4].

Note that, in either formulation, computing $A(x)$ for any given x, the machine M can make only finitely many queries to the oracle tape; in other words, it can access the value of $B(m)$ for at most finitely many m. This definition can be extended to computing values $f(n)$ of functions in the obvious way. We will also consider situations in which the oracle tape codes a finite string. In that case, if the machine attempts to make any queries beyond the length of the string, the computation automatically diverges. All the notation we introduce below for oracle tapes coding sets applies to strings as well. Note that $X \leqslant_T \emptyset$ iff X is computable. Indeed, if Y is computable, then $X \leqslant_T Y$ iff X is computable.

If $X \leqslant_T B$ then we say that X is B-computable, and X is computable *relative to* B.

It is not hard to check that Turing reducibility is transitive and reflexive, and thus is a preordering on the subsets of $\mathbb{N}$.

Definition 5.5.3.

1. The equivalence classes of the form $\deg(A) = \{B : B \equiv_T A\}$ code a notion of equicomputability and are called *(Turing) degrees (of unsolvability)*.
2. We always use boldface letters such as **a** for Turing degrees.
3. A Turing degree is *computably enumerable* if it contains a computably enumerable set.
4. It is important that the reader realize that a degree **a** being c.e. does **not** imply that all the sets in the degree are c.e.. For example, $\overline{K}$ is in the degree of the halting problem and is not c.e., lest the halting problem be algorithmically solvable.

The Turing degrees inherit a natural ordering from Turing reducibility: $\mathbf{a} \leqslant \mathbf{b}$ iff $A \leqslant_T B$ for some (or equivalently all) $A \in \mathbf{a}$ and $B \in \mathbf{b}$. We will

---

[4] Yet another way to formulate the idea above (and they are all the same in reality) is to separate out the queries into positive and negative parts. We regard a partial computable function as corresponding to a c.e set and have a suitable computably enumerable set with certain consistency properties you would expect for reductions $\Phi^B(x) = y \downarrow$ iff there exist $\langle x, y, u, v \rangle \in W$ and $D_u \subset B$ and $D_v \subset \overline{B}$ where D_z is the finite set with Gödel number z. See Rogers [Rog87], §9.2

relentlessly mix notation by writing, for example, $A <_T \mathbf{b}$, for a set A and a degree $\mathbf{b}$, to mean that $A <_T B$ for some (or equivalently all) $B \in \mathbf{b}$.

The Turing degrees form an uppersemilattice. The join operation is induced by $\oplus$, where $A \oplus B = \{2n : n \in A\} \cup \{2n + 1 : n \in B\}$. Clearly $A, B \leqslant_T A \oplus B$, and if $A, B \leqslant_T C$, then $A \oplus B \leqslant_T C$. Furthermore, if $A \equiv_T \widehat{A}$ and $B \equiv_T \widehat{B}$, then $A \oplus B \equiv_T \widehat{A} \oplus \widehat{B}$. Thus it makes sense to define the *join* $\mathbf{a} \vee \mathbf{b}$ of the degrees $\mathbf{a}$ and $\mathbf{b}$ to be the degree of $A \oplus B$ for some (or equivalently all) $A \in \mathbf{a}$ and $B \in \mathbf{b}$.

We let $\mathbf{0}$ denote the degree of the computable sets. Note that each degree is countable and has only countably many predecessors (since there are only countably many oracle machines), so there are continuum many degrees.

5.5.2 Universal Oracle Machines

For an oracle machine Φ, we write Φ^A for the function computed by Φ with oracle A (i.e., with A coded into its oracle tape). The analog of Proposition 5.1.6 holds for oracle machines.

Proposition 5.5.1 (Universal Oracle Turing Machine). *There is a computable list $\{\Phi_e : e \in \mathbb{N}\}$ of oracle Turing Machines, and a universal oracle machine Φ such that for all e, x, and oracles X,*

$$\Phi^X(e, x) = \Phi_e^X(x).$$

Proof. Proposition 5.5.1 is most easily seen to be true via the functional formulation of Turing reduction, since there is a computable listing of partial computable functions (acting on $\{0, 1\}^*$, without loss of generality, as the coding is immaterial), and a universal partial computable function guaranteed by Proposition 5.1.6. For any e we can regard Φ_e as initially undefined. As we find longer and longer τ's and σ's with $\varphi_e(\tau) = \sigma[s]$ and obeying the rule that we already have seen $\varphi(\widehat{\tau}) = \widehat{\sigma}$ for $\widehat{\sigma} \prec \sigma$. we can extend the definition of $\Phi_e^\tau(|\sigma|) \downarrow [s]$. Thus gives a dovetail enumeration of all the oracle computations, as required. $\square$

The general process of extending a definition or result in the non-oracle case to the oracle case is known as *relativization*. For example, there are corresponding notions of relativization for other models of computation. In the case of partial recursive functions, the act of relativization to an oracle B is particularly simple. The collection of B-partial recursive functions are obtained from the schema for the partial recursive functions and adding the characteristic function for B to the list. The point, of course, is that in any computation of a B partial recursive function $f(n)$, for argument n, we can only access a finite portion of χ_B during its computation of $f(n)$.

We have a version of the Church-Turing Thesis for relativized computation:

Relativized Church-Turing Thesis The class of intuitively computable partial functions computable with access to an oracle B coincides with exactly the B-Turing computable partial functions.

Note that we have replaced partial computable functions $\{\varphi_e : e \in \mathbb{N}\}$ by $\{\Phi_e : e \in \mathbb{N}\}$ to remind the reader that these are oracle machine, but of course φ_e is identical to some $\Phi_{j(e)}$ an oracle machine which ignores the oracle, and acts like φ_e.

Definition 5.5.4. When a set A has a computable approximation $\{A_s\}_{s \in \omega}$, we write $\Phi_e^A(n)[s]$ to mean the result of running Φ_e with oracle A_s on input n for s many steps, a relativized version of Kleene's T-predicate, given by the proof of Proposition 5.5.1.

One nice example of relativization is the following relativization of Proposition 5.4.2.

Proposition 5.5.2. *For all X-computable linear orderings $L = (L, <_L)$, there is a X-computable subset of $\mathbb{Q}$ such that L is X-computably isomorphic to $(A, \leqslant_{\mathbb{Q}})$.*

5.5.3 Use functions

The *use* of a converging oracle computation $\Phi^A(n)$ is $x + 1$ for the largest number x such that the value of $A(x)$ is queried during the computation. (If no such value is queried, then the use of the computation is 0.) We denote this use by $\varphi^A(n)$. In general, when we have an oracle computation represented by an uppercase Greek letter, its use function is represented by the corresponding lowercase Greek letter. Normally, we do not care about the exact value of the largest bit queried during a computation, and can replace the exact use function by any function that is at least as large. Furthermore, we can assume that an oracle machine cannot query its oracle's nth bit before stage n. So we typically adopt the following useful conventions on a use function φ^A.

1. The use function is strictly increasing where defined, that is, $\varphi^A(m) < \varphi^A(n)$ for all $m < n$ such that both these values are defined, and similarly, when A is being approximated, $\varphi^A(m)[s] < \varphi^A(n)[s]$ for all $m < n$ and s such that both these values are defined.

2. When A is being approximated, $\varphi^A(n)[s] \leqslant \varphi^A(n)[t]$ for all n and $s < t$ such that both these values are defined.

3. $\varphi^A(n)[s] \leqslant s$ for all n and s such that this value is defined.

Although in a sense trivial, the following principle is quite important.

Proposition 5.5.1 (Use Principle) *Let $\Phi^A(n)$ be a converging oracle computation, and let B be a set such that $B \upharpoonright \varphi^A(n) = A \upharpoonright \varphi^A(n)$. Then $\Phi^B(n) = \Phi^A(n)$.*

Proof. The sets A and B give the same answers to all questions asked during the relevant computations, so the results must be the same. $\square$

One important consequence of this result is that if A is c.e. and Φ^A is total, then A can compute the function f defined by letting $f(n)$ be the least s by which the computation of $\Phi^A(n)$ has settled, i.e., $\Phi^A(n)[t]\downarrow = \Phi^A(n)[s]\downarrow$ for all $t > s$. The reason is that $f(n)$ is the least s such that $\Phi^A(n)[s]\downarrow$ and $A \upharpoonright \varphi^A(n)[s] = A_s \upharpoonright \varphi^A(n)[s]$.

Remark 5.5.1. The reader should note that using lower case letters like φ to correspond to upper case letter like Φ when representing a use function has some potential for confusion, as here φ might not be a partial computable function. However, this convention is standard practice, and there should be no confusion because of the context.

5.5.4 The jump operator

For a set A, let

$$A' = \{e : \Phi_e^A(e)\downarrow\}.$$

The s-m-n-Theorem can be relativized[5] (put "A" on everything in the proof), so

$$A' \equiv_m \{\langle x, y \rangle : \Phi_x^A(y)\downarrow\}.$$

The set A' represents the halting problem *relativized* to A. For instance, Theorem 4.1.4 (the unsolvability of the halting problem) can be relativized with a completely analogous proof to show that $A' \not\leqslant_T A$ for all A.

Lemma 5.5.1.

1. $A <_T A'$
2. $A \leqslant_T B$ then $A' \leqslant_T B'$.

---

[5] The reader might think that this should be $\equiv_T$ as we'd need to relativize the definition of $\leqslant_m$. However, the proof of the s-m-n-Theorem (which is used in the proof that $K \equiv_m K_0$) a primitive recursive (recall, writing 'x' on the tape) is always the same and does not have anything to do with oracles.

Proof. 1. First $A \leqslant_T A'$ because there is an A-computable function, given
by the s-m-n-Theorem, such that $\Phi_{s(e)}(y) \downarrow$ iff $y \in A$. Then $y \in A$ iff
$\langle s(e), y \rangle \in K_0^A = \{\langle x, y \rangle : \Phi_x^A(y) \downarrow\}$. Now we simply relativize the proof
that the Halting Problem is unsolvable. Suppose that $A' \leqslant_T A$. Then
there is an A-computable Ψ such that

$$\Psi^A(x) = \begin{cases} 1 \text{ if } \Phi_x(x) \downarrow \\ 0 \text{ if } \Phi_x^A(x) \uparrow . \end{cases}$$

Define an A computable function g^A via

$$g^A(x) = \begin{cases} \Phi_x^A(x) + 1 \text{ if } \Psi^A(x) = 1 \\ 1 \text{ if } \Psi^A(x) = 0 \end{cases}$$

Then g^A is total and A-computable. Hence $g^A = \Phi_z^A$ for some z. But then

$$\Phi_z^A(z) \downarrow = g^A(z) = \Phi_z^A(z) + 1.$$

This is a contradiction.

2. Suppose that $A \leqslant_T B$. Let $\Psi^B = A$. We can construct an A-computable f
such that $\Phi_{f(e)}^B = \Phi_e^A$, by using B to simulate Φ_e^A computations using B as
an oracle, via Ψ^B. (That is, if $\Phi_e^A(x)$ queries "$y \in A$?" this is transferred to
B via Ψ, since from B we can compute $A(y)$. Then $\Phi_e^A(e) \downarrow$ iff $\Phi_{f(e)}(e) \downarrow$.
That is $e \in K^A$ iff $\langle f(e), e \rangle \in K_0^A$.
   □

Note that with this notation, we would write

$$\emptyset' = \{e : \Phi_e^\emptyset(e) \downarrow\}.$$

Clearly, this is just another version of the Halting Problem.

Convention 5.5.2 Henceforth, we will write $\emptyset'$ in place of K, to make notation consistent with A'. We will write $\mathbf{0}'$ for the Turing degree of $\emptyset'$.

Another important example of relativization is the concept of a set B being
computably enumerable in a set A, which means that $B = \text{dom}(\Phi_e^A)$ for some
e. Most results in computability theory can be relativized in a completely
straightforward way, and we freely use the relativized versions of theorems
proved below when needed.

Definition 5.5.5 (Jump Operator). We often refer to A' as the *jump*
of A. The *jump operator* is the function $A \mapsto A'$. The *nth jump of A*,
written as $A^{(n)}$, is the result of applying the jump operator n times to
A.

So, for example, $A^{(2)} = A''$ and $A^{(3)} = A'''$. If $\mathbf{a} = \deg(A)$ then we write $\mathbf{a}'$ for $\deg(A')$, and similarly for the nth jump notation. This definition makes sense because $A \equiv_T B$ implies $A' \equiv_T B'$. Note that we have a hierarchy of degrees $\mathbf{0} < \mathbf{0}' < \mathbf{0}'' < \cdots$. Next we explore this hierarchy.

5.6 The Arithmetic Hierarchy

We define the notions of Σ_n^0, Π_n^0, and Δ_n^0 sets as follows. A set A is Σ_n^0 if there is a computable relation $R(x_1, \ldots, x_n, y)$ such that $y \in A$ iff

$$\underbrace{\exists x_1 \, \forall x_2 \, \exists x_3 \, \forall x_4 \cdots Q_n x_n}_{n \text{ alternating quantifiers}} R(x_1, \ldots, x_n, y).$$

Since the quantifiers alternate, Q_n is $\exists$ if n is odd and $\forall$ if n is even. In this definition, we could have had n alternating quantifier *blocks*, instead of single quantifiers, but we can always collapse two successive existential or universal quantifiers into a single one by using pairing functions, so that would not make a difference.

The definition of A being Π_n^0 is the same, except that the leading quantifier is a $\forall$ (but there still are n alternating quantifiers in total). It is easy to see that A is Π_n^0 iff $\overline{A}$ is Σ_n^0.

Finally, we say a set is Δ_n^0 if it is both Σ_n^0 and Π_n^0 (or equivalently, if both it and its complement are Σ_n^0). Note that the Δ_0^0, Π_0^0, and Σ_0^0 sets are all exactly the computable sets. The same is true of the Δ_1^0 sets, as shown by Proposition 5.6.1 below.

These notions give rise to Kleene's *arithmetical hierarchy*, which can be pictured as follows.

$$\Pi_1^0 \qquad\qquad \Pi_2^0$$

$$\Delta_1^0 \qquad\qquad \Delta_2^0 \qquad\qquad \Delta_3^0 \quad \ldots$$

$$\Sigma_1^0 \qquad\qquad \Sigma_2^0$$

As we will see in the next section, there is a strong relationship between the arithmetical hierarchy and enumeration. The following is a simple example at the lowest level of the hierarchy

Proposition 5.6.1 (Kleene) *A set A is computably enumerable iff A is Σ_1^0.*

Proof. If A is c.e. then $A = \operatorname{dom} \varphi_e$. Now we can apply the Kleene Normal Form Theorem, Theorem 3.3.8.

Conversely, if $A \in \Sigma_1^0$, we have a computable R with $n \in A$ iff $\exists x R(x, n)$. We can define a partial computable φ with $\varphi(y) \downarrow [s]$ iff $s > n$ and $\exists x < s(R(x, n))$. Then $n \in A$ iff $n \in \operatorname{dom} \varphi$. $\quad\square$

5.6.1 The Limit Lemma

There is an important generalization of Proposition 5.6.1 due to Post which we will state and prove in §5.6.3. It ties in the arithmetical hierarchy with the degrees of unsolvability, and gives completeness properties of degrees of the form $\mathbf{0}^{(n)}$, highlighting their importance. In this section we prove some preliminary results and related characterizations, beginning with Shoenfield's Limit Lemma.

Saying that a set A is c.e. can be thought of as saying that A has a computable approximation that, for each n, starts out by saying that $n \notin A$, and then changes its mind at most once. More precisely, there is a computable binary function f such that for all n we have $A(n) = \lim_s f(n, s)$, with $f(n, 0) = 0$ and $f(n, s+1) \neq f(n, s)$ for at most one s. Generalizing this idea, Shoenfield's Limit Lemma [Sho59] characterizes the sets computable from the halting problem as being those that have computable approximations with *finitely many* mind changes for each argument. Hence sets computable from the halting problem are "effectively approximable". (In other words, the sets computable from $\emptyset'$ are exactly those that have computable approximations.)

Theorem 5.6.2 (Shoenfield's Limit Lemma [Sho59]).
For a set A, we have $A \leqslant_{\mathrm{T}} \emptyset'$ iff there is a computable binary function g such that, for all n,

(i) $\lim_s g(n, s)$ *exists (i.e., $|\{s : g(n, s) \neq g(n, s+1)\}| < \infty$), and*
(ii) $A(n) = \lim_s g(n, s)$.

Proof. ($\Rightarrow$) Suppose $A = \Phi^{\emptyset'}$. Define g by letting $g(n, s) = 0$ if either $\Phi^{\emptyset'}(n)[s] \uparrow^6$ or $\Phi^{\emptyset'}(n)[s] \downarrow \neq 1$, and letting $g(n, s) = 1$ otherwise. Fix n, and let s be a stage such that $\emptyset'_t \restriction \varphi^{\emptyset'}(n) = \emptyset' \restriction \varphi^{\emptyset'}(n)$ for all $t \geqslant s$. By the use principle (Proposition 5.5.1), $g(n, t) = \Phi^{\emptyset'}(n)[t] = \Phi^{\emptyset'}(n) = A(n)$ for all $t \geqslant s$. Thus g has the required properties.

($\Leftarrow$) Suppose such a function g exists. Without loss of generality, we can assume that $g(n, 0) = 0$ for all n. To show that $A \leqslant_{\mathrm{T}} \emptyset'$, it is enough to build a c.e. set B such that $A \leqslant_{\mathrm{T}} B$, since by $\emptyset' \equiv_m K_0 = \{\langle x, y \rangle : \varphi_x(y) \downarrow\}$, every c.e. set is computable from $\emptyset'$. We put $\langle n, k \rangle$ into B whenever we find that

$$|\{s : g(n, s) \neq g(n, s+1)\}| \geqslant k.$$

---

[6] Remember this notation means that everything is approximated at stage s, including the oracle.

Now define a Turing reduction Γ as follows. Given an oracle X, on input n, search for the least k such that $\langle n, k \rangle \notin X$, and if one is found, then output 0 if k is even and 1 is k is odd. Clearly, $\Gamma^B = A$. $\square$

Intuitively, the proof of the "if" direction of the limit lemma boils down to saying that, since (by Propositions 5.6.1) $\emptyset'$ can decide whether $\exists s > t\,(g(n,s) \neq g(n, s+1))$ for any n and t, it can also compute $\lim_s g(n, s)$.

As we have seen, we often want to relativize results, definitions, and proofs in computability theory. The limit lemma relativizes to show that $A \leq_T B'$ iff there is a B-computable binary function f such that $A(n) = \lim_s f(n, s)$ for all n. Combining this fact with induction, we have the following generalization of the limit lemma.

Corollary 5.6.1 (Limit Lemma, Strong Form). *Let $k \geq 1$. For a set A, we have $A \leq_T \emptyset^{(k)}$ iff there is a computable function g of $k+1$ variables such that $A(n) = \lim_{s_1} \lim_{s_2} \ldots \lim_{s_k} g(n, s_1, s_2, \ldots, s_k)$ for all n.*

5.6.2 Exercises

Exercise 5.6.3 A special kind of Turing reduction is called a *weak truth table* reduction, and is defined as $A \leq_{wtt} B$ iff there is a Turing procedure Φ, and a computable function φ with $\Phi^B = A$ with use function φ. In general there is no computable bound on the size of queries if $A \leq_T B$. Prove that $A \leq_{wtt} \emptyset'$ iff there are computable functions $f(\cdot, \cdot), g$ such that, for all n,

1. $A(n) = \lim_s f(n, s)$, and
2. $|\{s : f(n, s+1) \neq f(n, s)\}| < g(n)$

Exercise 5.6.4 A function $f : (0, 1) \to \mathbb{R}^+$ is called a *(type-2) computable function*[7] (really we should say "functional") if there is an oracle Turing machine M, such that for all $x \in (0, 1)$, given a (code C for a) fast Cauchy sequence $\{q_i : i \in \mathbb{N}\}$ for x, which the reader should recall from Exercise 5.4.6 means that $|q_n - x| < 2^{-n}$, $M^C(m)$ computes a code for a rational r_m such that

$$|f(x) - r_m| < 2^{-m}.$$

Here is some computable calculus[8]:

(i) Why does f being computable make f continuous?
(ii) Show that x^2, and $\sin(x)$ are computable functions.

---

[7] This a small sample from the area of *computable analysis* which is a fascinating area looking at computability on objects which are not discrete. A good introduction can be found in Pour-El and Richards [PER89]. A more recent book devoted to computable analysis for continuous algebraic structures is Downey and Melnikov [DMar].

[8] Compare with the comments by Borel [Bor12] we mentioned in the introduction.

(iii) (Computable Intermediate Value Theorem) Suppose that f and g are computable, with $a < b$ and with $a, b \in (0, 1)$, such that $f(a) < 0$ and $f(b) > 0$. Show that there is a *computable* real z with $a < z, b$ and $f(z) = 0$. (See Exercise 5.4.6 for the definition of a computable real. You may assume that a and b are computable reals, since the rational numbers are dense.)

(iv) Suppose that f is computable and bounded on $(0, 1)$. Show that $I(x) = \int_{z=0}^{x} f(z)$ is computable. Here $\int$ is the Riemann integral, though it really does not matter which integral is used.

Exercise 5.6.5 (Downey and Knight [DK92]) Let L be a $\emptyset'$-presented linear ordering. Thus we can consider L as a Δ_2^0 subset of $(\mathbb{Q}, <)$ (Here I have dropped the $\mathbb{Q}$ from $<$ as the meaning is clear.) Show that there is a computable presentation of $(\mathbb{Q} + 2 + \mathbb{Q}) \cdot L$. (Hint: If we knew L, then for each $a \in L$, we would turn it into $\mathbb{Q} + 2 + \mathbb{Q}$, and could do this by first making an adjacency (for the 2), then building a dense ordering each side of the 2. But we don't actually know $a \in L$. By the Limit Lemma, a might enter and leave L_s for finitely many s. Each time it enters, while it remains in L_t, we would perform this task. If $a \in L_s \setminus L_{s+1}$, then we would instead turn the 2 into a dense linear ordering. If a re-enters L_t at some $t > s + 1$, we'd pick two new points with large Gödel numbers for the adjacency to represent a. This process will stabilize at stage $s_0 = s_0(a)$, with either $a \in L_t$ for all $t > s_0$ (in which case we'd have a stable 2, with $a \mapsto 2$, or $a \notin L_t$ for all $t > s_0$, in which case nothing is built and all past mistakes have been densified.)

5.6.3 Post's Theorem

We now turn to Post's characterization of the levels of the arithmetical hierarchy. Let $\mathcal{C}$ be a class of sets (such as a level of the arithmetical hierarchy). A set A is $\mathcal{C}$-*complete* if $A \in \mathcal{C}$ and $B \leqslant_T A$ for all $B \in \mathcal{C}$. If in fact $B \leqslant_m A$ for all $B \in \mathcal{C}$, then we say that A is $\mathcal{C}$ *m-complete*, and similarly for other strong reducibilities.

Theorem 5.6.6 (Post's Theorem). *Let* $n \geqslant 0$. *Recall that* $\emptyset^{(n)}$ *denotes the n-th jump of* $\emptyset$

(i) *A set B is $\Sigma_{n+1}^0 \Leftrightarrow B$ is c.e. in some Σ_n^0 set $\Leftrightarrow B$ is c.e. in some Π_n^0 set.*

(ii) *The set $\emptyset^{(n)}$ is Σ_n^0 m-complete.*

(iii) *A set B is Σ_{n+1}^0 iff B is c.e. in $\emptyset^{(n)}$.*

(iv) *A set B is Δ_{n+1}^0 iff $B \leqslant_T \emptyset^{(n)}$.*

Proof. (i) First note that if B is c.e. in A then B is also c.e. in $\overline{A}$. Thus, being c.e. in a Σ_n^0 set is the same as being c.e. in a Π_n^0 set, so all we need to show is that B is Σ_{n+1}^0 iff B is c.e. in some Π_n^0 set.

The "only if" direction has the same proof as the corresponding part of Proposition 5.6.1, except that the computable relation R in that proof is now replaced by a Π_n^0 relation R.

For the "if" direction, let B be c.e. in some Π_n^0 set A. Then, by Proposition 5.6.1 relativized to A, there is an e such that $n \in B$ iff

$$\exists s\, \exists \sigma \prec A\, (\Phi_e^\sigma(n)[s]\downarrow). \tag{5.1}$$

The property in parentheses is computable, while the property $\sigma \prec A$ is a combination of a Π_n^0 statement (asserting that certain elements are in A) and a Σ_n^0 statement (asserting that certain elements are not in A), and hence is Δ_{n+1}^0. So (5.1) is a Σ_{n+1}^0 statement.

(ii) We proceed by induction. By Proposition 5.6.1, $\emptyset'$ is Σ_1^0 m-complete. Now assume by induction that $\emptyset^{(n)}$ is Σ_n^0 m-complete. Since $\emptyset^{(n+1)}$ is c.e. in $\emptyset^{(n)}$, it is Σ_{n+1}^0. Let C be Σ_{n+1}^0. By part (i), C is c.e. in some Σ_n^0 set, and hence it is c.e. in $\emptyset^{(n)}$. As in the unrelativized case, it is now easy to define a computable function f such that $n \in C$ iff $f(n) \in \emptyset^{(n+1)}$. (In more detail, let e be such that $C = W_e^{\emptyset^{(n)}}$, and define f so that for all oracles X and all n and x, we have $\Phi_{f(n)}^X(x)\downarrow$ iff $n \in W_e^X$.)

(iii) By (i) and (ii), and the fact that if X is c.e. in Y and $Y \leqslant_T Z$, then X is also c.e. in Z.

(iv) The set B is Δ_{n+1}^0 iff B and $\overline{B}$ are both Σ_{n+1}^0, and hence both c.e. in $\emptyset^{(n)}$ by (ii). But a set and its complement are both c.e. in X iff the set is computable in X. Thus B is Δ_{n+1}^0 iff $B \leqslant_T \emptyset^{(n)}$. □

There are many "natural" sets, such as certain index sets, that are complete for various levels of the arithmetical hierarchy. The following result gives a few examples.

Theorem 5.6.7.

(i) $Fin = \{e : W_e \text{ is finite}\}$ is Σ_2^0 m-complete.
(ii) $Tot = \{e : \varphi_e \text{ is total}\}$ and $Inf = \{e : W_e \text{ is infinite}\}$ are both Π_2^0 m-complete.

Proof. None of these are terribly difficult. We do (i) as an example. We know that $\emptyset''$ is Σ_2^0 m-complete by Post's Theorem, and it is easy to check that Fin is itself Σ_2^0, since $e \in$ Fin iff

$$\exists t \forall s > t \forall x > t[\varphi_e(x) \uparrow [s]].$$

We need to m-reduce $\emptyset''$ to Fin. Using the s-m-n theorem, we can define a computable function f such that for all s and e, we have $s \in W_{f(e)}$ iff there is a $t \geqslant s$ such that either $\Phi_e^{\emptyset'}(e)[t]\uparrow$ or $\emptyset_{t+1}' \restriction \varphi_e^{\emptyset'}(e)[t] \neq \emptyset_t' \restriction \varphi_e^{\emptyset'}(e)[t]$. Then $f(e) \in$ Fin iff $\Phi_e^{\emptyset'}(e)\downarrow$ iff $e \in \emptyset''$.

Part (ii) is similar, and is left to the reader in Exercise 5.6.9. □

The following result is more difficult, and can easily be skipped; but the reader who works through the proof will see some of the more dynamic methods used in classical computability theory.

Theorem 5.6.8. * $Cof = \{e : W_e \text{ is cofinite}\}$ *is* Σ_3^0 *m-complete.*

Proof. * First Cof is Σ_3^0 because

$$e \in \text{Cof iff } \exists v \forall t \forall y \geqslant v \exists s (\varphi_e(y) \downarrow [s]).$$

Showing Σ_3^0 m-hardness involves approximating the relevant relations. We wish to show Cof is Σ_3^0 m-complete. Let S be a Σ_3^0 relation, and hence $x \in S$ iff $\exists v \forall t \exists s R(x, v, t, s)$, with R computable. For the pair (x, v), we will say that the relation R *fires* for $t = 0$ at stage s if $R(x, v, 0, s)$ holds. More generally, we will say it fires for the t-th time if it has already fired for the $(t - 1)$-st time and $\exists s' \leqslant s$ with (x, v, t, s') holding.

The goal is to construct a partial computable $g = \varphi_{f(x)}$ whose index $f(x)$ is computable from x, and make sure that

if there is some v such $\forall t((x, v)$ fires t times$)$ then,

from some point m onwards, for all $z \geqslant m, g(z) \downarrow$.

The simplest method is to use the Recursion Theorem with parameters (actually we could do this with the s-m-n-Theorem, at a pinch). For each x, we will build a partial computable function $\varphi_{f(x)}$, with f computable, and such that $\text{dom}\,\varphi_{f(x)}$ is co-finite iff $x \in S$. The domain of $\varphi_{f(x)}$ is broken into (separated) *zones*, which will be denoted by $Z_{v,s}$ for the v-th zone at stage s. At each stage s we will look only at zones $Z_{i,s}$ for $i \leqslant s$. The zones have movement rules

1. If $|Z_{v,s+1}| > |Z_{v,s}|$ then we *initialize* $Z_{v',s+1}$ for all $v' > v$. Initialization means that for the smallest v with $|Z_{v,s+1}| > |Z_{v,s}|$, we will ensure that $Z_{v,s+1} \supseteq \cup_{s \geqslant v' \geqslant v} Z_{v',s}$, that is, $Z_{v,s+1}$ "eats" all the zones above it.
2. If $|Z_{v,s+1}| > |Z_{v,s}|$ with v least, then if b is the largest number yet used in the construction, in order of $s + 1 \geqslant v' > v$, we will define $Z_{v',s+1} = \{b + (v' - v)\}$. This action is sometimes called *kicking* into *fresh* numbers as we are kicking the zones into new areas. Note that $Z_{v',s+s} \cap Z_{v',s} = \emptyset$.

The construction runs as follows. For each zone $Z_{v,s}$ and $v \leqslant s + 1$, we will ensure that it has a largest number $z(v, s)$ with $\varphi_{f(x)}(z(v, s)) \uparrow [s]$. For every other $z \in Z_{v,s} \setminus \{z(v, s)\}$, we *declare* that $\varphi_{f(x)}(z) \downarrow [s]$. In the below, parameters do not change values unless specified.

Construction

Stage 0. $Z_{0,0} = \{0\}$, so that $z(0, 0) = 0$. Declare that $\varphi_{f(x)}(z(0, 0)) \uparrow [0]$.

Stage $s + 1$.

Case 1. No (x, v) with $v \leqslant s$ fires. In this case we will add a new Z and otherwise change nothing: For $v \leqslant s$, $Z_{v,s+1} = Z_{v,s}$, and we pick the least fresh number a, and define $Z_{s+1,s+1} = \{a\}$ so that $z(s+1, s+1) = a$ and declare that $\varphi_{f(x)}(a) \uparrow [s+1]$.

Case 2. Some (x, v) with v least fires at stage $s+1$. Then, as above, we let $Z_{v,s+1}$ eat $Z_{v',s}$ for $s \geqslant v' > v$, and for all elements $a \in Z_{v,s} \cup \bigcup_{s \geqslant v' > v} Z_{v',s}$, declare that $\varphi_{f(x)}(a) \downarrow [s+1]$, if currently $\varphi_{f(x)}(a) \uparrow$, that is, if necessary. We add the least number $z \notin Z_{v,s} \cup \bigcup_{s \geqslant v' > v} Z_{v',s}$ into $Z_{v,s+1}$ and declare that $\varphi_{f(x)}(z) \uparrow [s+1]$, so that $z = z(v, s+1)$. Then, as indicated by the zone rules, we define $Z_{v',s+1}$ each with single with fresh numbers for $s + 1 \geqslant v' > v$.

End of Construction

Now we verify that the construction works. Suppose that $x \in$ Cof. Then $\exists v \forall t \exists s R(x, v, t, s)$. We call v a witness if $\forall t \exists s R(x, v, t, s)$. There is some *least* witness, and we consider this v. Since for all $w < v$, $\exists t \forall s > t \neg R(x, w, t, s)$, there is some stage s_0 after which (x, w) will not fire for any $w < v$. If $s_0 - 1$ is the last stage any such (x, w) fires, then after s_0, $Z_{v,s}$ will never again be initialized, since zones are initialized only by smaller (x, w) firing.

Now, since $\forall t \exists s R(x, v, s, t)$, (x, v) must fire infinitely many times. Every stage s it fires it eats all bigger zones v' for $v \leqslant v' \leqslant s$ in existence and makes $\varphi_{f(x)}(a) \downarrow [s+1]$ for all $p \leqslant a \leqslant s$, where p is the least number in $Z_{v,s}$ (which does not change after stage s_0). Because it fires infinitely often, for *every* number $a > p$ there will be some stage where we define $\varphi_{f(x)}(a) \downarrow$. Thus $f(x) \in$ Cof.

Conversely, suppose that x is not in Σ_3^0. That is

$$\forall v \exists s \forall t \neg R(x, v, s, t).$$

Then for each v, we will reach a stage $s = s_v$ such that, for all $w \leqslant v$, (x, w) does not fire after stage s_v. It follows that $Z_{v,s_v} = Z_{v,s'}$ for all $s' \geqslant s_v$. Every zone at each stage has one element $z(v, s)$ with $\varphi_{f(x)}(z(v, s)) \uparrow [s]$, and this won't change after stage s_v. Therefore there are infinitely many elements, namely $z(v) = \lim_s z(v, s)$ with $\varphi_{f(x)}(z(v)) \uparrow$, so that $f(x) \notin$ Cof. $\square$

5.6.4 Exercises

Exercise 5.6.9 Prove that the following sets are are both Π_2^0 m-complete.

1. Tot $= \{e : \varphi_e$ is total$\}$.
2. Inf $= \{e : W_e$ is infinite$\}$.

Exercise 5.6.10 Show that the index sets $J = \{e : \varphi_e$ halts in exactly 3 places$\}$ and $Q = \{e : \varphi_e$ halts in at most 3 places$\}$ are both Σ_2^0 m-complete.

Exercise 5.6.11 Show that the index set $P = \{e : \varphi_e$ has infinite co-infinite domain (i.e. $\exists^\infty x \notin \text{dom}\, \varphi_e \wedge \exists^\infty y \in \text{dom}\, \varphi_e)\}$ is Π_2^0 m-complete.

Exercise 5.6.12 * (You will need to use the method of Theorem 5.6.8.) Prove that $\{\langle x, y\rangle : W_x =^* W_y\}$ is Σ_3^0 m-complete. Here $W_x =^* W_y$ means that they differ by only a finite number of elements.

Exercise 5.6.13 *[Lerman [Ler81]] (You will need to use the method of Theorem 5.6.8.) Let L be a linear ordering. A block of size n in L is a collection of elements $x_1 <_L \cdots <_L x_n$, such that (x_i, x_{i+1}) is an adjacency and there is no $y \in L$ with either (y, x_1) nor (x_n, y) an adjacency. (That is x_1 and x_n are limit points.) Consider a computable ordering L of order-type $\mathbb{Z} + n_1 + \mathbb{Z} + n_1 + \dots$. Then let $B(L) = \{n : L$ has a block of size $n\}$ Such and L is called an $\mathbb{Z}$-*representation* of S.

1. Show that $B(L)$ is Σ_3^0.
2. Show that if S is an Σ_3^0 set, then it has a $\mathbb{Z}$-representation; that is there is a computable ordering L (of the indicated order-type), with $B(L) = S$[9].

    (Hint: Without loss of generality, suppose that $2 \in S$, and is the smallest element of S. Let $x \in S$ iff $\exists v \forall s \exists t R(x, v, s, t)$. Break the ordering into $\mathbb{Z} + 2 + \mathbb{Z} + 2 + \dots$ at stage 0. Devote n-th copy of 2 to $\langle x, v\rangle = n$. First turn the 2-block into an x-block. The idea is to try to build around 2 by putting points y between the 2 block and the $\mathbb{Z}$'s on each side: $\mathbb{Z} + \text{(points in here)} + x + \text{(points in here)} + \mathbb{Z}$. Whilst $\langle x, v\rangle$ is not firing try to turn the x into a copy of $\mathbb{Z}$, by putting point on the *outside* of the $\mathbb{Z} + (yyyyyyxyyyyyy) + \mathbb{Z}$ (here y denotes single points) by putting one y more at each end Each time $\langle x, v\rangle$ fires also put a y on the right hand end of the first lot of y-points, and one of the left hand end of the second. If $\langle x, v\rangle$ fires infinitely often, it will isolate out the x-block, and we'll produce $\mathbb{Z} + \mathbb{Z} + x + \mathbb{Z} + \mathbb{Z}$. If it stops firing then we'll replace the x block by a copy of $\mathbb{Z}$, and we'll finish with $\mathbb{Z} + \mathbb{Z} + \mathbb{Z}$.)

5.7 The Structure of Turing Degrees and Post's Problem

It is consistent with our knowledge so far that the only Turing degrees are $\mathbf{0}, \mathbf{0}', \mathbf{0}'', \dots$ (Actually with "transfinite" extensions of this beginning with $\mathbf{0}^{(\omega)} = \deg_T(\emptyset^{(\omega)})$ where $\emptyset^{(\omega)} =_{\text{def}} \oplus_{n \in \mathbb{N}} \emptyset^{(n)}$, a uniform upper bound of all the $\mathbf{0}^{(n)}$'s.), but certainly the only *arithmetical* degrees we have seen have been the degrees of the iterated jumps of $\emptyset$, given by Post's Theorem, Theorem 5.6.6. Maybe those are the only Turing degrees; namely the natural degrees

---

[9] This result will relativize. For any X, and S a Σ_3^X set, there is an X-computable linear ordering $\mathbb{Z}$-representing S. By Post's Theorem there exist $S \in \Sigma_3^{\emptyset'} \setminus \Sigma_3^0$, such as $\emptyset''$. Since $B(L)$ is a classical invariant, it follows that there are $\emptyset'$-computable linear orderings not *classically* isomorphic to computable ones. The reason is that if $\widehat{L}$ was a computable ordering isomorphic to L, then $B(\widehat{L})$ is Σ_3^0, a contradiction. The existence of such an ordering is due to Feiner [Fei68] using slightly different methods.

generated by the jump operator[10]. That would make the structure of the ordering of the degrees a linear ordering $\mathbf{0} < \mathbf{0'} < \mathbf{0''} < \ldots$.

This picture is very far from being true. The Turing degrees, even below $\mathbf{0'}$ form a complex structure which is an upper semilattice and not a lattice. We won't have time to look at the complex structure as that is best left to a second course in computability theory as per the classic texts Rogers [Rog87], Lerman [Ler83], and Soare [Soa87, Soa16].

In this section we will have a first foray into the fascinating techniques developed to understand the structure of the Turing degrees. It is fair to say that the impetus for this analysis came from Post's classic paper [Pos44]. Rice's Theorem 5.2.1 shows that all index sets are of degree $\geqslant \mathbf{0'}$. In 1944, Post [Pos44] observed that all computably enumerable problems known at the time were either computable or of Turing degree $\mathbf{0'}$. He asked the following question.

Question 5.7.1 (Post's Problem). Does there exist a computably enumerable degree $\mathbf{a}$ with $\mathbf{0} < \mathbf{a} < \mathbf{0'}$?

As we will see in §5.7.2, Post's problem was finally given a positive answer by Friedberg [Fri57] and Muchnik [Muc56], using a new and ingenious method called the priority method. This method was an effectivization of an earlier method discovered by Kleene and Post [KP54]. The latter is called the finite extension method, and was used to prove the following result.

Theorem 5.7.1 (Kleene and Post [KP54]). *There are degrees $\mathbf{a}$ and $\mathbf{b}$, both below $\mathbf{0'}$, such that $\mathbf{a}|_T\mathbf{b}$. That is $\mathbf{a} \not\leqslant_T \mathbf{b}$ and $\mathbf{b} \not\leqslant_T \mathbf{a}$. In other words, there are $\emptyset'$-computable sets that are incomparable under Turing reducibility.*[11]

Proof. As with §5.4, we will break our task down into infinitely many *requirements*. We construct $A = \lim_s A_s$ and $B = \lim_s B_s$ in stages, to meet the following requirements for all $e \in \mathbb{N}$.

$$\mathcal{R}_{2e} : \Phi_e^A \neq B.$$
$$\mathcal{R}_{2e+1} : \Phi_e^B \neq A.$$

Note that if $A \leqslant_T B$ then there must be some procedure Φ_e with $\Phi_e^B = A$. Hence, if we meet all our requirements then $A \not\leqslant_T B$, and similarly $B \not\leqslant_T A$,

---

[10] There is a longstanding question of whether the jump operator and its iterates are the only "degree invariant" (think natural) operators. *Martin's Conjecture* states that these are the only such operators. See Montalbán [Mon19] for more on this problem

[11] The difference between the Kleene-Post Theorem and the solution to Post's Problem is that the degrees constructed in the proof of Theorem 5.7.1 are not necessarily computably enumerable, but merely Δ_2^0.

so that A and B have incomparable Turing degrees. The fact that $A, B \leqslant_T \emptyset'$ will come from the construction and will be observed at the end.

The argument is by finite extensions, in the sense that at each stage s we specify a finite portion A_s of A and a finite portion B_s of B. These finite portions A_s and B_s will be specified as binary strings. The key invariant that we need to maintain throughout the construction is that $A_s \preccurlyeq A_u$ and $B_s \preccurlyeq B_u$ for all stages $u \geqslant s$. Thus, after stage s we can only *extend* the portions of A and B that we have specified by stage s, which is a hallmark of the finite extension method.

Construction.
Stage 0. Let $A_0 = B_0 = \lambda$ (the empty string).

Stage $2e + 1$. (Attend to $\mathcal{R}_{2e}$.) We will have specified A_{2e} and B_{2e} at stage $2e$. Pick some number x, called a *witness*, with $x \geqslant |B_{2e}|$, and ask whether there is a string σ properly extending A_{2e} such that $\Phi_e^\sigma(x)\downarrow$.

If such a σ exists, then let A_{2e+1} be the length-lexicographically least such σ. Let B_{2e+1} be the string of length $x + 1$ extending B_{2e} such that $B_{2e+1}(n) = 0$ for all n with $|B_{2e}| \leqslant n < x$ and $B_{2e+1}(x) = 1 - \Phi_e^\sigma(x)$.

If no such σ exists, then let $A_{2e+1} = A_{2e}0$ and $B_{2e+1} = B_{2e}0$.

Stage $2e + 2$. (Attend to $\mathcal{R}_{2e+1}$.) Define A_{2e+2} and B_{2e+2} by proceeding in the same way as at stage $2e + 1$, but with the roles of A and B reversed.
End of Construction.

Verification. First note that we have $A_0 \prec A_1 \prec \cdots$ and $B_0 \prec B_1 \prec \cdots$, so A and B are well-defined.

We now prove that we meet the requirement $\mathcal{R}_n$ for each n; in fact, we show that we meet $\mathcal{R}_n$ at stage $n + 1$. Suppose that $n = 2e$ (the case where n is odd being completely analogous). At stage $n + 1$, there are two cases to consider. Let x be as defined at that stage.

If there is a σ properly extending A_n with $\Phi_e^\sigma(x)\downarrow$, then our action is to adopt such a σ as A_{n+1} and define B_{n+1} so that $\Phi_e^{A_{n+1}}(x) \neq B_{n+1}(x)$. Since A extends A_{n+1} and $\Phi_e^{A_{n+1}}(x)\downarrow$, it follows that A and A_{n+1} agree on the use of this computation, and hence $\Phi_e^A(x) = \Phi_e^{A_{n+1}}$. Since B extends B_{n+1}, we also have $B(x) = B_{n+1}(x)$. Thus $\Phi_e^A(x) \neq B(x)$, and $\mathcal{R}_n$ is met.

If there is no σ extending A_n with $\Phi_e^\sigma(x)\downarrow$, then since A is an extension of A_n, it must be the case $\Phi^A(x)\uparrow$, and hence $\mathcal{R}_n$ is again met.

Finally we argue that $A, B \leqslant_T \emptyset'$. Notice that the construction is in fact fully computable except for the decision as to which case we are in at a given stage. There we must decide whether there is a convergent computation of a particular kind. For instance, at stage $2e + 1$ we must decide whether the following holds:

$$\exists \tau \, \exists s \, [\tau \succ A_{2e} \wedge \Phi_e^\tau(x)[s]\downarrow]. \tag{5.2}$$

This is a Σ_1^0 question, uniformly in x, and hence can be decided by $\emptyset'$.[12] □

The reasoning at the end of the above proof is quite common: we often make use of the fact that $\emptyset'$ can answer any Δ_2^0 question, and hence any Σ_1^0 or Π_1^0 question.

A key ingredient of the proof of Theorem 5.7.1 is the use principle (Proposition 5.5.1). In constructions of this sort, where we build objects to defeat certain oracle computations, a typical requirement will say something like "the reduction Γ is not a witness to $A \leqslant_\mathrm{T} B$." If we have a converging computation $\Gamma^B(n)[s] \neq A(n)[s]$ and we "preserve the use" of this computation by not changing B after stage s on the use $\gamma^B(n)[s]$ (and similarly preserve $A(n)$), then we will preserve this disagreement. But this use is only a finite portion of B, so we still have all the numbers bigger than it to meet other requirements. In the finite extension method, this use preservation is automatic, since once we define $B(x)$ we never redefine it, but in other constructions we will introduce below, this may not be the case, because we may have occasion to redefine certain values of B. In that case, to ensure that $\Gamma^B \neq A$, we will have to structure the construction so that, if Γ^B is total, then there are n and s such that $\Gamma^B(n)[s] \neq A(n)[s]$ and, from stage s on, we preserve both $A(n)$ and $B \restriction \gamma^B(n)[s]$.

5.7.1 Exercises

Exercise 5.7.2 A set A is called *autoreducible* if there is a Turing functional Φ such that for all x,

$$A(x) = \Phi^{(A \setminus \{x\})}(x) = A(x).$$

That is, for each x, A can determine if $x \in A$, *without* using x in any query to itself. For example, a complete theory T has this property since, for each (code of) a sentence x, $x \in T$ iff $\neg x \notin T$. Using the finite extension method construct a set $A \leqslant_\mathrm{T} \emptyset'$ which is *not* autoreducible.

Exercise 5.7.3 (Jockusch and Posner [JP78])

1. A set A is called *1-generic* if for all c.e. sets of strings V (that is, $V = \{\sigma : \text{the code of } \sigma \in W_e\}$ for some e), one of the following holds.

   a. There is a $\sigma \in V$ with $\sigma \prec A$.
   b. There is a $\sigma \prec A$, such that for all $\tau \in V$, $\sigma \not\prec \tau$.

---

[12] More precisely, we use the *s-m-n* theorem to construct a computable ternary function f such that for all e, σ, x, and z, we have $\Phi_{f(e,\sigma,x)}(z)\downarrow$ iff (5.2) holds. Then (5.2) holds iff $f(e,\sigma,x) \in \emptyset'$.

We say that A *meets or avoids* V. Use the finite extension method to
construct a 1-generic set $A \leqslant \emptyset'$. (Hint: Meet requirements of the form R_e saying
A meets or avoids V_e, the e-th c.e. set of strings.)

2. Show that if A is 1-generic, then it is not computably enumerable.

Exercise 5.7.4 A pair of Turing degrees $\mathbf{a}, \mathbf{b}$ are said to form a *minimal pair*
if they are both nonzero and for all $\mathbf{c}$ if $\mathbf{c} \leqslant \mathbf{a}, \mathbf{b}$, then $\mathbf{c} = \mathbf{0}$. That is, there
is no common information between $\mathbf{a}$ and $\mathbf{b}$. Use the finite extension method
to construct a minimal pair below $\mathbf{0}'$. (Hint: Build $A = \lim_s A_s$, $B = \lim_s B_s$ to
meet the requirements

$$R_{i,j} : \Phi_i^A = \Phi_j^B = f \text{ total implies } f \text{ is computable,}$$

as well as

$$P_e^A : A \neq W_e,$$
$$P_e^B : B \neq W_e.$$

The P_e^A and P_e^B are met by diagonalization with witnesses. The more complex requirements $R_{i,j}$ requirements are met as follows. Suppose that we have A_s and B_s and are now
dealing with $R_{i,j}$ at step $s + 1$. We seek extensions σ and τ, and a witness n such that
$A_s \preccurlyeq \sigma$, $B_s \preccurlyeq \tau$ and $t \geqslant s$ with

$$\Phi_i^\sigma(n) \downarrow \neq \Phi_j^\tau(n) \downarrow [t].$$

If $\langle \sigma, \tau, n, t \rangle$ is found set $A_{s+1} = \sigma, B_{s+1} = \tau$.)

5.7.2 *Post's Problem and the Finite Injury Priority Method*

A more subtle generalization of the finite extension method is the *priority
method*. We begin by looking at the simplest incarnation of this elegant technique, the *finite injury priority method*. This method is somewhat like the
finite extension method, but with backtracking. It also resembles the wait
and see arguments we have met in §5.4, again considered with backtracking.

The idea behind it is the following. Suppose we must again satisfy require-
ments $\mathcal{R}_0, \mathcal{R}_1, \ldots$, but this time we are constrained to some sort of effective
construction, so we are not allowed questions of a noncomputable oracle dur-
ing the construction. As an illustration, let us reconsider Post's Problem
(Question 5.7.1). Post's Problem asks us to find a c.e. degree strictly between $\mathbf{0}$ and $\mathbf{0}'$. It is clearly enough to construct c.e. sets A and B with
incomparable Turing degrees. The Kleene-Post method does allow us to con-
struct sets with incomparable degrees below $\mathbf{0}'$, using a $\emptyset'$ oracle question
at each stage, but there is no reason to expect these sets to be computably
enumerable. To make A and B c.e., we must have a *computable* (rather than
merely $\emptyset'$-computable) construction where elements go into the sets A and
B but never leave them. As we will see, doing so requires giving up on satisfying our requirements in order. The key idea, discovered independently by

Friedberg [Fri57] and Muchnik [Muc56], is to pursue multiple strategies for each requirement, in the following sense.

In the proof of the Kleene-Post Theorem, it appears that, in satisfying the requirement $\mathcal{R}_{2e}$, we need to know whether or not there is a σ extending A_{2e} such that $\Phi_e^\sigma(x) \downarrow$, where x is our chosen witness. Now our idea is to first *guess* that no such σ exists, which means that we do nothing for $\mathcal{R}_{2e}$ other than keep x out of B. If at some point we find an appropriate σ, we then make A extend σ and put x into B if necessary, as in the Kleene-Post construction.

The only problem is that putting x into B may well upset the action of other requirements of the form $\mathcal{R}_{2i+1}$, because such a requirement might need B to extend some string τ (for the same reason that $\mathcal{R}_{2e}$ needs A to extend σ), which may no longer be possible. If we nevertheless put x into B, we say that we have *injured* $\mathcal{R}_{2i+1}$. Of course, $\mathcal{R}_{2i+1}$ can now choose a new witness and start over from scratch, but perhaps another requirement may injure it again later. So we need to somehow ensure that, for each requirement, there is a stage after which it is never injured.

To make sure that this is the case, we put a *priority ordering* on our requirements, by stating that $\mathcal{R}_j$ has stronger priority than $\mathcal{R}_i$ if $j < i$, and allow $\mathcal{R}_j$ to injure $\mathcal{R}_i$ only if $\mathcal{R}_j$ has stronger priority than $\mathcal{R}_i$. Thus $\mathcal{R}_0$ is never injured. The requirement $\mathcal{R}_1$ may be injured by the action of $\mathcal{R}_0$. However, once this happens $\mathcal{R}_0$ will never act again, so if $\mathcal{R}_1$ is allowed to start over at this point, it will succeed. This process of starting over is called *initialization*. Initializing $\mathcal{R}_1$ means that we restart its action with a new witness, chosen to be larger than any number previously seen in the construction, and hence larger than any number $\mathcal{R}_0$ cares about. This new incarnation of $\mathcal{R}_1$ is guaranteed never to be injured. It should now be clear that, by induction, each requirement will eventually reach a point, following a finite number of initializations, after which it will never be injured and hence will succeed in reaching its goal.

We take the following description of this idea from my book with Hirschfeldt [DH10].

"We may think of this kind of construction as a game between a team of industrialists (each possibly trying to erect a factory) and a team of environmentalists (each possibly trying to build a park). In the end we want the world to be happy. In other words, we want all desired factories and parks to be built. However, some of the players may get distracted by other activities and never decide to build anything, so we cannot simply let one player build, then the next, and so on, because we might then get permanently stuck waiting for a player who never decides to build. Members of the two teams have their own places in the pecking order. For instance, industrialist 6 has stronger priority than all environmentalists except the first six, and therefore can build anywhere except on parks built by the first six environmentalists. So industrialist 6 may choose to build on land already demarcated by environmentalist 10, say, who would then need to find another place to build a park. Of course, even if this event happens, a higher ranked environmentalist, such as number 3, for instance, could later lay claim to that same land, forcing industrialist 6 to find another place to build a factory. Whether the highest ranked industrialist has priority over the highest ranked environmentalist or vice-versa is irrelevant to the construction, so we leave that detail to each reader's political leanings.

For each player, there are only finitely many other players with stronger priority, and once all of these have finished building what they desire, the given player has free pick of the remaining land (which is infinite), and can build on it without later being forced out."

In general, in a finite injury priority argument, we have a list of requirements in some priority ordering. There are several different ways to meet each individual requirement. Exactly which way will be possible to implement depends upon information that is not initially available to us but is "revealed" to us during the construction. The problem is that a requirement cannot wait for others to act, and hence must risk having its work destroyed by the actions of other requirements. We must arrange things so that only requirements of stronger priority can injure ones of weaker priority, and we can always restart the ones of weaker priority once they are injured. In the wait and see arguments of Section 5.4, we would pick a single witness for a requirement, wait for the bait to be snatched by the opponent, and win the requirement once and for all. If the opponent does not snatch the bait, we would win by default. In a finite injury argument, we try to do the same thing, but the difference is that after the opponent snatches the bait, it could undo all the good work we have done for other requirements. We would need to restart the other requirements with new bait, as the are *injured*.

In a finite injury argument, any requirement *requires attention* only finitely often, and we argue by induction that each requirement eventually gets an environment wherein it can be met. Beyond the scope of a text like this, there are much more complex *infinite injury* (or even *monstrous injury* ([Lac76])!) arguments where one requirement might injure another infinitely often, but the key there is that the injury is somehow controlled so that it is still the case that each requirement eventually gets an environment wherein it can be met. Of course, imposing this *coherence criterion* on our constructions means that each requirement must ensure that its action does not prevent weaker requirements from finding appropriate environments where they can succeed in being satisfied. (A principle known as Harrington's "golden rule".) For a more thorough account of these beautiful techniques and their uses in modern computability theory, see Soare [Soa87].

We now turn to the formal description of the solution to Post's Problem by Friedberg and Muchnik, which was the first use of the priority method.

Theorem 5.7.5 (Friedberg [Fri57], Muchnik [Muc56]). *There exist computably enumerable sets A and B such that A and B have incomparable Turing degrees.*

Proof. We build $A = \bigcup_s A_s$ and $B = \bigcup_s B_s$ in stages to satisfy the same requirements as in the proof of the Kleene-Post Theorem. That is, we make A and B c.e. while meeting the following requirements for all $e \in \mathbb{N}$.

$$\mathcal{R}_{2e} : \Phi_e^A \neq B.$$
$$\mathcal{R}_{2e+1} : \Phi_e^B \neq A.$$

The strategy for a single requirement. We begin by looking at the strategy for a single requirement $\mathcal{R}_{2e}$. We first pick a witness x to *follow* $\mathcal{R}_{2e}$. This number is targeted for B, and, of course, we initially keep x out of B. We then wait for a stage s such that $\Phi_e^A(x)[s]\downarrow = 0$. If such a stage does not occur, then either $\Phi_e^A(x)\uparrow$ or $\Phi_e^A(x)\downarrow \neq 0$. In either case, since we keep x out of B, we have $\Phi_e^A(x) \neq 0 = B(x)$, and hence $\mathcal{R}_{2e}$ is satisfied.

If a stage s as above occurs, then we put x into B and *protect A_s*. That is, we try to ensure that any number entering A from now on is greater than any number seen in the construction thus far, and hence in particular greater than $\varphi_e^A(x)[s]$. If we succeed then, by the use principle, $\Phi_e^A(x) = \Phi_e^A(x)[s] = 0 \neq B(x)$, and hence again $\mathcal{R}_{2e}$ is satisfied.

Note that when we take this action, we might injure a requirement $\mathcal{R}_{2i+1}$ that is trying to preserve the use of a computation $\Phi_i^B(x')$, since x may be below this use. As explained above, the priority mechanism will ensure that this can happen only if $2i + 1 > 2e$.

We now proceed with the full construction. We will denote by A_s and B_s the sets of elements enumerated into A and B, respectively, by the end of stage s.

Construction.

Stage 0. Declare that no requirement currently has a follower.

Stage $s+1$. Say that $\mathcal{R}_j$ *requires attention* at this stage if one of the following holds.

(i) $\mathcal{R}_j$ currently has no follower.
(ii) $\mathcal{R}_j$ has a follower x and, for some e, either

    (a) $j = 2e$ and $\Phi_e^A(x)[s]\downarrow = 0 = B_s(x)$ or
    (b) $j = 2e + 1$ and $\Phi_e^B(x)[s]\downarrow = 0 = A_s(x)$.

Find the least j with $\mathcal{R}_j$ requiring attention. (If there is none, then proceed to the next stage.) We suppose that $j = 2e$, the odd case being symmetric. If $\mathcal{R}_{2e}$ has no follower, then let x be a *fresh large* number (that is, one larger than all numbers seen in the construction so far) and appoint x as $\mathcal{R}_{2e}$'s follower.

If $\mathcal{R}_{2e}$ has a follower x, then it must be the case that $\Phi_e^A(x)[s]\downarrow = 0 = B_s(x)$. In this case, enumerate x into B and *initialize* all $\mathcal{R}_k$ with $k > 2e$ by canceling all their followers.

In either case, we say that $\mathcal{R}_{2e}$ *receives attention* at stage s.
End of Construction.

Verification. We prove by induction that, for each j,

(i) $\mathcal{R}_j$ receives attention only finitely often, and
(ii) $\mathcal{R}_j$ is met.

Suppose that (i) holds for all $k < j$. Suppose that $j = 2e$ for some e, the odd case being symmetric. Let s be the least stage such that for all $k < j$, the requirement $\mathcal{R}_k$ does not require attention after stage s. By the minimality of s, some requirement $\mathcal{R}_k$ with $k < j$ received attention at stage s (or $s = 0$), and hence $\mathcal{R}_j$ does not have a follower at the beginning of stage $s + 1$. Thus, $\mathcal{R}_j$ requires attention at stage $s + 1$, and is appointed a follower x. Since $\mathcal{R}_j$ cannot have its follower canceled unless some $\mathcal{R}_k$ with $k < j$ receives attention, x is $\mathcal{R}_j$'s permanent follower.

It is clear by the way followers are chosen that x is never any other requirement's follower, so x will not enter B unless $\mathcal{R}_j$ acts to put it into B. So if $\mathcal{R}_j$ never requires attention after stage $s + 1$, then $x \notin B$, and we never have $\Phi_e^A(x)[t]\downarrow = 0$ for $t > s$, which implies that either $\Phi_e^A(x)\uparrow$ or $\Phi_e^A(x)\downarrow \neq 0$. In either case, $\mathcal{R}_j$ is met.

On the other hand, if $\mathcal{R}_j$ requires attention at a stage $t + 1 > s + 1$, then $x \in B$ and $\Phi_e^A(x)[t]\downarrow = 0$. The only requirements that put numbers into A after stage $t + 1$ are ones weaker than $\mathcal{R}_j$ (i.e., requirements $\mathcal{R}_k$ for $k > j$). Each such strategy is initialized at stage $t + 1$, which means that, when it is later appointed a follower, that follower will be bigger than $\varphi_e^A(x)[t]$. Thus no number less than $\varphi_e^A(x)[t]$ will ever enter A after stage $t + 1$, which implies, by the use principle, that $\Phi^A(x)\downarrow = \Phi_e^A(x)[t] = 0 \neq B(x)$. So in this case also, $\mathcal{R}_j$ is met. Since $x \in B_{t+2}$ and x is $\mathcal{R}_j$'s permanent follower, $\mathcal{R}_j$ never requires attention after stage $t + 1$. $\square$

The reader may wonder if all this complexity is necessary for solving Post's Problem. There *are* solutions to Post's Problem *not using the priority method* such as Kučera [Kuč86] and Downey, Hirschfeldt, Nies and Stephan [DHNS03]. These proofs are *much* more difficult than the proof above. In fact using metamathematical techniques, they have been shown to be *provably* more complicated in a certain technical sense! Also the non-priority techniques used have not yet seen many further applications. The priority technique has seen many applications throughout computability theory, computable structure theory, and also in areas such as algorithmic randomness and descriptive set theory. Such applications are beyond the scope of the present text, but we refer the reader to Downey and Hirschfeldt [DH10], Ash and Knight [AK00], Downey and Melnikov [DMar], and Moschovakis [Mos09].

The above proof is an example of the simplest kind of finite injury argument, what is called a *bounded injury* construction. That is, we can put a computable bound *in advance* on the number of times that a given requirement $\mathcal{R}_j$ will be injured. In this case, the bound is $2^j - 1$.

We give another example of this kind of construction, connected with the important concept of lowness. It is natural to ask what can be said about the jump operator beyond the basic facts we have seen so far. The next theorem proves that the jump operator on degrees is not injective. Indeed, injectivity fails in the first place it can, in the sense that there are noncomputable sets that the jump operator cannot distinguish from $\emptyset$. Recall that X is low if $X' \equiv_T \emptyset'$.

Theorem 5.7.6 (Friedberg). *There is a noncomputable c.e. low set.*

Proof. We construct our set A in stages. To make A noncomputable we need to meet the requirements

$$\mathcal{P}_e : \overline{A} \neq W_e.$$

To make A low we meet the requirements

$$\mathcal{N}_e : \forall n \left[(\exists^\infty s \, \Phi_e^A(n)[s]\downarrow) \implies \Phi^A(n)\downarrow \right].$$

To see that such requirements suffice, suppose they are met and define the computable binary function g by letting $g(x,s) = 1$ if $\Phi_e^A(x)[s]\downarrow$ and $g(x,s) = 0$ otherwise. Then $g(e) = \lim_s g(e,s)$ is well-defined, and by the limit lemma, $A' = \{e : g(e) = 1\} \leqslant_T \emptyset'$.

The strategy for $\mathcal{P}_e$ is simple. We pick a fresh large follower x, and keep it out of A. If x enters W_e, then we put x into A. We meet $\mathcal{N}_e$ by an equally simple conservation strategy. If we see $\Phi_e^A(n)[s] \downarrow$ then we simply try to ensure that $A \upharpoonright \varphi_e^A(n)[s] = A_s \upharpoonright \varphi_e^A(n)[s]$ by initializing all weaker priority requirements, which forces them to choose fresh large numbers as followers. These numbers will be too big to injure the $\Phi_e^A(n)[s]$ computation after stage s. The priority method sorts the actions of the various strategies out. Since $\mathcal{P}_e$ picks a fresh large follower each time it is initialized, it cannot injure any $\mathcal{N}_j$ for $j < e$. It is easy to see that any $\mathcal{N}_e$ can be injured at most e many times, and that each $\mathcal{P}_e$ is met, since it is initialized at most 2^e many times. $\square$

We finish this section with a classical application of the bounded injury method to computable linear orderings. The conflicts are very clear between the two teams of requirements in this proof. The presentation of this proof is taken from [DMNar].

Theorem 5.7.7 (Tennenbaum, unpublished). *There is a computable copy of $\omega^* + \omega$ with no infinite computable ascending or descending suborderings.*

Proof. We will build the ordering $(A, \leqslant_A)$ in stages. The domain of A will be $\mathbb{N}$ so that suborderings correspond to subsets of $\mathbb{N}$. Recall that W_e denotes the e-th computably enumerable set. We meet the requirements

$$R_{2e} : |W_e| = \infty \to W_e \text{ is not an ascending subordering of } A.$$

$$R_{2e+1} : |W_e| = \infty \to W_e \text{ is not a descending subordering of } A.$$

Additionally, we must made sure that the order type of $(A, \leqslant_A)$ is $\omega^* + \omega$. To this end, we will build A as $B + C$, where we will refer to the members of B as *blue* and those in C as *red*. At each stage s we will let

$$B_s = b_{0,s} <_A b_{1,s} <_A \ldots i <_A b_{n,s}$$

and similarly

$$C_s = c_{m,s} <_A c_{m-1,s} <_A \cdots <_A c_{0,s}.$$

Thus we will need to ensure that for all i, $\lim_i c_{i,s} = c_i$ exists and similarly $\lim_s b_{i,s} = b_i$ exists. (We could explicitly add this as new requirements, but it is unnecessary as we see below.)

Now the blue part is of course the ω^* part and the red part the ω part of A. At any stage $s+1$ red element can become blue and vice versa. If $b_{i,s}$ becomes red, say, then every element $x \in B_s$ with $x \geqslant_A b_{i,s}$ will also become red, so that if $b_{i,s}$ is the $\leqslant_A$-least element that becomes red, and k elements become red, then at stage $s + 1$ we'd have the blue elements now $b_{0,s}, \ldots, b_{i-1,s}$, and the red ones now $c_{m+k,s}, c_{m+k-1,s}, \ldots, c_{0,s}$. That is $b_{j,s+1} = b_{j,s}$ for $j \leqslant i-1$ and $c_{j,s+1}$ are the same for $j \leqslant m$ and for $j > m$ are defined as the erstwhile blue elements.

Why would we do such a thing? We wish to make sure that W_e is not an ascending (blue) ω sequence. If W_e contains a red element then it cannot be such a sequence. So the obvious strategy for R_{2e} if to wait till we see some blue $b_{i,s} \in W_{e,s}$, and make it red, as indicated. To make sure that we don't do this for all elements, we will only look at $b_{i,s} > 2e$, so that R_{2e} has no authority to recolour elements $\{0, 1, \ldots, 2e\}$. On the other hand, R_{2q+1} is trying to stop W_q from being an infinite ω^* sequence, and by the same token wants to make red elements blue. If we allowed it to undo the work we just did for R_{2e} by making the erstwhile $b_{i,s}$ blue again, we would undo the work we have done to meet R_{2e}. Thus, when we act to R_k, we will do so with priority k, and make some element the colour demanded by R_k, unless $R_{k'}$ for $k' < k$ wishes to change its colour.

So we would say R_{2e} requires attention at stage s if R_{2e} not currently declared satisfied and we see some $b_{i,s} \in W_{e,s}$ not protected by any R_k for $k < 2e$ and $b_{i,s} > 2e$, and similarly R_{2q+1} with $c_{i,s}$ in place of $b_{i,s}$. The construction if to take the smallest k if any, and perform the demanded re-colouring demanded by R_k which requires attention. The only other part of the construction if that at ever stage, we will add one more blue element to the right of B_{s+1} and one more red element to the left of C_{s+1}. The remaining details are to let the requirements fight it out by priorities.

A routine induction on k shows that we meet R_k. Once the higher priority requirements have ceased activity, if R_k requires attention via some $n \in W_{d,s}$ the whatever colour R_k gives n will thereafter be fixed, as R_k has priority. Finally, if W_d is infinite the such an n will occur. $\square$

Note that R_k can only be injured at most $O(2^k)$ many times.

5.7.3 Exercises

Exercise 5.7.8 Construct an infinite collection $\{A_i : i \in \mathbb{N}\}$ of c.e. sets such that for all $i \neq j$, $A_i |_T A_j$.

Exercise 5.7.9 A disjoint pair is c.e. sets A_1, A_2 is called a *maximal pair* if $|\overline{A_1 \sqcup A_2}| = \infty$, and for all disjoint pairs of c.e. sets W_1, W_2 if $W_1 \supset A_1$ and $W_2 \supset A_2$, then for $i \in \{1, 2\}$, $W_i \setminus A_i$ is finite. Use the bounded injury method to construct a maximal pair of c.e. sets. (Hint: look at requirements of the form $R_{\langle i,j \rangle} : W_i \supseteq A_1 \wedge W_j \supseteq A_2) \rightarrow |W_i \setminus A_1| < \infty \wedge |W_j \setminus A_2| < \infty$, requirements of the form $N_e : \lim_s a_{i,s} = a_i$ exists, where $\{a_{i,s} : i \in \mathbb{N}\}$ lists $\overline{A_{1,s} \sqcup A_{2,s}}$ in order of magnitude. Try to negate the hypothesis of the R_e.)

Exercise 5.7.10 [Ladner [Lad73]]

1. Use the finite injury method to construct a c.e. set which is not autoreducible (see Ex. 5.7.2).
2. (This is not easy.) Show that a c.e. set A is autoreducible iff there exist (disjoint) c.e. sets $A_1 \sqcup A_2 = A$ with $A \equiv_T A_1 \equiv_T A_2$. (Such sets are called *mitotic*.)

Exercise 5.7.11 Given a linear ordering L, a *self-embedding* is a map $f : L \rightarrow L$ such that for all $x <_L y$, $f(x) <_L f(y)$. It is called nontrivial of it is not the identity. Dushnik and Miller [DM40] showed that every infinite countable linear ordering has a nontrivial self embedding. Construct a computable copy of $\mathbb{N}$ with no nontrivial computable self embedding. (Hint: The requirements are $R_e : \varphi_e$ is not a computable nontrivial self-embedding. If φ_e is nontrivial, then there must be some x with $x <_L \varphi_e(x) <_L \varphi_e(\varphi_e(x)) <_L \ldots$. Now consider the cardinalities of $[\varphi_e^{(k)}(x), \varphi_e^{(k+1)}(x)]$ and $[\varphi_e^{(k+1)}(x), \varphi_e^{(k+2)}(x)]$. It is within our power in a construction to put more elements into the former than are in the latter.)

5.7.4 Sacks' Splitting Theorem*

There are priority arguments in which the number of injuries to each requirement, while finite, is not bounded by any computable function. One example is the following proof of Sacks' Splitting Theorem [Sac63]. We write $A = A_0 \sqcup A_1$ to mean that $A = A_0 \cup A_1$ and $A_0 \cap A_1 = \emptyset$.

Theorem 5.7.12 (Sacks [Sac63]). *Let A be a noncomputable c.e. set. Then there exist Turing incomparable c.e. sets A_0 and A_1 such that $A = A_0 \sqcup A_1$.*

Proof. We build $A_i = \bigcup_s A_{i,s}$ in stages by a priority argument to meet the following requirements for all $e \in \mathbb{N}$ and $i = 0, 1$, while ensuring that $A = A_0 \sqcup A_1$.

$$\mathcal{R}_{e,i} : \quad \Phi_e^{A_i} \neq A.$$

These requirements suffice because if $A_{1-i} \leqslant_T A_i$ then $A \leqslant_T A_i$.

Without loss of generality, we will assume that we are given an enumeration of A so that exactly one number enters A at each stage. At each stage we must put this number $x \in A_{s+1} \setminus A_s$ into exactly one of A_0 or A_1, to ensure that $A = A_0 \sqcup A_1$.

To meet $\mathcal{R}_{e,i}$, we define the *length of agreement function*

$$l^i(e,s) = \max\{n : \forall k < n \, (\Phi_e^{A_i}(k)[s] = A(k)[s])\},$$

and the *maximum length of agreement function*

$$m^i(e,s) := \max\{l^i(e,t) : t \leqslant s\},$$

which can be thought of as a high water mark for the length of agreements seen so far. Associated with this maximum length of agreement function is a use function

$$u(e,i,s) = \max\{\varphi_e^{A_i}i(z) : z \leqslant l^i(e,t)[t] : t \leqslant s\},^{13}$$

using the convention that use functions are monotone increasing where defined.

The main idea of the proof is perhaps initially counterintuitive. Let us consider a single requirement $\mathcal{R}_{e,i}$ in isolation. At each stage s, although we want $\Phi_e^{A_i} \neq A$, instead of trying to destroy the agreement between $\Phi_e^{A_i}(k)[s]$ and $A(k)[s]$ represented by $l^i(e,s)$, we try to *preserve* it. (A method sometimes called the *Sacks preservation strategy*.) The way we do this preservation is to put numbers entering $A \upharpoonright u(e,i,s)$ after stage s into A_{1-i} and *not* into A_i. By the use principle, since this action freezes the A_i side of the computations involved in the definition of $l^i(e,s)$, it ensures that for all $k < l^i(e,s)$, we have $\Phi_e^{A_i}(k) = \Phi_e^{A_i}(k)[s]$.

Now suppose that $\liminf_s l^i(e,s) = \infty$, so for each k we can find infinitely many stages s at which $k < l^i(e,s)$. For each such stage, $\Phi_e^{A_i}(k) = \Phi_e^{A_i}(k)[s] = A(k)[s]$. Thus $A(k) = A(k)[s]$ for any such s. So we can compute $A(k)$ simply by finding such an s, which contradicts the noncomputability of A. Thus $\liminf_s l^i(e,s) < \infty$, which clearly implies that $\mathcal{R}_{e,i}$ is met.

In the full construction, of course, we have competing requirements, which we sort out by using priorities. That is, we establish a priority list of our requirements (for instance, saying that $\mathcal{R}_{e,i}$ is stronger than $\mathcal{R}_{e',i'}$ iff $\langle e,i \rangle < \langle e',i' \rangle$). At stage s, for the single element x_s entering A at stage s, we find the strongest priority $\mathcal{R}_{e,i}$ with $\langle e,i \rangle < s$ such that $x_s < u(e,i,s)$ and put x_s into A_{1-i}. We say that $\mathcal{R}_{e,i}$ *acts* at stage s. (If there is no such requirement, then we put x_s into A_0.)

---

[13] Of course, in defining this set we ignore t's such that $l^i(e,t) = 0$. We will do the same without further comment below. Here and below, we take the maximum of the empty set to be 0.

To verify that this construction works, we argue by induction that each requirement eventually stops acting and is met. Suppose that all requirements stronger than $\mathcal{R}_{e,i}$ eventually stop acting, say by a stage $s > \langle e, i \rangle$. At any stage $t > s$, if $x_t < u(e, i, t)$, then x_t is put into A_{1-i}. The same argument as in the one requirement case now shows that if $\liminf_s l^i(e, s) = \infty$ then A is computable, so $\liminf_s l^i(e, s) < \infty$, which implies that $\mathcal{R}_{e,i}$ eventually stops acting and is met. $\square$

Corollary 5.7.1. [14] *There are no minimal computably enumerable degrees: That is, if $\mathbf{a} > \mathbf{0}$ is c.e., then there is a c.e. degree $\mathbf{0} <_T \mathbf{b} <_T \mathbf{a}$.*

In the above construction, injury to a requirement $\mathcal{R}_{e,i}$ happens whenever $x_s < u(e, i, s)$ but x_s is nonetheless put into A_i, at the behest of a stronger priority requirement. How often $\mathcal{R}_{e,i}$ is injured depends on the lengths of agreement attached to stronger priority requirements, and thus cannot be computably bounded[15].

5.7.5 Exercises

Exercise 5.7.13 Show that Sacks method can be used to prove the following. Given a c.e. set $C \neq_T \emptyset$, there is a c.e. noncomputable set A with $C \not\leq_T A$. (Hint: The requirements are $R_e : A \neq W_e$, and $N_e : \Phi_e^A \neq C$. Use the length of agreement function $\ell(e, s) = \max\{x : \forall y < x[\Phi_e^A(y) = C(y)[s]]\}$, where $C = \cup_s C_s$ is a computable enumeration of C.)

Exercise 5.7.14

1. Consider the proof of Sacks Splitting Theorem. Prove that for any non-computable c.e. set C, we can add requirements of the form $\Phi_e^{A_i} \neq C$ to the above construction, satisfying them in the same way that we did for the $\mathcal{R}_{e,i}$ via the method of Exercise 5.7.13. Thus, as shown by Sacks [Sac63], in addition to making $A_0 \mid_T A_1$, we can also ensure that $A_i \not\geq_T C$ for $i = 0, 1$.

2. Show that C being c.e. is not essential. Such an argument shows that if C is a noncomputable Δ_2^0 set, then there exist Turing incomparable c.e. sets A_0 and A_1 such that $A = A_0 \sqcup A_1$ and $A_i \not\geq_T C$ for $i = 0, 1$. Here you would use the Limit Lemma to begin with a Δ_2^0 approximation $C = \lim_s C_s$.

---

[14] A much more difficult result is that the computably enumerable degrees are dense. That is if $\mathbf{a} > \mathbf{c}$ are c.e., then there is a c.e. degree $\mathbf{c} <_T \mathbf{b} <_T \mathbf{a}$. This beautiful result due to Sacks [Sac64], and needs the infinite injury method to prove it.

[15] In fact, it is possible to *calibrate* how many times the injury must occur using more complex techniques, and we refer the reader to [DASMar]

Part III
Computational Complexity Theory

Chapter 6
Computational Complexity

Abstract This chapter looks at the basics of computational complexity theory. We examine how to calibrate computation by measuring the amount of time and space a machine uses. We introduce polynomial time and polynomial space. We prove the hierarchy theorems and Blum's speedup theorem.

Key words: computational complexity theory, time, space, hierarchy theorems, speedup theorem, union theorem, polynomial time, polynomial space

6.1 Introduction

In the previous chapters, we have developed tools to distinguish what is computable from what is not computable. But suppose we have some process which is computable, how should we measure how hard it is to compute? Certainly some tasks seem more difficult than others. For computability theory, mathematicians had an intuitive idea of computation. In the same way, people had an idea that certain tasks were hard and certain ones were easy.

The formal development of a framework for what we now call *computational complexity theory* was initiated in the two papers Hartmanis and Sterns [HS65] and Edmonds [Edm65]. By mathematical standards, these papers occurred "only yesterday". Computational complexity theory is still a subject that is rapidly developing. It is also a subject where, as we see, some of the central problems are completely open. One of these problems is called "$P \neq NP$?" and is regarded by many as the most important problem in mathematics and computer science. It is one of the Clay Millennium Prizes[1], and hence solving this prize would win the solver one million US Dollars. Later we will discuss this problem. We will see why "$P \neq NP$?" is

---

[1] https://www.claymath.org/millennium-problems

© The Author(s), under exclusive license to Springer Nature Switzerland AG 2024   159
R. Downey, *Computability and Complexity*, Undergraduate Topics
in Computer Science, https://doi.org/10.1007/978-3-031-53744-8_6

important, especially as we are in the computer age. But first we will need
to deal with some preliminaries. We'll need to develop the framework.

Turing showed that we could break down computation into elementary
steps, and computable functions were built from combinations of these steps.
The most intuitive idea for trying to understand how hard some computation
is would be to try to understand the resources we need to perform these steps.
Classically, the resources which have been the most important arc:

- *time*. How much time will it cost us to perform a computation?
- *space*. How much memory or space will the computation require?

These two measures will be clarified once we have developed models for
them. Once we have developed models, we will then develop measures of
hardness which we will use to calibrate the intrinsic difficulty of the problems.

The Dream of Complexity Theory Once we have clarified how to
measure resources, then the dream of computational complexity theory
is to develop techniques which, given some algorithm for some computation process, will enable us to prove that the algorithm at hand is the
best we can ever hope to find for the problem, at least in terms of the
relevant resources.

It's good to have dreams.

This dream is *provably impossible*, in that there are some functions which
provably don't have fastest algorithms in any reasonable sense. We will see
this when we prove the *Speedup Theorem* in §6.2. Nonetheless, as will later
see in Chapter 9, §9.6, modulo some *reasonable* complexity assumptions, we
have broadly achieved this dream for a wide class of functions *met in practice*.
But first we move to formalize our concepts.

6.1.1 Size matters

Suppose that we have some function f which is computed by some model
of computation. Presumably, as the input becomes larger and larger, the
resources needed for the computation will increase. Thus the first idea of
computational complexity is the following:

We will measure the complexity of a process according to the *growth
rate* of the resource use as a function of the size of the input.

But what do we want to use as "size"?

So far, our Turing machines, register machines, and the like, have dealt with functions computing their argument in unary notation. Recall that for Turing machines, n was n ones on an otherwise blank tape. But imagine you wished to compute, for example, $f(x) = 2x$. In unary we have seen that this is a block copying exercise. But imagine you had Turing Machines where in the input data was in *binary*. We would be starting on the leftmost (or rightmost) bit of the string representing x and outputting $2x$. In binary, this equates to adding a 0 to the right end of x. The unary simulation would take about $O(n^2)$ many steps whilst the binary one would take the number of steps needed to find the right end of the input, supposing that we began on the leftmost nonzero bit. At first glance, this tradeoff would not seem too bad, unless you cared deeply about very fine grained analysis of some resource on a particular device. In the complexity theory, we want the following:

> As with the theory of computation, whatever we mean by computational complexity should broadly be *model independent*. It is, at present, not altogether clear that this should be possible.

It seems that what takes about n steps on a Turing machine with binary input seems to take about n^2 many steps on unary.

Wrong!!. If I give M the number 2^{100} in binary the representation would be a string of 1 followed by 99 many 0's. In unary, the input would be 2^{100} many 1's, exponentially longer. For the input $n = 2^{50,000,000}$, in unary the input itself, would be larger than the estimated size of the universe! It would have *galactic size*.

What's the correct way to measure the difference between the cost for unary and for binary? We want to measure the performance in terms of the *natural size* of the input.

The following notation/definition is important:

Given (computable) functions f and g, we will say that f is

- $O(g)$ if there is a constant c such that $\forall n(f(n) < c \cdot g(n))$
- $o(g)$ if $\limsup_{n\to\infty} \frac{f(n)}{g(n)} = 0$
- $\Omega(g)$ if there is a constant c such that $\exists n^\infty(c \cdot f(n) > g(n))$

So if $f(n) = 3n, g(n) = 27n+2$, and $h(n) = \frac{2^n}{5}$, then f is $O(g)$, g is $O(f)$, f is $o(h)$, and certainly h is $\Omega(f)$. The number of steps for $f(n) = 2n$ with the input in binary is $o(g)$ where g is the number of steps to even read the input in unary.

Why base 2? The issue above is that representing objects in unary means that the input is exponentially longer than in binary. But this is not true between bases binary and ternary for instance. Since we are only interested in asymptotic behaviour and gross differences such as exponential vs polynomial, we see that base 2 is representative for integers as inputs.

Keep in mind: If we want to have an upper bound on running times we use big O, if we want to show something takes more time than a lower bound we use Ω. If we want to say g is growing much faster than f we use little o.

Suppose we are interested in the running times of algorithms for problems on graphs, such as the HAMILTON CYCLE problem for deciding if there is a cycle through all the vertices of a graph, then we would need to similarly think about what a good measure of size is for a graph. To specify a graph we need to specify the vertices as a list, and then for each i, j whether i is incident with j. For a graph (or directed graph) with n vertices, we specify the labels of the graph in binary each needs $O(\log n)$ many bits, but since we need to say whether they are connected, the specification of the graph takes $O(n^2)$ many bits of information, where n is the number of vertices of the graph.

It seems that all natural problems have natural representations and natural sizes as input to algorithms.

Thus we see that the theory we will be looking at does depend on the way that we specify the input; but that the specification is natural.

Pairing. We will also have to be careful with how we used various computable functions from the previous chapters. Consider the pairing function: $\langle x, y \rangle$. Now supposing that $x, y \in \{0,1\}^*$, presumably we want the size of the code representing the pair as not being too much larger. One way to do this is locally double x, put 01 and then write y. For example, if $x = 10001$ and $y = 1101$, then we can represent $\langle x, y \rangle$ as 1100000011011101. Then $|\langle x, y \rangle| = 2|x| + |y| + 2$. This is "small" relative to $|x| + |y|$.

The reader might wonder how much we can compress the information in (x, y). The answer comes from an area of research called *Kolmogorov Complexity*, and we refer the reader to, for example, Li and Vitanyi [LV93] or Downey and Hirschfeldt [DH10].

6.1.2 The Basic Model

For our purposes we will deal with languages $L \subseteq \{0,1\}^*$. The length of a string will be the size of the input. For simplicity, we will be concerned with the following form, at least at present:

MEMBERSHIP OF L
Input: $x \in \Sigma^*$.
Question: Is $x \in L$?

The idealized computation device is a Turing machine. To rid ourselves of problematical worries about having to copy blocks across parts of the tape, it is convenient to use multi-tape Turing machines where there are a fixed k number of tapes each with their own heads, and an output tape. It is not difficult to show that a k-tape machine running in time t can be simulated by a 2-tape machine in time $O(t^2)$ and with some effort this can be done in time $O(t \log t)$ (Hennie and Sterns [HS66]). (See Exercise 6.1.4.)

- *For Time.* We count the invocation of a quadruple and applying the quadruple as being one step. If we have an algorithm A deciding membership of L, then we will measure its behaviour asymptotically. That is $R(A(x))$ as $|x| \to \infty$, where $R(A(x))$ is the running time, the number of steps, needed to compute $A(x)$ *as a function of* $|x|$.
- *For Space.* We will consider the number of squares we need to use to perform the computation. There is a slight problem here in that for an input x to even read x takes $O(|x|)$ many squares. We are concerned with how much memory is needed for the actual computation. Thus, what we do is to measure space by augmenting the Turing Machine model by adding a work tape, or more precisely k work tapes, and only use the input tape as a read only input. Then we measure space by counting the amount of work tape we need for the computation[2].

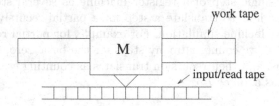

Fig. 6.1: The Basic Model; There Can Be k Work Tapes.

We will examine running times and space considerations using Turing Machines as a model. Now the reader might argue that Turing Machines seem a silly model for measuring running times for computations. For example, it seems unlikely that we would choose a Turing Machine to run a big Machine

---

[2] For the reader familiar with the various sorting algorithms in computing, the bubble sort is one that uses very small space, since it needs only 3 registers.

Learning algorithm. *But for the general theory, we are considering the* coarse *complexity theoretical behaviour such as* $|x|^2$ *vs* $2^{|x|}$, *and hence for coarse behaviour we have the following variation of the Church-Turing Thesis*:

Church-Turing Thesis for Complexity   Turing Machines are a universal model for computational complexity. For any process running in $\leqslant f(n)$ many steps on a universal computational model for inputs x of size (length) $n = |x|$, or using $\leqslant g(n)$ much space deciding $x \in L$, there is a k such that can be simulated by a Turing Machine running in time $O(f(n)^k)$ and space $O(g(n)^k)$.

We have seen as far as computability is concerned, we know that we can translate between Turing Machines and all of the other models. The Church-Turing Thesis for complexity says that these translations have at worst a *polynomial overhead*.

The following is a *fact* we will not prove here as it is horribly tedious and will detract from the narrative:

Theorem 6.1.1. *All of the models* which we have discussed in this book *can be translated from one to another with at worst a cubic overhead.*

That is, not only is there a polynomial overhead, but the polynomial has low degree (3). In the case of space these translations only take *constantly extra* space. The proof comes from very carefully counting how we translate, for example, a single step of a register machine as several steps in a Turing Machine. Or how to translate a step for a partial recursive function as steps in Turing Machine simulation. For example, for partial recursive functions, on input x, we count $|x|$ many steps for the basic zero, successor and monus functions, and then recursion translates as counting etc. We will take Theorem 6.1.1 as a *Black Box*.

Feasible Computations. The reader should keep in mind is that we will be interested in what we mean by being *feasible*. By feasible we mean a process that is able to be computed in "real life instances". That is, a computation being able to be run in time or space without involving galactic numbers; at least for reasonable input sizes. We also wish to examine this notion of feasibility more or less independently or the model. The following can be thought of as the underlying *heuristic principle*:

If a computation is feasible, then we can run it using a polynomial number of steps on a Turing Machine. Moreover, the coefficients and degree of the polynomial will be small.

6.1.3 Discussion and Basic Definitions

Whilst we tend to believe the Church-Turing Thesis, the belief that

> feasible implies "can be simulated using only a polynomial number of steps on a Turing Machine"

is somewhat harder to believe. In some sense it must be viewed as a hypothesis in physics, which makes it more of a statement about *physical devices*. This hypothesis is supported by the fact that currently all classical computation models have the mutual simulation property expressed in Theorem 6.1.1. However, the introduction of quantum computers *could* result in a class of functions which don't have a polynomial translation into a Turing Machine computation. But the picture remains murky, as

- nobody knows how to build a large quantum computer, and
- for the algorithmic tasks (like factoring) which are theoretically (apparently) faster on a quantum computer, we also don't know how to establish that they are *not* of the same speed on a Turing Machine.

Perhaps there are other models created from physics yet to be discovered which could run more than polynomially faster on some problem than a Turing Machine can.

The upshot is that much of computational complexity theory is somewhat *pragmatic*. The basic definitions are the following.

Definition 6.1.1 (Basic Definitions-Deterministic). Let $L \subseteq \{0,1\}^*$.

1. Let g be a function. We sat that $L \in \mathrm{DTIME}(g)$ iff there is a deterministic (multi-tape) Turing Machine M which accepts L, such that, for all $x \in \{0,1\}^*$, $M(x)$ uses only $O(g(|x|))$ many steps.
2. $L \in \mathrm{DSPACE}(g)$ iff there is a deterministic (multi-tape) Turing Machine M which accepts L, such that, for all $x \in \{0,1\}^*$, $M(x)$ uses only $O(g(|x|))$ many squares on the work tape(s).

As with automata, we may use nondeterministic Turing Machines, and arrive at analogous definitions for the above. As with automata, for a nondeterministic automaton, M *accepts* x means that some computation path accepts x.

Definition 6.1.2 (Basic Definitions-Nondeterministic).
Let $L \subseteq \{0,1\}^*$.

1. Let g be a function. We sat that $L \in \mathrm{NTIME}(g)$ iff there is a nondeterministic Turing Machine M which accepts L, such that, for all $x \in \{0,1\}^*$, $M(x)$ uses only $O(g(|x|))$ many steps (on all computation paths).
2. $L \in \mathrm{NSPACE}(g)$ iff there is a nondeterministic Turing Machine M which accepts L, such that, for all $x \in \{0,1\}^*$, $M(x)$ uses only $O(g(|x|))$ many squares on the work tape during all computations.

Various special g's are given names. So if $g(n) = O(n)$, then $L \in \mathrm{DTIME}(g)$ is called *linear time on a Turing Machine*, and similarly quadratic time, for $g(n) = O(n^2)$. Notice that, for such small time classes, we are careful to specify the computation device. Linear time on a Turing Machine would not be the same as linear time on a Register Machine. There is a branch of computational complexity theory concerned with finding, for example, the fastest algorithm for processes such as matrix multiplication[3]. For such questions, we would not work with a general model such as a Turing Machine. Rather, we would be concerned with a model which more closely resembles a modern computer; for example a Register Machine or something similar such as a "random access machine". For a fixed machine model it does make sense to discuss questions such as time n^2 vs time $n^{2.5}$.

But keep in mind we are more concerned with coarser delineations such as $O(n)$ vs $O(2^n)$, *feasible* vs *infeasible*. In that case, the model plays no part, at least with the current state of human knowledge.

6.1.4 Exercises

Exercise 6.1.2 The following are Blum's Axioms for measuring resources used in computation. We say that a collection of partial computable functions $\{\Gamma_e : e \in \mathbb{N}\}$ are a *complexity measure* iff the following two axioms are satisfied.

(i) $\varphi_e(x) \downarrow$ iff $\Gamma_e(x) \downarrow$.
(ii) The question "Does $\Gamma_e(x) = s$?" is algorithmically decidable.

---

[3] As of 2023, this is $O(n^{2.371552})$ Williams, Xu, Xu, and Zhou [WXXZ23].

For instance, $\Gamma_e(x) = s$ might mean that $\varphi_e(x) \downarrow$ in exactly s steps of a Turing machine.

1. Prove that $\Gamma_e(x) = s$ meaning that $\varphi_e(x)$ uses exactly s work tape squares (i.e. space) is a complexity measure in the sense of Blum.
2. Prove that the two axioms above are independent.

Exercise 6.1.3 Prove that if L can be accepted in time $f(n)$ on a two tape Turing machine, there is a one tape Turing machine accepting L in time polynomial in $f(n)$.

Exercise 6.1.4 prove that a k-tape Turing Machine running in time $O(t)$ can be simulated by a 2-tape one running in time $O(t^2)$.

6.1.5 Polynomial Time and Space

We now introduce the main players of our story.

Definition 6.1.3 (P and NP). We can now define

$$\mathbf{P} = \bigcup_c \mathrm{DTIME}(|x|^c)$$

$$\mathbf{NP} = \bigcup_c \mathrm{NTIME}(|x|^c)$$

$$\mathbf{PSPACE} = \bigcup_c \mathrm{DPACE}(|x|^c)$$

$$\mathbf{NPSPACE} = \bigcup_c \mathrm{NSPACE}(|x|^c)$$

P is the complexity class *polynomial time*, NP is the complexity class *non-deterministic polynomial time*, PSPACE *polynomial space*, and NPSPACE is *nondeterministic polynomial space*.

The *pragmatic idealization* used throughout computational complexity theory is to make the following identification:

The class of languages in P are the feasible languages.

This idealization is *plainly false*. Consider polynomials of the form

$$g(n) = n^{2^{2^{2^{2^{100}}}}}, \text{ or,}$$

$$\widehat{g}(n) = 2^{2^{2^{2^{100}}}} \cdot n.$$

Consider DTIME(g) and DTIME($\widehat{g}$). The first has an exponent which is far to large to ever be run on an computer even for inputs of length 2. The second is in linear time, but similarly could never be run as the constant is of galactic size.

What are some properties we'd like the class of "feasible" functions to have? Here are two:

(i) If $f(x)$ and $g(x)$ are both feasible, then $f(g(x))$ and $c \cdot g(x)$ for any constant c, should also be feasible.

(ii) Problems that can be computed in time n^2 should be considered feasible, since we actually do run quadratic algorithms in practice.

These two properties have the unfortunate consequence that algorithms that take time n^{1000} must be considered "feasible". So we are faced with two choices:

- jettison one of the nice properties above, or
- keep the nice properties, and see if the resulting theory is useful.

The resulting theory has been *very* useful, if we are trying to understand what problems are *not* feasible. We just need to understand that we don't really mean that everything in P is really "feasible". Another reason is *pragmatism*, and reflects the nature of "real-life" problems. Especially in the early days of computational complexity theory, the following general principle seemed true:

If a natural problem is in P, then there is an algorithm for it which runs in polynomial time with small constants and small exponent.

In Chapter 9, we will see that this heuristic principle seems no longer to be true. Indeed, we will see that there are certain natural problems which lie in P and we can prove that they *cannot* have feasible algorithms unless certain complexity assumptions we believe to be true are false. More on this later.

But at present we will concentrate on P. P has very nice *closure properties* in that it is closed under union, concatenation, composition, etc, as these are true of polynomials. Such closure is not true of, for example, linear time. Thus P admits *mathematical analysis*.

The reader should think of P as being an *outer boundary* of "feasible", and if something is *not* in P then likely it is *not feasible* for large inputs.

The P vs NP Problem.

Evidently, P $\subseteq$ NP, since every deterministic machine is a non-deterministic one that happens to not make any choices. It is not known whether P = NP; this is arguably the most important open problem in computer science. It is

not even known if P = PSPACE! We remark that later we will show that
PSPACE = NPSPACE. We will also prove that NP $\subseteq$ PSPACE, and we also
don't know if NP = PSPACE. Lots of basic open questions!

6.1.6 Universal Enumerations

One of the key results in classical computability theory is Theorem 3.2.13,
which states that there is a universal Turing Machine. This result was the key
to the definition of the halting problem, and more generally, diagonalization
in the context of computation. What happens in complexity theory?

The key to such an enumeration may seem a strange concept at first.

Definition 6.1.4.

1. A computable function $f : \mathbb{N} \to \mathbb{N}$ is called *time constructible* iff
   there is a Turing Machine M which on input 1^n runs in exactly time
   $f(n)$.
2. A computable function $f : \mathbb{N} \to \mathbb{N}$ is called *space constructible* iff
   there is a Turing Machine M which on input 1^n uses exactly space
   $f(n)$.

6.1.7 Using Clocks

Now the reader might wonder what kinds of functions are not, for example,
time constructible. We will construct some examples later when we look at
the Union Theorem, Theorem 6.3. However, for our purposes, constructibility
allows us to define "clocks" for time (and space).

Namely suppose that g is time constructible via machine M. We can enumerate

$$\{\langle \varphi_e, c \rangle : \varphi_e \text{ runs in } c \text{ times (i.e. multiplied) the time constructed by } M, c \geqslant 1\}.$$

That is, on input x we would run $\varphi_e(x)$ only for $c \cdot M(1^{|x|})$ many steps. If it
does not halt in time, we output 0. This would computably list DTIME(g).
We have an analog of Theorem 3.2.13, an enumeration theorem. Strictly
speaking, this takes about $2 \cdot c \cdot g(|x|)$ many steps, but we only really care
up to an O factor for the definition of a time (or space) class. (There is a
technical *linear speed-up theorem*, which says that, by changing the number

of states in the Turing Machine, if you can do something in time $cg(|x|)$ and g is superlinear with $c \geqslant 1$, then you can do it in time $g(|x|)$. We will not prove this result as it detracts from the story.)

Lemma 6.1.1. *The following are time and space constructible.*

1. n^k *for any fixed k.*
2. $n!$
3. $2^{c \cdot n}$ *for any fixed c.*
4. 2^{n^k} *for any fixed k.*
5. $n \cdot \lceil \log n \rceil^k$ *for any fixed $k \geqslant 1$.*

Proof. We sketch the proof of 1, but they are all similar. For simplicity we will show $O(n^k)$ is time constructible since we are only caring about $O(n^k)$ anyway. First we build a machine calculating n^k on input 1^n. This works by writing out n in binary and then multiplying it by itself k times, using at worst, the method (once?) learned in school. Now we simply start subtracting 1 until nothing remains. This whole activity takes $O(n^k)$ many steps. $\square$

For more details on time and space constructibility we refer the reader to Balcázar, Dáz and Gabarró [BDG90].

6.1.8 Hierarchy Theorems

How can we use time constructibility? Take a computable machine A which lists DTIME(g). Suppose that we wished to simulate the $\langle e, c \rangle$'th member of the list on some input x. Then we can run a machine $\widehat{A}$ which does the following. It takes as input $\langle e, c, x \rangle$. It then runs $A(\langle e, c \rangle)(x)$ and for a *fixed* $\langle e, c \rangle$, this will take $O(c \cdot g(|x|))$ many steps. We will use this method in the proof of following which is a basic hierarchy theorem. There are more complex ones, but it will suffice for our purposes.

Theorem 6.1.5 (Hierarchy Theorem-Hartmanis and Sterns [HS65]).

1. *Suppose that f and g are time constructible and superlinear[4], that is $f(|x|), g(|x|) > O(|x|)$. Suppose that f is $o(g)$. Then DTIME(f) $\subset$ DTIME($g \log g$)[5]. Here $\subset$ means proper subset.*

---

[4] This also holds for f linear.

[5] The reader might wonder about the presence of log here. The point is that we are able to use a multi-tape Turing Machine, but note that a k tape machine (for a fixed k) running in time $O(h(n))$ can be simulated by a 2-tape machine in time $O(h(n) \log(h(n)))$. This is a result of Hennie and Sterns [HS66]. The proof would detract from our narrative flow. The simpler simulation using time $O(h(n)^2)$ is assigned to you in Exercise 6.1.4.

2. Also, if $f(n+1)$ is $o(g(n))$ then $NTIME(f) \subset NTIME(g)$.
3. The analogous results hold for space.

Proof. We prove 1. 2 requires somewhat more complex techniques which are rather technical and would detract from our main line of development.

It is clear that $\mathrm{DTIME}(f) \subseteq \mathrm{DTIME}(g \log g)$. The nontrivial part of the proof is by a wait and see argument. Since f and g are time constructible, we can have a computable list $\{\langle \varphi_e, c \rangle : c, e \in \mathbb{N}\}$ of functions computable running in time $O(f)$. We also have some machine N which runs in time $g(1^n)$. This can be simulated on one of the tapes of the Turing Machine, and we can check running times against it. Since f is $o(g)$, for each c, there is an n such that for all $m \geqslant n$

$$c \cdot f(1^m) < g(1^m).$$

We will build a language $L \in \mathrm{DTIME}(g) \setminus \mathrm{DTIME}(f)$. We must meet requirements of the form

$$R_{e,c} : \langle \varphi_e, c \rangle \text{ does not witness } L \in \mathrm{DTIME}(f).$$

To keep $L \in \mathrm{DTIME}(g)$ we will use g's clock N and on inputs of size n will only allow the construction to run for $g(n)$ many steps. Since the only number we will be interested in will be of the form 1^d for some d, we declare that $L(x) = 0$ for all $x \neq 1^d$.

Construction To satisfy $R_{e,c}$, we will assume that the construction has entered *state* $\langle e, c \rangle$, and this happened at step s of the construction.

At step $s+1$, we will choose 1^{s+1} as a potential witness and do the following. Try to compute $\varphi_e(1^{s+1})$, but only do this for $g(s+1)$ many steps. To do this you will be fixing $\langle e, c \rangle$ until we satisfy the requirement and then simulating the algorithm $\widehat{A}$ mentioned above. Now, if $\varphi_e(1^{s+1})$ returns before the clock N's alarm goes off to say stop, then we can diagonalize: We can put

$$1^{s+1} \in L \text{ iff } \varphi_e(1^{s+1}) = 0.$$

In this case, we would move the construction into state $\langle e', c' \rangle$ where $\langle e', c' \rangle = \langle e, c \rangle + 1$. Declare $R_{e,c}$ as met.

If the clock tells N tells us to stop, then put nothing of length $s+1$ into L, keep the state of the construction as $\langle e, c \rangle$ and move to step $s+2$.
End of construction

We show the construction succeeds by induction. Suppose that we have met all $R_{e'',c''}$ for $\langle e'', c'' \rangle < \langle e, c \rangle$ and we did so by the end of step s. Then by construction, at step s we would have put the construction in state $\langle e, c \rangle$, as this is the next least number. Since f is $o(g)$, for each d, there is an n such that for all $m \geqslant n$, $d \cdot f(1^m) < g(1^m)$. If we choose $d = 2c$, say, then there will be some step $n \geqslant s+1$, where for all $m \geqslant n \ \varphi_e(1^m)$ will return before the clock alarm goes off. When this happens, we will diagonalize, making sure

$\varphi_e(1^m) \neq L(1^m)$, and then moving on to the next requirement in the list. Every step of the construction is kept $O(g(s))$ for some O independent of the construction. So $L \in \mathrm{DTIME}(g)$. $\quad\square$

Several corollaries can be obtained from this:

Corollary 6.1.1.

1. $DTIME(n^k) \subset DTIME(n^{k+1})$.
2. $NTIME(n^k) \subset NTIME(n^{k+1})$.
3. $DSPACE(n^k) \subset DSPACE(n^{k+1})$.
4. $NSPACE(n^k) \subset NSPACE(n^{k+1})$.

At this stage, we introduce further complexity classes.

Definition 6.1.5.

1. $\mathrm{LOG} = \bigcup_{c \geqslant 1} \mathrm{DSPACE}(c \cdot \log n)$
2. $\mathrm{NLOG} = \bigcup_{c \geqslant 1} \mathrm{NSPACE}(c \cdot \log n)$
3. $\mathrm{E} = \mathrm{DEXT} = \bigcup_{c \geqslant 0} \mathrm{DTIME}(2^{c \cdot n})$
4. $\mathrm{NE} = \mathrm{NEXT} = \bigcup_{c \geqslant 0} \mathrm{NTIME}(2^{c \cdot n})$
5. $\mathrm{ESPACE} = \bigcup_{c \geqslant 0} \mathrm{DSPACE}(2^{c \cdot n})$
6. $\mathrm{EXPTIME} = \bigcup_{c \geqslant 0} \mathrm{DTIME}(2^{n^c})$
7. $\mathrm{NEXPTIME} = \bigcup_{c \geqslant 0} \mathrm{NTIME}(2^{n^c})$

The following separations use the same method as the Hierarchy Theorem.

Theorem 6.1.6.

1. $P \subset E \subset EXPTIME$.
2. $LOG \subset PSPACE$.

Proof. They are all similar. We do[6] $P \subset E$. To establish proper containment, we need a language $L \in E$ diagonalizing against $\langle e, k, c \rangle$ saying that membership of L cannot be computed by φ_e in time $c \cdot |x|^k$. But 2^n is time constructible and can be used as a clock. Also for any c and k, $c \cdot n^k$ is $o(2^n)$. So the same simulation and diagonalization will succeed. $\quad\square$

---

[6] Another proof is the following: Let $f = n^{\log n}$ and $g = 2^n$. Both are time-constructible. By the Hierarchy Theorem $\mathrm{DTIME}(f)$ is properly contained in $\mathrm{DTIME}(g)$. The latter is contained in E, and the former contains P.

6.1.9 Exercises

Exercise 6.1.7 Show that $P \subset DTIME(n^{\log n})$.

Exercise 6.1.8 Show that if f and g are time constructible and $\exists^{\infty} n(g(n) > f(n))$, then $DTIME(g) \nsubseteq DTIME(f)$.

6.2 Blum's Speedup Theorem

In this section, we will look at a theorem which is important in that it shows that there is no hope of ever associating a fastest *algorithm*, even up to a cost factor, to a computable *function*. Strangely, it has become unfashionable for courses in computational complexity theory. The reason seems to be that the functions we construct to prove this theorem are synthetic functions achieved by diagonalization. It is unclear if there are *natural* problems exhibiting the speedup phenomenon. (Although it is conjectured by some that there is no optimal algorithm for matrix multiplication.) Also the proof of the Speedup Theorem uses a priority argument, and hence is often out of the scope of many first courses.

The Speedup Theorem below should be thought of in the same context as Rice's Theorem. Rice's Theorem says says that machine descriptions yield nothing in terms of algorithms in that, for example, to show a regular language is nonempty we need much more than just a machine description. The Speedup Theorem says that we can compare running times of *algorithms*, but not of *functions*. I believe that it is conceptually important to see Blum's Speedup Theorem.

Why? Consider the question: What is the real complexity of a *function*. The *guess* would be *the running time of the fastest algorithm for* g, even for any amplification by another function. But the theorem below says that there may be no such algorithm. The proof uses the quantifier "a.a. x" which means "for all but finitely many x".

Theorem 6.2.1 (Blum's Speedup Theorem-Blum [Blu67]). *Let Ψ_e denote the running time of φ_e. Let f be a monotone computable function with $f(n) \geqslant n^2$ for all n. Then there is a computable set R such that, for any φ_i accepting R, there is a j with φ_j also accepting R, and $f(\Psi_j(x)) < \Psi_i(x)$ for almost all x.*

Proof. In the following proof f^n is interpreted to mean

$$\underbrace{f \circ f \circ f \circ \cdots \circ f}_{n\text{-times}}.$$

That is, f composed with itself n times. We will meet the requirements:

$S_i : (\varphi_i$ accepts R in time $\Psi_i) \to \exists j (\varphi_j$ accepts $R \wedge (a.a.x) f(\Psi_j(x)) < \Psi_i(x))$.

Thus we build a computable set R by diagonalization against all φ_i. At each stage s, for each $i \leqslant s$, we will designate i as possibly being *active* or *inactive*. Once declared inactive, it never becomes active. At stage s, for each x with $|x| = s$, we try to find an active $i \leqslant |x|$ such that $\varphi_i(x)$ converges in fewer than $f^{|x|-i}(|x|)$ steps (that is, $\Psi_i(x) < f^{|x|-i}(|x|)$). If such an i is found, for the first such i, set $R(x) = 1 - \varphi_i(x)$ diagonalizing $\varphi_i(x)$, and meeting S_i forever. Declare i inactive from that point on since we have already ensured φ_i does not accept R, we no longer need to deal with φ_i. If no such i is found, set $R(x) = 0$. By construction, R is computable, since f is a computable function so that all the searches are bounded.

Note that any φ_i accepting R must satisfy $\Psi_i(x) \geqslant f^{|x|-i}(|x|)$ for almost all $|x| \geqslant i$. Otherwise, φ_i would be diagonalized at some point, because for any x such that $\Psi_i(x) < f^{|x|-i}(|x|)$ is a candidate for diagonalization. The only way φ_i can escape diagonalization is when some other φ_j with $j < i$ is diagonalized as it has higher priority. But this situation can happen at most i times. Now, suppose φ_e is a functional that computes R. We need to show that there is another functional which computes R much faster than φ_e.

Fix a large k and assume we are given as 'advice' the finite list σ_k of indices $i < k$ such that φ_i eventually gets diagonalized against (and therefore i becomes inactive) in the construction of R. Now, we can compute R via the following procedure. In a first phase, simply follow the construction of R as described above, until we reach a point where all $i \in \sigma_k$ have become inactive. At this point, we know (only because we know σ_k!) that none of the $\{\varphi_i \mid i < k\}$ are relevant for the construction of R on future x. Thus, we enter a second phase where to compute each $R(x)$, we only need to simulate, for $k \leqslant j \leqslant |x|$, $\varphi_j(x)$ during $f^{|x|-j}(|x|)$ steps of computation. By dovetailing, this can be done in time polynomial in $(|x| \cdot f^{|x|-k}(|x|))$ (the polynomial being independent of k). The polynomial being independent of k means that iterating f enough times will see us dominating it. If f is fast growing enough and k large enough compared to e, $(|x| \cdot f^{|x|-k}(|x|)) < f(f^{|x|-e}(|x|))$ (which in turn is $< f(\Psi_e, x))$) for almost all x. Note that such a k can be computed uniformly given e. $\square$

The presentation of the proof given here is adapted from a recent paper by Bienvenu and Downey [BD20]. In that paper the authors used the Speedup Theorem to show that access to a randomness source (as an oracle in the sense of Chapter 8) will *always* accelerate the computation of *some* computable functions by more than a polynomial amount. More on this in §8.1.1.

6.3 The Union Theorem*

In this section, we will prove another classic theorem not normally covered in complexity courses today. A corollary to this result is that there is a single computable function f such that $P = \text{DTIME}(f)$. It follows that f cannot be time constructible; for otherwise the machine M running in time $O(f)$ would accept (or could be modified to accept) a language in $L \in \text{DTIME}(f)$, meaning that it would be in P. But that would contradict the Hierarchy Theorem, since f would then need to be dominated by a polynomial $O(|x|^k)$ and the Hierarchy Theorem says that $\text{DTIME}(f) \subseteq \text{DTIME}(|x|^k) \subset \text{DITME}(|x|^{k+1})$.

Theorem 6.3.1 (Union Theorem-Meyer and McCreight [MM69]).
Suppose that $\{t_n : n \in \mathbb{N}\}$ is a uniformly computable sequence of computable functions with $t_n : \mathbb{N} \to \mathbb{N}$, for each n. (That is, there is a computable function g with $g(n)$ an index for t_n.) Suppose also that

$$\forall n \forall m (t_n(m) \leqslant t_{n+1}(m)).$$

Then there is a computable function t such that

$$DTIME(t) = \bigcup_n DTIME(t_n).$$

Proof. For convenience for each n, and for each $c \in \mathbb{N}$, there is some t_m for $m \geqslant n$, such that $t_m(k) \geqslant c \cdot t_n(k)$ for almost all k. This makes the construction simpler as then we don't need to worry about the constants in the $\text{DTIME}(t_n)$.

We build t in stages using a priority argument. To ensure that

$$\bigcup_n \text{DTIME}(t_n) \subset \text{DTIME}(t),$$

we will meet the requirements

$$P_e: \quad \text{(a.a. m)} t_e(m) \leqslant t(m).$$

Here "a.a. m" means "for all but finitely many." It is convenient to define the *diagonal* function $d(m) = t_m(m)$. Note that d is computable as the sequence t_n is uniformly computable. (In the case of $t_n = |x|^n$, then $d(m) = m^m$.)

Note that if $e \leqslant m$ then $t_e(m) \leqslant d(m)$, by construction. But we can't use $d = t$ since in the case of $d(m) = m^m$, this is definitely not a polynomial, and would be in $\text{DTIME}(d)$. So we also have negative requirements

$$N_e: \quad (a.a.x)\varphi_e(x) \text{ runs in time } \leqslant t(|x|) \text{ then } \varphi_e \in \bigcup_n \text{DTIME}(t_n).$$

By this we mean the language L_e accepted by φ_e is in $\bigcup_n \mathrm{DTIME}(t_n)$. We denote the running time of φ_e by Ψ_e, to simplify notation.

We satisfy this as

$$N_e : L_e \notin \bigcup_n \mathrm{DTIME}(t_n) \to \exists^\infty x[t(|x|) < \Psi_e(x)].$$

For all m, we will force $t(m) \leqslant d(m)$. Now, if $L_e \notin \bigcup_n \mathrm{DTIME}(t_n)$, then

$$\forall n \exists^\infty x(t_n(|x|) < \Psi_e(x)).$$

Then as the construction proceeds we ask that the pair $\langle e, n \rangle$ requires attention when we see some x with

$$t_x(|x|) < \Psi_e(x) \leqslant d(|x|).$$

If we make the value of $t(|x|)$ smaller than $\Psi_e(x)$, we would be working towards meeting N_e.

Thus we will take a pairing $\langle \cdot, \cdot \rangle$, and at stage n, see if there x with $|x| = n$ and for which there is an $\langle e, z \rangle$ not yet cancelled, such that

(i) $\langle e, z \rangle \leqslant |x|$.
(ii) $t_{\langle e,z \rangle}(|x|) < \Psi_e(|x|) \leqslant d(|x|)$.

If there is no such pair, set $t(|x|) = d(|x|)$.

If there is such a pair, choose the least one and define $t(|x|) = \Psi_e(|x|) - 1$. Then cancel $\langle e, z \rangle$.

Clearly t is computable by the hypotheses of the theorem, and because we can see if Ψ_e has halted yet; if not it's running time is too large.

First we argue that we meet P_e. Note that $t(|x|) < t_e(|x|)$ only when we defined $t(|x|) = \Psi_p(x) - 1$ in the construction, when we would need $t_{\langle f,p \rangle}(|x|) < \Psi_p(x)$. But this can only happen if $\langle f, p \rangle < e$, and after it happens we cancel $\langle f, p \rangle$. So, from some point onwards $t(|x|) \geqslant t_e(|x|)$.

Second, we argue that N_e is met. Suppose that $L_e \notin \bigcup_n \mathrm{DTIME}(t_n)$. Then for all z,

$$\exists^\infty x(t_{\langle e,x \rangle}(|x|) < \Psi_e(x)).$$

If there are infinitely many x with $d(|x|) < \Psi_e(x)$, and since $t(|x|) < d(|x|)$, we are done. Otherwise, for almost all x, $\Psi_e(x) \leqslant d(|x|)$, and hence

$$\forall z \exists^\infty x(t_{\langle e,z \rangle}(|x|) < \Psi_e(x) \leqslant d(|x|)).$$

Then $\langle e, z \rangle$ will eventually receive attention, and we set $t(|x|) < \Psi_e(x)$ by construction. $\square$

Chapter 7
NP- and PSPACE-Completeness

Abstract We introduce NP-completeness. We prove the Cook-Levin Theorem. Using it we prove many natural problems are NP-complete, and using similar ideas show QBF is PSPACE complete, and then show several natural problems are PSPACE complete. We prove Savitch's Theorem showing that NPSPACE=PSPACE. We finish by looking at advice classes, BPP and randomization.

Key words: NP, P, PSPACE, Cook-Levin Theorem, NP-completeness, PSPACE completeness, Savitch's Theorem, advice, BPP, Valiant-Vazirani theorem

7.1 The Polynomial Time Hierarchy

In this Chapter, we will look at the class NP in more detail. We will see that it relates to natural problems in way that was discovered in the 1970's, and is a widespread and important concept. There is a vague analog of c.e. vs NP. By Kleene's Normal Form Theorem, Theorem 3.3.8, a language L is c.e. iff there is a computable (primitive recursive) R such that

$$x \in L \text{ iff } \exists y R(x, y).$$

Suppose that L is in the class NP. Then there is a c, and a nondeterministic Turing Machine M such that $x \in L$ iff *some computation path of M accepts x in $\leqslant |x|^c$ many steps*. With this in mind, we can define an analog of the polynomial hierarchy, but in terms of time bounded classes.

Definition 7.1.1.

1. We say that a relation $R(\cdot, \cdot)$ is polynomial time if there is a (deterministic) Turing Machine M and a constant c such that $R(x, y)$ holds iff $M(\langle x, y \rangle) = 1$ and $M(z)$ runs in $|z|^c$ many steps.
2. We say that L is Σ_1^P iff there is a constant d and a polynomial time computable relation R, such that

$$x \in L \text{ iff } \exists y(|y| \leqslant |y|^d \wedge R(x, y)).$$

3. We say that L is Π_1^P if $\overline{L}$ is Σ_1^P.
4. L is Σ_{n+1}^P if there is a Π_n^P relation R and a constant d such that

$$x \in L \text{ iff } \exists y(|y| \leqslant |y|^d \wedge R(x, y)).$$

5. Similarly, L is Π_n^P iff $\overline{L}$ is Σ_n^P.

Again we get the following containments:

$$
\begin{array}{ccccc}
\Pi_1^P & & \Pi_2^P & & \\
\subseteq \quad \subseteq & & \subseteq \quad \subseteq & & \subseteq \\
\Sigma_1^P \cap \Pi_1^P & \Sigma_2^P \cap \Pi_2^P & & \Sigma_3^P \cap \Pi_3^P \ \ldots \\
\subseteq \quad \subseteq & & \subseteq \quad \subseteq & & \subseteq \\
& \Sigma_1^P & & \Sigma_2^P &
\end{array}
$$

This hierarchy is called the *Polynomial (Time) Hierarchy, PH*, a name first given by Meyer and Stockmeyer [MS72]. In the case of the arithmetical hierarchy, we know that all containments are *proper*, but alas for the time bounded versions, we don't know if *any* are proper!

The reader might note that the we have not denoted $\Sigma_k^P \cap \Pi_k^P$ by Δ_k^P. There is a class Δ_k^P, but it won't play any part in this book. It will be defined when we look at polynomial time Turing reducibility in §7.2.4.

The analog of Kleene Normal Form is the following:

Theorem 7.1.1. $L \in NP$ *iff* $L \in \Sigma_1^P$. *That is,* $NP = \Sigma_1^P$.

Proof. Suppose $L \in NP$ accepted by M in time $|x|^c$.
So $x \in L$ iff $\exists y \ |y| \leqslant |x|^c$, and where y is a computation path in M accepting x. So we define $R(x, y)$ to mean that y is an accepting computation path for x. Then $R(x, y)$ is computable in deterministic polynomial time from the quadruples of M as the choices are coded into the path.

Conversely, if L is Σ_1^P i.e., $x \in L$ iff $\exists y(|y| \leqslant |x|^d \wedge R(y,x))$ then we design a nondeterministic Turing Machine which, on input x, begins by guessing y of length $\leqslant |x|^d$ and then checks if $R(x,y)$ holds. $\square$

The theorem above says that NP models "guess and check" and that all the nondeterminism can be *guessed at the beginning.* Theorem 7.1.1 is analogous to Kleene's Theorem (Lemma 3.3.1) which says that we can code c.e. sets with a single search over a primitive recursive relation.

Definition 7.1.2. We say that Π_1^P is *co-NP.*

7.2 NP Completeness

7.2.1 Machine NP-Complete Problems

If we continue to pursue the analog of our actions with computably enumerable sets and undecidability, recall that the first step was to prove that the halting problem is not decidable. Using reductions, we then showed that many decision problems could have the halting problem coded into them. Hence such problems are undecidable too.

For complexity theory, the analogous approach would be to

1. prove P$\neq$NP and then
2. take a problem $L \in$ NP\P and code L into many problems.

At present, *nobody has any idea how to do* 1.

But we are able to do do 2. What we will do is identify a core problem (analogous to the halting problem) in NP, and prove it to be NP-complete. That is, we prove this core problem to be as hard as any other problem in the class NP. Next we show that there are many natural problems also in NP which have the same complexity as this core problem. This strategy says that NP-completeness is widespread, and says that *none of these problems are in* P, *unless* P = NP, which we don't think is true. Certainly, these results say that none of the problems have a polynomial time algorithm to the best of our knowledge. As before, we need a notion of reduction. Before the reduction was constrained to preserve *decidability.* Now the reduction must also preserve *complexity.*

Definition 7.2.1. Let L_1, L_2 be languages. We say that L_1 is *polynomially many-one reducible to* L_2, in symbols $L_1 \leqslant_m^P L_2$, iff there is a polynomial time computable function f, such that, for all x,

$$x \in L_1 \text{ iff } f(x) \in L_2.$$

The following result is immediate.

Proposition 7.2.1. *If* $L_2 \in P$ *and* $L_1 \leqslant_m^P L_2$, *then* $L_1 \in P$.

Proof. Suppose that $L_1 \leqslant_m^P L_2$ with function f running in time $|x|^c$, and M is accepting L_2 in time $|x|^d$. Given an instance x of the problem L_1, compute $f(x)$ and ask whether $f(x) \in L_2$. The algorithm for L_2 runs in polynomial time in the size of its input z, which might be bigger than $|x|^c$ but is certainly has size $\leqslant O(|x|^c)^d$, since f is computable in polynomial time in the size of x. Then the composition of two polynomials is still a polynomial, so the overall algorithm is polynomial in the size of x. $\square$

Definition 7.2.2. We say A is *NP-hard* if $B \leqslant_m^P A$ for all $B \in$NP.

The following is also immediate.

Proposition 7.2.2. *If* A *is NP-hard and* $A \in P$, *then* $P=NP$.

Proof. For any $B \in$NP, compose the polynomial-time algorithm for A with the polynomial-time function reducing B to A to get a polynomial-time algorithm for B. $\square$

The following is the core definition:

Definition 7.2.3. We say A is *NP-complete* if A is NP-hard and $A \in$ NP.

Essentially, an NP-complete problem is a hardest problem in the class NP. Okay, so we have a great definition, but are there any NP-complete problems? Let $\mathcal{L} = \{N_e : e \in \mathbb{N}\}$ list all nondeterministic Turing Machines.

Theorem 7.2.1.

1. *The following language is NP-complete.*
   $L = \{\langle e, x, 1^n \rangle : \text{some computation of } N_e \text{ accepts } x \text{ in } \leqslant n \text{ steps}\}$.
2. *The following language is co-NP-complete.*
   $L_2 = \{\langle e, x, 1^n \rangle : \text{all computation paths of } N_e \text{ accept } x \text{ in } \leqslant n \text{ steps}\}$.

The languages in Theorem 7.2.1 are *unnatural*, but are useful machine-based ones.

Proof. We do 1., 2. being more or less identical.

First L is NP-hard. Let $A \in$ NP. Then there is a N_e and a constant c such that $x \in A$ iff some computation of N_e accepts x in $\leqslant |x|^c$ many steps. Thus, we can reduce as follows. Define a polynomial time computable function f via $f(x) = \langle e, x, 1^{|x|^c} \rangle$. For a fixed e and c, this is a polynomial time computable function. Note that

$$x \in A \text{ iff } N_e(x) \text{ accepts iff } \langle e, x, 1^{|x|^c} \rangle \in L.$$

That is, $x \in A$ iff $f(x) \in L$.

Second, $L \in NP$. On input z see if $z = \langle e, x, 1^n \rangle$. This takes $O(|z|)$ many steps. If the answer is no, reject. If the answer is yes, the answer is yes iff some computation of N_e accepts in $\leqslant n$ many steps, and this whole analysis is $O(|z|^2)$ at worst, nondeterministically. □

7.2.2 Exercises

Exercise 7.2.2 Prove that a language L is in NP iff there is a polynomial time computable language B and a polynomial p such that: for all x, $x \in L$ iff $\exists z[|z| = p(|x|) \wedge \langle x, z \rangle \in B]$.

Exercise 7.2.3 Suppose that $L \in$ NP. Suppose that $\widehat{L} \leqslant_m^P L$. Show that $\widehat{L} \in$ NP.

Exercise 7.2.4 Show that if $\Sigma_n^P = \Sigma_{n+1}^P$ then for all $m \geqslant n$, $\Sigma_n^P = \Sigma_m^P$.

Exercise 7.2.5 Show that the following problems are in the class NP.

(i) NON-PRIME[1]
   Input : A number $n \in \mathbb{N}$ in binary form.
   Question : Is n not prime?

---

[1] It is known to be in P, quite famously by Agrawal, and Kayal, and Saxena [AKS04]. It is not known, however, if *finding* the factors of a composite number is in P, and conjecturally it is not. If factorization is in P, then one of the principal algorithms used in Public Key Cryptography, the RSA scheme, would be insecure. There is *no evidence* that factorization is not in P, save the fact that we have no idea how to do it.

(ii) GRAPH ISOMORPHISM[2]
    Input : Two Graphs G_1, G_2.
    Question : Is $G_1 \cong G_2$?

7.2.3 The Cook-Levin Theorem

By Theorem 7.2.1, we know that there are some abstract languages taking the role of the halting problem. They are NP-complete. But are there any *interesting, natural* languages which are NP-complete? The answer is most definitely yes. Cook [Coo71] and Levin [Lev73] both realized that there were natural NP-complete problems. In the original papers, Cook showed that SATISFIABILITY, below, is NP-complete as well as SUBGRAPH ISOMORPHISM (see CLIQUE below), and Levin proved a version of SATISFIABILITY, as well as SET COVER and a tiling problem akin to that of Exercise 4.3.4 (see also Exercise 7.2.9 below). We will begin with the Cook-Levin Theorem which is concerned with satisfiability in propositional logic; that is, finding out of the truth table of a propositional formula has a line that evaluates to be *true*. We can view this example in the context of the Undecidability of First Order Logic. We know that Propositional Logic is decidable, since the decision process is to build a truth table for the formula. We see below that we can use satisfiability of a formula of propositional logic to emulate instances of an NP-complete problem, and hence the satisfiability problem is NP-complete. Both Cook and Levin had the idea of using a *generic* reduction, meaning that we will perform a direct simulation of nondeterministic Turing Machines.

SAT or SATISFIABILITY
Input: A formula X of propositional logic.
Question: Does X have a satisfying assignment of boolean variables? That is, is there at least one assignment of the boolean variables which makes X true?

Theorem 7.2.6 (Cook-Levin Theorem [Coo71, Lev73]). SAT *a NP-complete. That is:*

(1). SAT $\in$ NP
(2). SAT *is NP-hard*

Proof. (1) Clearly, SAT $\in$ NP. Simply take the machine M which on input X, guesses values for the variables of X and then evaluates X on those values.

   (2) Take any $A \in$ NP, for example, $A = L$ from Theorem 7.2.1. We need to show that $A \leqslant_m^p$ SAT.

---

[2] We do *not* think that this problem is NP-complete.

Now, A is accepted by some nondeterministic Turing Machine M in NTIME($|x|^k$). We will describe a function σ that will generate a formula $\varphi = \sigma(x)$ such that $x \in A$ iff φ is satisfiable.

Recall the definition of configuration in Definition 3.3.3. The possible executions of M on input x form a branching tree of configurations, where each configuration gives information on the current instantaneous state of the computation. It includes all relevant information such as tape contents, head position, current state etc. M is polynomially time bounded by running time $|x|^k$, so we can assume the depth of this tree is at most $N = |x|^k$ for some fixed k (not dependent on x).

A valid computation sequence of length N can use no more than N tape cells, since at the very worst M moves right one tape cell in each step. Thus there are at most N time units and N tape cells to consider. (Here $N = |x|^k$.) We can think of the tape as potentially travelling left to position $-N$, or right to position N, so the possible scope of operation is the interval $[-N, N]$, so $2N + 1$ many tape squares.

We will encode computations of M on input x as truth assignments to various arrays of Boolean variables, which describe things such as which square the read head is at time i, which symbol is occupying cell j at time i, what state M is in at time i, and the like, things necessary for a machine configuration.

We write down clauses involving these variables that will describe legal moves of the machine and legal starting and accepting configurations of M on x. A truth assignment will simultaneously satisfy all these clauses iff it describes a valid computation sequence of M on input x. In the end, we take $\varphi = \sigma(x)$ to be the conjunction of all these clauses. Then the satisfying truth assignments of φ correspond one-to-one to the accepting computations of M on x. We suppose that M has $R + 1$ many states beginning with q_0 and with q_y the accept state (and the machine accepts it it ever moves in to state q_y), and $P + 1$ many symbols starting with S_0.

We list the relevant propositional variables for our formula φ listing the intended meaning of the relevant variable;

1. $Q[i, k]$, $0 \leqslant i \leqslant N$, $0 \leqslant k < Q$. This is intended to mean that "At time i, M is in state q_k." Note that gives $O(N \cdot R)$ many propositional variables.

2. $H[i, j]$, $0 \leqslant i \leqslant N$ and $-N \leqslant j \leqslant N$. This is intended to mean "At time i, M is scanning tape cell j." Note this gives $O(N^2)$ many propositional variables.

3. $S[i, j, k]$, $0 \leqslant i \leqslant N$ and $-N \leqslant j \leqslant N$. This is intended to mean "At time i, tape cell j contains symbol S_k." Note that this gives $O(N^2 \cdot P)$ many propositional variables.

4. We also have B for the blank symbol as we wish to distinguish this.

The quadruples of M and the mechanics of Turing Machines give restrictions on that the possible values the above variables can have and how interrelate. We construct these as six clauses groups, where the reader should recall a *clause* is a disjunction of literals. We break the clauses into 6 groups, starting each with their intended meaning.

G_1 At each time i, M is in exactly one state

$$\bigwedge_{0\leqslant i\leqslant N}\left[(Q[i,0]\vee\cdots\vee Q[i,Q])\wedge\bigwedge_{j_1\neq j_2}(\neg Q[i,j]\vee\neg Q[i,j_2])\right]$$

G_2 At each time i, M is scanning exactly one square

$$\bigwedge_{0\leqslant i\leqslant N}\left[\left(\bigvee_{-N\leqslant j\leqslant N}H[i,j]\right)\wedge\left(\bigwedge_{-N\leqslant j<k\leqslant N}(\neg H[i,j]\vee\neg H[i,k])\right)\right]$$

G_3 At each time i, each square contains exactly one symbol

$$\bigwedge_{\substack{0\leqslant i\leqslant N\\-N\leqslant j\leqslant N}}\left[(S[i,j,0]\wedge\cdots\wedge S[i,j,P])\wedge\bigwedge_{k_1\neq k_2}(S[i,j,k_1]\vee\neg S[i,j,k_2])\right]$$

G_4 At time 0, computation is in the initial configuration of reading input $x=S_{i_0}S_{i_1}\cdots S_{i_{n-1}}$

$$Q[0,0]\wedge H[0,0]\wedge\bigwedge_{0\leqslant j\leqslant|x|-1}S[0,j,i_j]\wedge\bigwedge_{\substack{|x|\leqslant j\leqslant N\\-N\leqslant j<0}}S[0,j,B]$$

G_5 By time N, M has entered state q_y (the accept state) and accepted x, $\bigwedge_{0\leqslant i\leqslant N}Q[i,y]$

G_6 For each i, configuration of M at time $i+1$ follows by a single application of transition function from the configuration at time i. i.e., the action is faithful.

The transition function of M is the set Δ of quadruples for M.

(i) Suppose that the quadruple is $\langle q_d,S_k,S_{k'},q_{d'}\rangle$. The the formula would be

$$\bigwedge_{\substack{0\leqslant i\leqslant N\\-N\leqslant j\leqslant N}}\left[(Q(i,d)\wedge H(i,j)\wedge S(i,j,k))\to\right.$$

$$(Q(i+1,d')\wedge H(i+1,j)\wedge S(i,j,k')\wedge\bigwedge_{\substack{j'\neq j\\0\leqslant p\leqslant P}}(S(i,j',p)\to S(i+1,j',p))\bigg].$$

(ii) If the quadruple is $\langle q_d,S_k,L,q_{d'}\rangle$.

$$\bigwedge_{\substack{0\leqslant i\leqslant N\\-N\leqslant j\leqslant N}}\left[(Q(i,d)\wedge H(i,j)\wedge S(i,j,k))\to\right.$$

$$\left[Q(i+1,d') \wedge H(i+1,j-1) \wedge \bigwedge_{\substack{-N \leqslant j' \leqslant N \\ 0 \leqslant p \leqslant P}} (S(i,j',p) \rightarrow S(i+1,j',p).) \right].$$

(iii) If the quadruple is $\langle q_d, S_k, R, q_{d'} \rangle$.

$$\bigwedge_{0 \leqslant i \leqslant N; -N \leqslant j \leqslant N} \left[(Q(i,d) \wedge H(i,j) \wedge S(i,j,k)) \rightarrow \right.$$

$$\left[Q(i+1,d') \wedge H(i+1,j+1) \wedge \bigwedge_{\substack{-N \leqslant j' \leqslant N \\ 0 \leqslant p \leqslant P}} (S(i,j',p) \rightarrow S(i+1,j',p)) \right].$$

Note we have one subformula corresponding to each quadruple, and hence possibly many subformulae for each $\langle q_i, S_j, \cdot, \cdot \rangle$ since the machine is nondeterministic. Thus we would have a disjunction (or) for these subformulae as the computation path would choose one to implement on any run. That is, if there was a quadruple of type (i) and type (ii) for the same $\langle q_i, S_j, \cdot, \cdot \rangle$, we'd have the formula for (i) $\vee$ the formula for (ii).

The formula φ is the conjunction of all these formulae. Its length is polynomial in $|x|$, and can be constructed from x in polynomial time.

Every satisfying truth assignment to φ gives rise to an accepting computation of M, and vice-versa. $\square$

The proof also shows that CNFSAT, below, is NP-complete as clauses G_1-G_5 are in CNF and G_6 can easily be transformed into CNF using the distributive and de Morgan laws.

CNFSAT
Input: A boolean formula X in conjunctive normal form
Question: Is X satisfiable?

In many ways SAT and CNFSAT are the bases for most practical NP-completeness results. As mentioned above, this proof is an example of a generic reduction, where we emulate a nondeterministic Turing Machine step-by-step. Usually, if we want to show B is NP-complete we show for some known NP-complete problem A that $A \leqslant_m^P B$. One particularly useful variation of SAT is 3-SAT, which is exactly the same problem as CNFSAT but the clause size is at most 3.

Corollary 7.2.1. 3-SAT *is NP-complete.*

Proof. We can transform any function φ to ψ where ψ has clauses of size $\leqslant 3$, so φ is satisfiable iff ψ is satisfiable. This is done by adding extra variables. If a clause C_j of φ is of the form $(Y_1 \vee Y_2 \vee \cdots \vee Y_n)$, where Y_i is a literal (i.e. propositional variable, or a negated propositional variable), and $n > 3$ add $n-3$ new propositional variables (specific to the clause) $\{Z_{j,k} : 1 \leqslant k \leqslant n-3\}$.

Then replace the clause by new clauses $(Y_1 \vee Y_n \vee Z_{j,1}) \wedge (\neg Z_{j,1} \vee Y_2 \vee Z_{j,2}) \wedge$ $(\neg Z_{j,2} \vee \ldots) \wedge \ldots$. For example,

$$\varphi = (p_1 \vee p_2 \vee \neg p_3 \vee p_4)$$
$$\psi = (p_1 \vee p_2 \vee z) \wedge (\neg z \vee \neg p_3 \vee p_4).$$

The reader should note that you cannot use the $Z_{j,k}$'s to make the conjunction of the clauses evaluate to true. You must make one of the literals C_j evaluate to true. Moreover, if you can make C_j true, then this assignment can use the $Z_{j,k}$'s to make the conjunction of the clauses in ψ to be true. That is φ is satisfiable iff $f(\varphi) = \psi$ is satisfiable. $\square$

The reader should note that the polynomial time m-reduction used to reduce CNFSAT to 3-SAT is *not* a direct emulation as we have used in machine to machine reductions. They are simply equi-satisfiable formulae. What about formulae where we want every line of the truth table to evaluate to be true?

BOOLEAN TAUTOLOGY
Input: A formula X of propositional logic.
Question: Is X valid? That is, does *every* assignment of values to the boolean variables evaluate to true?

The proof of SAT being NP-complete is so generic that it yields the following corollary.

Corollary 7.2.2. BOOLEAN TAUTOLOGY *is co-NP complete.*

The reader might also wonder if we can do better than 3-SAT. What about clause size 2?

Theorem 7.2.7. *2*-SAT $\in P$. *In fact, it is solvable in time which is low level polynomial for reasonable models of computation.*

Proof. Take some φ in 2-CNF. Then the clauses in φ contain at most two literals, and without loss of generality, exactly two. Think of every clause $(l \vee l')$ as a pair of implications:

$$(\neg l \rightarrow l') \text{ and } (\neg l' \rightarrow l).$$

Construct a directed graph $G = (V, E)$ with a vertex for every literal (positive and negated) and directed edges corresponding to the implications as above.

Recall a subset of the nodes of a directed graph is a *strongly connected component* if there is a path between any pair of nodes in the subset. We claim that φ is satisfiable iff no pair of complementary literals both appear in the same strongly connected component of G.

Under any satisfying truth assignment, all literals in a strong component of G must have the same truth value. So if any variable x appears both positively and negatively in the same strong component of G, φ is not satisfiable.

Conversely, suppose that no pair of complementary literals both appear in the same strong component of G. We can produce a quotient G' by collapsing the strong components of G so that G' is acyclic and hence induces a partial order on its vertices. This partial order extends to a total order. Assign $x :=$ false if the strong component of x occurs before the strong component of $\neg x$ in this total order, and $x :=$ true otherwise. This gives a satisfying assignment.

We can find strong components of G in quadratic time[3], giving a linear-time algorithm test for 2-SAT. Having done this we can then produce a satisfying assignment in linear time, if one exists. □

7.2.4 Search vs Decision Problems: Polynomial Time Turing Reductions

Given a problem such as SAT, presumably we wish to *find* the satisfying assignment if there is one. Does the knowledge that X is satisfiable tell us how to find the satisfying assignment? In terms of computational complexity, the search and decision problems are of the same polynomial time complexity.

Theorem 7.2.8. *Suppose that there is a polynomial-time procedure A for deciding whether a given boolean formula Y has a satisfying assignment. Then, given a boolean formula X, in polynomial time in $|X|$, we can either say that X has no satisfying assignment, or find one if there is one.*

Proof. Given X first ask A if X has a satisfying assignment. If *no* say *no*.

If *yes*, let $\{x_1, \ldots, x_n\}$ list the boolean variables in X. Let $\hat{X}$ be the result of setting x_1 in X; that is $\hat{X} = X \wedge x_1$. Ask A is $\hat{X}$ is satisfiable. If *no*, then we know that in any satisfying assignment for X, we need to set x_1 to be false. So we would let $X_1 = X \wedge \neg x_1$. If the answer is *yes*, then we know there is a satisfying assignment with x_1 assigned to be true in X. In this case, we set $X_1 = \hat{X}$. Notice that since X is satisfiable, at least one of these two options must apply.

Now we repeat this process with X_1 in place of X and x_2 in place of x_1. After n steps we will have discovered a satisfying assignment of the variables $\{x_1, \ldots, x_n\}$. □

The reader should note that the reduction of this process is *not* a polynomial m-reduction, as it involves many questions to the decision problem

---

[3] This is a fact from graph theory. It is not hard to prove. To prove the problem in P it would be enough to compute the transitive closure of the directed graph, and algorithms such as Warshall's Algorithm do this in polynomial time.

oracle. This is a polynomial time Turing reduction. As we did in §5.5, we can also define the notion of an oracle Turing machine:

Definition 7.2.4 (Polynomial Time Turing Reduction). We say that $L \leqslant^P_T B$ iff there is an oracle Turing Machine M, and a constant c such that for all $x \in \Sigma^*$, $\Phi^B(x) = L(x)$ in time $\leqslant |x|^c$, and using only $\leqslant |x|^c$ many queries of length $\leqslant |x|^c$ to B. It does not matter whether we regard a query "$z \in B$?" as cost 1 or cost $|z|$.

The kind of Turing reduction reducing the search to the decision problem is known in the literature as a *self-reduction*. The following is yet another open problem in computational complexity theory:

Question 7.2.1. Do all NP-complete problems have self-reductions from their search versions to their decision versions?

If P=NP the answer is definitely yes. But assuming P≠NP, we don't know. All *natural* problems have this property.

We also remark that we can use polynomial time Turing reductions to define the class Δ^P_k for $k \in \mathbb{N}^+$. Recall from the Limit Lemma (Theorem 5.6.2) and Post's Theorem (Theorem 5.6.6) that A is Δ^0_2 iff $A \in \Sigma^0_2 \cap \Pi^0_2$ iff $A \leqslant_T \emptyset'$. So Δ^0_2 is the same as being $\leqslant_T$ the halting problem. Complexity theorists defined Δ^P_k as those languages polynomial time Turing reducible to Σ^P_{k-1}. Note that $\Delta^P_k \subseteq \Sigma^P_k \cap \Pi^P_k$. (See Exercise 7.2.12.) We won't pursue this further as it plays no part in our story, save to point out that all kinds of analogs we might expect from computability theory do not hold in computational complexity theory, at least with the current state of knowledge.

7.2.5 Exercises

Exercise 7.2.9 (Levin [Lev73]) Recall from Exercise 4.3.4 that a TILING SYSTEM T is a finite collection of unit square tiles with coloured edges and a set of rules saying which colours can be next to which. Using the ideas from Exercise 4.3.4, show that the following problem is NP-complete.

TILING
Input: A Tiling System T and an integer N.
Question: Can T colour the $N \times N$ square from a given starting position?

Exercise 7.2.10 Suppose that I have a polynomial time oracle procedure deciding whether a given graph G has a HAMILTON CYCLE. Construct self-reduction showing that we can find Hamilton cycles in polynomial time.

Exercise 7.2.11 EXACT k-CNFSAT is CNFSAT where the clause size is exactly k. Prove that for $k \geqslant 3$, EXACT-k-CNFSAT is NP-complete.

Exercise 7.2.12 * Prove that $\Delta_2^P \subseteq \Sigma_2^P \cap \Pi_2^P$.

7.2.6 Some Natural NP-Complete Problems

In this section, we will discover that *many* problems are NP-complete. There are tens of thousands more. Imagine you are a worker in, say, computational biology, or linguistics or some other area where discrete computational problems naturally occur. You have some problem and it seems hard to find an efficient algorithm for it. Should you wish to show the problem is NP-hard, then likely one of the problems in the current section will be helpful to reduce from. Moreover, the techniques we use such as *local replacement* and *component design* are standard in applications of this hardness theory.

We start by generating a collection of NP-complete graphical problems. We will begin by reducing from SAT. We recall the following definition.

Definition 7.2.5. Given a graph G with set of vertices V, G is a k-*clique* iff there exists $U \subseteq V$, $|U| = k$ and such that for all $x, y \in U$, $xy \in E(G)$.

CLIQUE
Input : A graph $G = (V, E)$, integer k
Question : Does G have a k-clique?

Lemma 7.2.1 (Karp [Kar73]). CNF-SAT $\leqslant_m^p$ CLIQUE.

Proof. We construct a graph G and integer k from a formula φ in CNF such that φ is satisfiable iff G contains K_k, the complete graph on k vertices.

Take k to be the number of clauses in φ. The vertices of G are all the occurrences of literals in φ.
We put an edge of G between two such vertices if they are in different clauses and the two literals are not complementary; i.e., we can assign both literals the value true without conflict.
So a k-clique means there is a set containing one literal from each clause such that each can be assigned true without conflict, hence φ is satisfiable.
Formally we can prove that φ is satisfiable iff G has a k-clique:
Suppose φ is satisfiable. Then there is some satisfying truth assignment r. At least one literal in each clause must be assigned true under r.
Then in G all these literals will be connected as they are in different clauses

and are not complementary. There are k clauses, hence those literals form a k-clique in G.

Conversely, suppose G has a k-clique.

G is k-partite[4], with partition elements corresponding to clauses, so the k-clique must have one element in each clause.

Now, assign true to the literals corresponding to the vertices in the k-clique. This is possible since no pair of complementary literals are connected, hence cannot be in the clique. The remaining literals can have values assigned arbitrarily.

The resulting truth assignment assigns true to at least one literal in each clause, hence φ is satisfiable. □

The following lemma is immediate by the NP-completeness of CNFSAT, since CLIQUE is clearly in NP. However, we will give a direct proof to show the translation of the problem directly into boolean satisfiability. This translation exhibits a general method used in practice to translate a problem directly into a satisfiability problem.

Lemma 7.2.2. * CLIQUE $\leqslant_m^p$ CNFSAT

Proof. * Given an undirected graph $G = (V, E)$ and a number k, we construct a propositional formula φ in CNF such that G has a k-clique iff φ is satisfiable.

Take as variables x_i^u for $u \in V$ and $1 \leqslant i \leqslant k$. Consider x_i^u as saying "u is the ith element of the clique".

Let $\varphi = C \wedge D \wedge E$, such that

C: "For every i, $1 \leqslant i \leqslant k$, there is at least one $u \in V$ such that u is the ith element of the clique".

$$C = \bigwedge_{1 \leqslant i \leqslant k} \left(\bigvee_{u \in V} x_i^u \right)$$

D: "For every i, $1 \leqslant i \leqslant k$, no distinct vertices are both the ith element of the clique".

$$D = \bigwedge_{1 \leqslant i \leqslant k} \left(\bigwedge_{\substack{u,v \in V \\ u \neq v}} (\neg x_i^u \vee \neg x_i^v) \right)$$

E: "If (u, v) is not an edge of G, then either u is not in the clique or v is not in the clique"

$$E = \bigwedge_{(u,v) \notin E} \left(\bigwedge_{1 \leqslant i,j \leqslant k} (\neg x_i^u \vee \neg x_j^v) \right)$$

---

[4] Meaning that the vertices of G can be split into k sets $S_1, \ldots, S_k$ of vertices all disjoint from the others, and if xy is an edge of G for some $i \neq j$, $x \in S_i$ and $y \in S_j$.

Suppose there is a satisfying assignment for $C \wedge D$. Then we can pick out a set of k vertices, namely those u such that $x_i^u = \text{true}$, for some i, in this truth assignment.

If the assignment also satisfies E then, these k vertices are a clique.

Conversely, if $u_1, \ldots, u_k$ is a k-clique in G, set $x_i^{u_i}$ true for all $1 \leqslant i \leqslant k$. This satisfies φ. □

Hence, we have shown that CNFSAT $\equiv_m^p$ CLIQUE, and so we know that CLIQUE is NP-complete.

Definition 7.2.6. An *independent set* in an undirected graph $G = (V, E)$ is a subset U of V such that no two vertices in U are connected by an edge in E.

INDEPENDENT SET
Input : A graph $G = (V, E)$, integer k.
Question does G have an independent set of size k?

Theorem 7.2.13. INDEPENDENT SET

Proof. We reduce from CLIQUE. Let $G = (V, E)$, k be an instance of CLIQUE. Consider the complementary graph $\bar{G} = (V, \bar{E})$, where

$$\bar{E} = \{(u, v) | u \neq v, (u, v)\ E\}.$$

Then G has a k-clique iff $\bar{G}$ has an independent set of size k. This simple one-to-one correspondence gives polynomial reductions in both directions. □

Definition 7.2.7. A *vertex cover* in an undirected graph $G = (V, E)$ is a set of vertices $U \subseteq V$ such that every edge in E is adjacent to some vertex in U.

VERTEX COVER
Input : A graph $G = (V, E)$, and a positive integer k.
Question : Is there a vertex cover in G of cardinality $\leqslant k$?

Theorem 7.2.14 (Karp [Kar73]). VERTEX COVER *is NP-complete.*

Proof. We reduce from INDEPENDENT SET. $U \subseteq V$ is a vertex cover iff $V - U$ is an independent set. This gives simple polynomial reductions in both directions. □

Definition 7.2.8. Let C be a finite set of colours and $G = (V, E)$ an undirected graph. A *colouring* is a map $\chi : V \to C$ such that $\chi(u) \neq \chi(v)$ for all $(u, v) \in E$.

GRAPH k-COLOURABILITY; k-COL
Input : A graph G, integer k.
Question : Is there a colouring of G using $\leqslant k$ colours?

Theorem 7.2.15 (Karp [Kar73]). *For $k \geqslant 3$, GRAPH k-COLOURABILITY is NP-complete.*

Proof. We reduce from CNFSAT. In fact, we can assume that we are reducing from an instance of EXACT 3-SAT, by Exercise 7.2.11. (In the diagram we have indicated the minor change needed to make the reduction work for CNFSAT.) We can trivially reduce k-COL to 3-COL for $k > 3$ by appending a $(k-3)$-clique and edges from every vertex of the $k-3$-clique to every other vertex. So we only need to show CNF-SAT$\leqslant_m^p$ 3-COL.

Let φ be a propositional formula in CNF. We will construct a graph G that is 3-colourable iff φ is satisfiable.

Create three central vertices, R, G, B, connected in a triangle. These letters indicate labels for us but are not part of the graph. We can think of these as aligning to the colours red, green, and blue. Also create a vertex for each literal in φ, connecting each to both its complement and B.

So one of x and $\bar{x}$ must be green and the other must be red.

Intuitively, consider green literals as true and red literals as false. For each clause $(x \lor y \lor z)$ in φ we add the following subgraph.

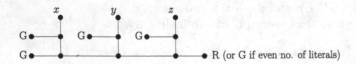

Now, if every vertex along the top is coloured red this corresponds to φ being unsatisfied.

But we easily see that this cannot give a legal 3-colouring, as the middle vertices must then all be blue so the bottom row alternates red and green, clashing with the final vertex in the bottom row.

Suppose however that one of the vertices on the top row is coloured green, which corresponds to φ being satisfied.

Then we can colour the vertex below it red and all other vertices on the middle row blue. This forces a colouring for the bottom row which is successful alternation of red and green, with one blue vertex under the green top row vertex. This 3-colouring extends to the whole graph G.

Thus, we have that if there is a legal 3-colouring then the subgraph corresponding to each cause must have at least one green literal, hence we have a satisfying truth assignment for φ.

Conversely, if there is a satisfying assignment then simply colour the true variables green and the false ones red. There is a green literal in each clause, so the colouring can be extended to a 3-colouring of the whole graph.

Hence, φ is satisfiable iff G is 3-colourable, and G can be constructed in polynomial time. □

The technique in the proof above is called *component design*.

Often in problems with a parameter k, like k-Col and k-CNF-SAT, larger values of k make the problem harder. However, this is not always true. Consider the problem of trying to determine whether a *planar graph*[5] has a k-colouring. The problem is easy for $k = 1$, $k = 2$, and trivial ("just say yes") for $k \geqslant 4$ (by the famous Four Colour Theorem [HA77, AHK77]).

Theorem 7.2.16 (Karp [Kar73]). 3-COL $\leqslant_m^p$ PLANAR 3-COL, *so* PLANAR 3-COL *is NP-complete.*

Proof. This proof uses *widget design* or *gadget design*. We will use a *crossing widget.*

Given an undirected graph $G = (V, E)$, embed it in the plane arbitrarily, letting edges cross if necessary. Replace each such cross with the following edge crossing gadget:

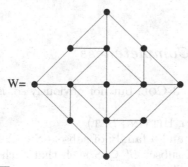

---

[5] That is, a graph that can be drawn on the plane with no edges crossing.

W is a planar graph with the following properties: (Exercise 7.2.17)

(i). In any legal 3-colouring of W, the opposite corners are forced to have the
same colour

(ii). Any assignment of colours to the corners such that opposite corners have
the same colour extends to a 3-colouring of all of W.

For each edge (u, v) in E, replace each point at which another edge crosses
(u, v) with a copy of W and identify the other corners of the extremal copies
with u and V, all except for one pair, which are connected by an edge:

The resulting graph $G' = (V', E')$ is planar.

If $\chi : V' \to \{$red, green, blue$\}$ is a 3-colouring of G' then property (i) of
W implies that $\chi|V$ is a 3-colouring of G.

Conversely, if $\chi : V \to \{$red, green, blue$\}$ is a 3-colouring of G then
property (ii) of W allows χ to be extended to a 3-colouring of G'.

Hence, we can find a 3-colouring of G iff we can find a 3-colouring of G'.
□

7.2.7 Exercises

Exercise 7.2.17 Prove that the widget of Theorem 7.2.16 does what is
claimed. This is basically case analysis.

Exercise 7.2.18 Prove that 2-COL $\in$ P (for any graph).

7.2.8 More NP-Completeness

One problem related to 3-COL, but not obviously so, is the following.

EXACT COVER (EXACT HITTING SET)
Input : A finite set X and a family of subsets S of X.
Question : Is there a subset $S' \subseteq S$ such that each element of X lies in
exactly one element of S'?

Theorem 7.2.19 (Karp [Kar73]). EXACT COVER *is NP-complete.*

Proof. We reduce from 3-COL. Let $G = (V, E)$ be the given graph. For each $v \in V$ we will add 4 elements to X, v together with elements labelled R_v, G_v, B_v. Then for each edge $uv \in E$, add three elements to X, R_{uv}, B_{uv}, G_{uv}. Then the total size of X is $O(|G|)$.

Now we need to specify the family of subsets. For each $v \in V$, construct a subset S_v^R which contains v and R_{uv} for each $vu \in E$. Now do the same to define S_v^B and S_v^G. To complete the construction, add sets of the form $\{p\}$ for $p \in X$ and $p \neq v$ for $v \in V \cup \bigcup_{uv \in E} \{R_{uv}, B_{uv}, G_{uv}\}$. Again the size is $O(|G|)$.

Now, suppose that EXACT COVER has a solution. Then, because of the sets of the form $\{p\}$, each of these must be chosen. To cover each $v \in V$, we must choose *exactly* one of R_v^B, R_v^G, or R_v^B. This is a 3-colouring. The reasoning reverses, giving the result. $\square$

7.2.9 Exercises

Exercise 7.2.20 Consider the following problem:

DOMINATING SET
Input : A graph G and an integer k.
Question : Does G have a *Dominating Set* of size k? That is, is there a collection of vertices D (of size k) of G such that for all $v \in V(G)$, either $v \in D$ or there is some $w \in D$ with $vw \in E(G)$.

Prove that DOMINATING SET is NP-complete. (Hint: reduce from VERTEX COVER.)

Exercise 7.2.21 Consider the following problem.

DOMINATING SET REPAIR[6].
Input : A graph G, and a dominating set D of G and an integer k.
Question: If I remove one element of D as well as all the edges from it, can I add at most k elements of G to repair the dominating set for the resulting graph $\widehat{G}$? That is, can we add at most k elements $\{v_1, \ldots, v_k\}$ such that $D \cup \{v_1, \ldots, v_k\}$ is a dominating set for $\widehat{G}$?

The question is the following. Is DOMINATING SET REPAIR NP-complete or in P?

Exercise 7.2.22 Consider the following scheduling problem. You are given positive integers m and t and an undirected graph G. The vertices of G specify unit time jobs and the edges specify that two jobs cannot be scheduled

---

[6] You can think of the dominating sets as placements of e.g. radar or cell phone towers. One of these gets destroyed and is no longer available for placement of towers. Maybe the land is now a designated park.

simultaneously. The question is whether the jobs can be scheduled for m identical processors so that all jobs can be completed within t time units.

(i) Prove that this problem is in P for $t = 2$.
(ii) Prove that it is NP complete for $t \geqslant 3$.

Exercise 7.2.23 Consider the restricted version of CNFSAT in which formulae can contain at most k occurrences of the same literal (either negated or unnegated), with k fixed.

(i) Prove that the problem is NP complete for $k \geqslant 3$. (Hint: Reduce from SAT.)
(ii) Prove the problem is in P if $k \leqslant 2$. (This might require some thought.)

7.2.10 Even More NP-Completeness

We will establish a few more NP-completeness results, especially some which are not graph problems. We turn to integer problems.

SUBSET SUM
Input : A finite set S, integer weight function $w : S \to \mathbb{N}$, and a target integer B.
Question : Is there a subset $S' \subseteq S$ such that $\sum_{a \in S'} w(a) = B$?

Theorem 7.2.24 (Karp [Kar73]). SUBSET SUM *is NP-complete.*

Proof. We reduce from EXACT COVER. Take $X = \{0, 1, \ldots, m - 1\}$ in the instance (X, S) of exact cover. For each $x \in X$, define $\#x = |\{A \in S : x \in A\}|$ i.e., the number of elements of S containing x.

Let p be a number of exceeding all $\#x$, $0 \leqslant x \leqslant m - 1$.

Encode $A \in S$ as the number $w(A) = \sum_{x \in A} p^x$, and take

$$B = \sum_{x=0}^{m-1} p^x$$
$$= \frac{p^m - 1}{p - 1}.$$

In p-ary notation, $w(A)$ looks like a string of 0's and 1's with a 1 in position x for each $x \in A$ and 0 elsewhere. B looks like a string of 1's of length m.

Adding the numbers $w(A)$ simulates the union of the sets A. The number p was chosen big enough so that we don't get into trouble with carries.

Asking whether there is a subset sum that gives B is the same as asking for an exact cover of X. $\square$

Related here are three similar problems:

PARTITION
Input : A finite set S, integer weight function $w : S \to \mathbb{N}$.
QUESTION : Is there a subset $S' \subseteq S$ such that $\sum_{a \in S'} w(a) = \sum_{a \in S - S'} w(a)$?

KNAPSACK
Input : A finite set S, integer weight function $w : S \to \mathbb{N}$, benefit function $b : S \to \mathbb{N}$, weight limit $W \in \mathbb{N}$, and desired benefit $B \in \mathbb{N}$.
Question : Is there a subset $S' \subseteq S$ such that $\sum_{a \in S'} w(a) \leqslant W$ and $\sum_{a \in S'} b(a) \geqslant B$?

BIN PACKING
Input : A finite set S, volumes $w : S \to \mathbb{N}$, bin size $B \in \mathbb{N}$, $k \in \mathbb{N}$.
Question : Is there a way to fit all the elements of S into k or fewer bins?

Theorem 7.2.25 (Karp [Kar73]). PARTITION, KNAPSACK *and* BIN PACK-ING *are all NP-complete.*

Proof. For PARTITION, we reduce from SUBSET SUM. We introduce two new elements of weight $N - B$ and $N - (Q - B)$, where we have $Q - \sum_{a \in S} w(a)$ and N is sufficiently large (with $N > Q$).

Choose N large enough so that both new elements cannot go in the same partition element, because together they outweigh all the other elements. Now ask whether this new set of elements can be partitioned into two sets of equal weight (which must be N).

By leaving out the new elements, this gives a partition of the original set into two sets of weight B and $Q - B$.

To reduce PARTITION to KNAPSACK take $b = w$ and $W = B = \frac{1}{2}Q$.

For BIN PACKING, take an instance of KNAPSACK. Then choose B to be half the total weight of all elements of S and $k = 2$. $\square$

The following problem is important in optimization theory.

INTEGER PROGRAMMING
Input : Positive rationals a_{ij}, c_j, b_i, $1 \leqslant i \leqslant m$, $1 \leqslant j \leqslant n$ and threshold B.
Question : Can we find the integers $x_1, \ldots, x_n$ such that $\sum_{j=1}^{n} c_j x_j \geqslant B$, subject to constraints $\sum_{j=1}^{n} a_{ij} x_j \leqslant b_i$, $1 \leqslant i \leqslant m$?

Theorem 7.2.26 (Karp [Kar73]). INTEGER PROGRAMMING *is NP-complete.*

Proof. Hardness: We reduce from SUBSET SUM. The subset sum instance, consisting of a set S with weights $w : S \to \mathbb{N}$ and threshold B, has a positive solution iff the integer programming instance: $0 \leqslant x_n \leqslant 1$, $a \in S$, $\sum_{a \in S} w(a) x_a = B$, as an integer solution.

Membership: INTEGER PROGRAMMING is in NP. We show that if there is an integer solution, then there is one with only polynomially many bits as a function of the size of the input. The integer solution can then be guessed and verified in polynomial time. Considering constraints $\sum_{j=1}^{n} a_{ij}x_j \leqslant b_i$, $1 \leqslant i \leqslant m$, we can convert to all the entries being integers by multiplying by the least common multiple of the denominators. Thus we can consider the coefficients as all integers. The fact that all entries are positive means that the b_i will be an upper-bound (after we turn everything into integers) for the values of the x_i and hence the solutions can only have at most a polynomial number of bits. $\square$

There is a related and very important optimization problem called LINEAR PROGRAMMING (where we optimize the sum over the real or complex numbers) which is is in solvable in polynomial time. The existence of a polynomial time algorithm for LINEAR PROGRAMMING was a celebrated result proven by Kachiyan [Kha79] in 1979. Kachiyan's algorithm is not practical, but there is another polynomial time algorithm which is is due to Karmarkar [Kar84] which is is "reasonably" practical. However, LINEAR PROGRAMMING shows the difference between theory and practice. As we will discuss in §10.5, the most commonly used algorithm for LINEAR PROGRAMMING, is one called the *simplex method* which *provably does not run in polynomial time*. However, the simplex method works well "mostly" in practice. It was, and is, a longstanding question to give a satisfactory explanation for this phenomenon. (More in §10.5.)

In practice, often INTEGER PROGRAMMING problems are approximated by LINEAR PROGRAMMING ones, hoping to get a reasonable approximate solution quickly. We discuss this idea in the proof of Theorem 10.2.8, where we are looking at approximation algorithms.

We finish with one of our original motivating examples.

Theorem 7.2.27 (Karp [Kar73]).

(i) HAMILTON CYCLE *is NP-complete.*
(ii) DIRECTED HAMILTON CYCLE *(in directed graphs, where to be a cycle all the edges must go in the same direction) is NP-complete.*

The proof below involves some relatively complicated gadgetry.

Proof. We show VERTEX COVER $\leqslant_m^p$ HAMILTONIAN CYCLE. We build a graph H which has a Hamilton cycle iff a given graph $G = (V, E)$ has a vertex cover of size k. We begin by considering the directed version.

For a directed graph we use a four vertex widget:

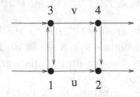

For an undirected graph we use a twelve vertex widget:

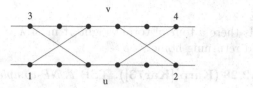

There is one widget for each edge $(u, v) \in E$, one side corresponds to the vertex u and the other to the vertex v.

The widgets have the property that any Hamiltonian cycle entering at vertex 1 must leave at vertex 2, and likewise with vertex 3 and 4. There are two ways for this to happen: either straight through, and so no vertices on the other side are visited, or zigzagging so that every vertex on both sides are visited. Any other path through will leave some vertex stranded and so cannot be part of a Hamiltonian cycle. This is proven by case analysis.

We form H as follows.

For each vertex u, string together all the u sides of all the widgets corresponding to edges in E incident to u. Call this the u loop.

Also, H has a set K of k extra vertices, where k is the parameter of vertex cover as given.

There is an edge from each vertex in K to the first vertex in the u loop, and an edge from the last vertex in the u loop to each vertex in K.

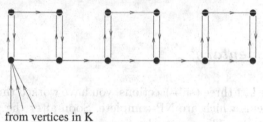

from vertices in K

Suppose there is a vertex cover $\{u_1, \ldots, u_k\}$ of G. Then H has a Hamiltonian cycle. Starting from the first vertex of K, go to the u_1 loop. When passing through the (u_1, v) widget, take the straight path if v is in the vertex cover (i.e., $v = u_j$ for some j - the other side will be picked up later when we traverse the u_j loop), and the zigzag path otherwise. When leaving the u_1 loop go to the second vertex of K and do the same, etc. until you get back to the first vertex of K.

Conversely, if H has a Hamiltonian cycle, it must traverse exactly k u loops, and that set of vertices u forms a vertex cover of G.

Hence, there is a vertex cover of size k of G iff H has a Hamiltonian cycle. (This argument holds for both directed and undirected graphs.) $\square$

An important variation of HAMILTON CYCLE is to put weights on the edges. You are planning a road trip and seek to minimize costs.

TRAVELLING SALESMAN PROBLEM (TSP)
Input : A number $k \geqslant 0$, directed graph $G = (V, E)$ with non-negative edge

weights $w : E \to \mathbb{N}$.

Question : Is there a tour of total weight at most k visiting each vertex at least once and returning home?

Theorem 7.2.28 (Karp [Kar73]). TSP *is NP-complete.*

Proof. We reduce from HAMILTON CYCLE. The reduction is simple. Let $G = (V, E)$ be an instance of HAMILTON CYCLE. We construct an instance G' of TRAVELLING SALESMAN. To do this $V(G') = V(G)$, $E(G') = E(G)$. Then in G' give all edges unit weight and ask for a TSP tour of weight $n = |V(G)|$.

TSP is in NP provided we can argue that optimal tours are short enough to guess and verify them in polynomial time. Each vertex can be visited at most n times in an optimal tour, otherwise we could cut out a loop and still visit all vertices. Thus we can guess a tour of length at most n^2 and verify that its total weight is at most k. $\square$

We summarize the NP-completeness results of this section:

Theorem 7.2.29 (Karp [Kar73]). *All of the following are NP-complete:* KNAPSACK, PARTITION, SUBSET SUM, EXACT COVER, BIN PACKING, INTEGER PROGRAMMING, (DIRECTED AND UNDIRECTED) HAMILTONIAN CYCLE, TSP.

7.2.11 Commentary*

By the end of the last three (sub-)sections, you have worked through a number of natural problems which are NP-complete. Soon after the papers of Cook and Levin, Karp [Kar73] was to the first to realize the *widespread nature of NP-completeness*. His original paper listed 21 problems, including all of those we have analysed. Many, many more can be found in the classic text by Garey and Johnson [GJ79]. Since then, thousands of problems have been shown to be NP-complete.

In a first course such as this, you gain the impression that proving NP-completeness of some problem is roughly like "bouldering" compared to the "mountain climbing" of theorem proving[7]. That is, the proofs are clever but not that difficult, and some elementary local replacement or gadget design will work.

However, there are many exceptionally complex and deep NP-completeness results requiring techniques *far* too advanced for this book, and involving substantial amounts of deep mathematical tools. Arguably, one of the high points of computational complexity theory is something called the PCP *Theorem*, which is fundamentally *a reduction*. This remarkable theorem states roughly that for some universal constant C, for every n, *any mathematical proof for a*

---

[7] Thanks to my friend Mike Fellows for this analogy.

statement of length n can be rewritten as a different proof of length polynomial in n, that is formally verifiable with 99% accuracy by a randomized algorithm that only looks at C many bits in the proof! The PCP Theorem was proven by Arora et. al. [ALM+98] in 1998. The proof involves algebraic coding theory and probability theory to build a reduction, complicated mathematical analysis to show that it works, and takes nearly 50 pages of proof. Even the simplified proof by Dinur [Din07] uses highly sophisticated techniques such as algebraic coding, expander graphs, and nontrivial probability theory.

But Does It All Matter? Beyond the question of P vs NP, one of the striking developments in recent years is the developing realization of the importance of the question: *"Does P $\neq$ NP really matter?"*. What do we mean by this? There are now many engineering applications using SAT SOLVERS. In the proof of CLIQUE being NP-complete we explicitly gave a reduction from CLIQUE to SAT (in Lemma 7.2.2). We gave this explicit reduction in spite of the fact that we could have simply argued that CLIQUE was in NP. Therefore, CLIQUE must have such a translation via the Cook-Levin Theorem. But we wanted to show that that detour was unnecessary and there was an easy reduction re-writing the CLIQUE problem into an instance of SAT. Many problems have easy and natural translations into instances of SAT.

So What? Isn't SAT supposedly hard? Is that not the point of NP-completeness? Well, not quite. We believe that such an NP-completeness result shows that if *L* is NP-complete, then there are *instances* of "$x \in L$?" not solvable in in polynomial time, assuming P $\neq$ NP. *What about instances that occur in practice?* It turns out that many *natural engineering* problems, when translated into instances of SAT, *and* can be solved with a class of algorithms called SAT SOLVERS. These are algorithms specifically engineered to solve SATISFIABILITY. Moreover, this methodology seems to work well and quickly for the versions of the problems *encountered in "real life"*. NASA, for example, uses SAT SOLVERS for robot navigation. When the author was young, showing something to be NP-complete meant that the problem looked scary and unapproachable; like they who must not be named in Harry Potter. But now many engineers regard a problem coming from practice as "solved" if it can be efficiently translated into an instance of SAT.

Of course, this translation does not always work, such as in certain biological problems and in cryptography (the latter being a *very good thing*, since otherwise all banking would be insecure).

A really good open question is *why*? What is it about these instances of SAT, coming from engineering problems, which allows us to solve them efficiently? We really have no idea.

7.2.12 Exercises

Exercise 7.2.30 Show that SET COVER below is NP-complete.

SET COVER
Input : A ground set $U = \{x_1, \ldots, x_n\}$, a collection of subsets $S_i \subseteq U$, $\{S_1, \ldots, S_k\}$ and an integer k.
Question : Is there a sub-collection of at most k S_i whose union is U?
  (Hint. Reduce from VERTEX COVER.)

Exercise 7.2.31 Prove that the following is NP-complete.

LONG PATH.
Input : A graph G with n vertices.
Question : Does G have a path with exactly $n - 1$ edges and no vertices repeated?
  (This might need a bit of thought.)

Exercise 7.2.32 A graph H is called a *topological minor* of a graph G if there is an injection $f : H \to G$, such that edges of H are represented by disjoint paths in G. That is, if $xy \in E(H)$ then there is a path $P(xy)$ from $f(x)$ to $f(y)$ in G, for all xy, if $P(xy)$ and $P(qr)$ has a vertex in common it is only $f(x)$ or $f(y)$.
  Prove that the problem below is NP-complete.

TOPOLOGICAL MINOR
Input : Graphs H and G.
Question : Is H a topological minor of G?

Exercise 7.2.33 BINARY INTEGER PROGRAMMING asks that the integer solutions $x_i \in \{0, 1\}$. Show that BINARY INTEGER PROGRAMMING is NP-complete.

7.3 Space

The goal of the present section is to look at calibrating computable functions according to the memory used. We recall that DSPACE (f) is the class of sets A for which there is a machine M such that $x \in A$ iff M accepts x and uses at most $f(|x|)$ squares of work tape. This uses the model of a Turing machine with extra work tapes whose used squares we count; the model we used in Figure 6.1, and similarly for NPSPACE(f).

  Certain natural inclusions come from the definition. The class L of languages accepted in deterministic logspace on the work tape, is evidently a subset of P, since the work tape is size $O(\log |x|)$ and thus for some fixed c, there are $\leq c^{\log |x|}$ many possible configurations. Working base 2, this gives a worst case running time of $O(2^{c \log |x|})$. But this is $O(|x|^c)$ and hence polynomial time. The upshot is $L \subseteq P$.

Similar simulations show that $NP \subseteq PSPACE$, since we can check every computation path of a nondeterministic machine running in nondeterministic time $|x|^k$, in space $O(|x|^k)$ by cycling through the computation paths in lexicographic order, one at a time. Furthermore, by the Hierarchy Theorem, Theorem 6.1.5, $L \subset PSPACE$. Thus, we get the following picture:

$$L \subseteq P \subseteq NP \subseteq PSPACE \subseteq NPSPACE \text{ and } L \neq PSPACE.$$

We also see that co-PSPACE = PSPACE, since we can cycle through all configurations of a machine M to see if we get acceptance.

7.3.1 Savitch's Theorem

The following theorem shows that space seems more pliable than time, and also shows that nondeterminism is often less interesting for space.

Theorem 7.3.1 (Savitch [Sav70]). *If t is space constructible and $t(|x|) > \log |x|$ for all x, then $NSPACE(t) \subseteq DSPACE(t^2)$.*

Proof. We use a method of "divide and conquer", where we keep guessing and testing halfway points of a computation.

Suppose $A \in NSPACE(t)$ i.e., M accepts A nondeterministically in space $t(|x|)$.

So we know there are finitely many quadruples, states, etc. and hence there is some $d \in \mathbb{N}$ such that the number of possible configurations is $d^{t(|x|)}$. Let $n = (\log d)t(|x|)$, so $2^n = 2^{(\log d)t(|x|)} = d^{t(|x|)}$.

Now, if C_1 and C_2 are configurations we will write $C_1 \overset{i}{\vdash} C_2$ if we can obtain C_2 from C_1 in $\leq 2^i$ steps. If $i \geq 1$ then $C_1 \overset{i}{\vdash} C_2$ iff $\exists C \ (C_1 \overset{i-1}{\vdash} C$ and $C \overset{i-1}{\vdash} C_2)$, where C is a halfway point. Take C_0 to be the initial configuration of M on input x. Then $x \in A$ iff there is some accepting configuration C_e such that $C_0 \overset{n}{\vdash} C_e$.

To test if $C_0 \overset{n}{\vdash} C_e$ for each C_e in a small space we proceed inductively;

(i). For $i = 0$, $C_0 \overset{i}{\vdash} C_e$ iff $C_0 = C_e$ or can be reached in one step.

(ii). For $i \geq 1$, $C_0 \overset{i}{\vdash} C_e$ iff $C_0 \overset{i-1}{\vdash} C$ and $C \overset{i-1}{\vdash} C_e$ for some C of length $t(|x|)$.

We call $\boxed{C_1|C_2|i|C}$ an i module.

Now C_1, C_2 have length $< t(|x|)$ and $i \leqslant n$, then we guess so that the tape at any stage appears as:

$C_1 \mid C_2 \mid i \mid C$	$C_1 \mid C \mid i-1 \mid C'$	$C_1 \mid C' \mid i-2 \mid C''$
i module	$i-1$ module	$i-2$ module

So we start with C_1, C_2, i, C and in the next module cycle through all the possible C' to see if C_1, C, $i-1$, C' is a legal $i-1$ module. If we show that $C_1 \overset{i-1}{\vdash} C$ via C' we then need to show $C' \overset{i-1}{\vdash} C_2$ in the same way, by cycling through all possible configurations recursively. There are n modules with $n = (\log d)t(|x|)$, and each module of length $\leqslant O((\log d)t(|x|))$ so the total length is $O(t(|x|))^2$.

Hence, NSPACE$(t) \subseteq$ DSPACE(t^2). $\square$

We remark that the $t(|x|) > \log|x|$ is needed for this theorem. It is unknown if NL, the nondeterministic version of L, is different from L. An example of a problem in NL is whether x and y are path connected in a directed graph G, since you only need to guess the vertices of the path one at a time, making sure that each pair are connected and stopping if you hit the target one. This connectivity problem is in fact complete for the class NL, under the appropriate reductions. It seems that to determine such connectivity deterministically we need the square of L. But again L $\neq$ NL is a longstanding open question. (See Papadimitriou [Pap94], Chapter 16 for a thorough discussion of this issue.)

7.3.2 Time vs Space

We return to the polynomial hierarchy, PH. It is not hard to use an analogue of the proof of Cook's Theorem to establish that there is an analogue of SATISFIABILITY for Σ_n^P. For example, the language which consists of propositional formulae $\varphi(y_1, \ldots, y_n, z_1, \ldots, z_m)$, with variables partitioned into two sets given by the $\overline{y_i}$ and $\overline{z_{ji}}$, is Σ_2^P-complete when we ask if there is an assignment for the literals y_i which satisfies every assignment for the z_j. (Exercise 7.3.3.) We would write this as $\exists^P \overline{y} \forall^P \overline{z} \varphi(\overline{y}, \overline{z})$.

To save on notation, if it is obvious we are using tuples rather than single variables we will write y for $\overline{y}$, so this expression would appear as $\exists^P y \forall^P z \varphi(y, z)$. The meaning will be clear from context. Since polynomial space is big enough to simulate all of the relevant searches, we have the following.

Proposition 7.3.1 (Folklore). PH $\subseteq$ PSPACE.

Less obvious, but also true, is that the "diagonal" of PH is complete for PSPACE. That is, we can define QBFSAT as the collection of codes for true quantified Boolean formulae.

QBF (QBFSAT)
Input : A quantified boolean formula $\varphi = \exists^P x_1 \forall^P x_2 \ldots \varphi(x_1, x_2, \ldots x_n)$.
Question : Is φ valid?

The difference between QBF and PH is that PH is only concerned with $\cup_n \Sigma_n^P$, the union of *fixed levels*, whereas in the case of QBFSAT we are concerned with all the levels at once. The following result is not altogether obvious. It is the analog of the Cook-Levin Theorem for space. The proof uses the recycling space idea of Savitch's Theorem.

Theorem 7.3.2 (Meyer and Stockmeyer [MS72]). QBF *is* PSPACE-*complete.*

Proof. We only sketch the proof as it uses the same ideas as Savitch's Theorem. Namely that to get from one configuration to another requires a half-way point, and the idea here is that this can be found by quantifying, with an additional layer of quantification as the computations get longer.

Fix a PSPACE bounded machine M with bound n^c. The proof uses the formula $\text{access}_j(\alpha, \beta)$ where α and β are free variables representing M-configurations, and the formula asserts that we can produce β from α in $\leqslant 2^j$ steps.

The key idea is that the formula is re-usable: for $j = 1$, access_j can be read off from the transition diagram, and for $j > 1$, you can express access_j as

$$\exists \gamma(\text{config}_M(\gamma) \wedge (\forall \delta \forall \lambda [(\delta = \alpha \wedge \lambda = \gamma) \vee (\delta = \gamma \wedge \lambda = \beta)] \rightarrow \text{access}_{j-1}(\delta, \lambda))).$$

Here $\text{config}_M(\gamma)$ asserts that γ is a configuration of M. The formula asserts that there is an accepting β such that on input α, $\text{access}_{n^c}(\alpha, \beta)$. □

7.3.3 Exercises

Exercise 7.3.3 Prove an analog of the Cook-Levin Theorem for Σ_2^P.

7.4 Natural PSPACE-Complete Problems

As with time, there is a machine based problem $\leqslant_m^P$ complete for PSPACE.

Proposition 7.4.1. *The language*

$$L = \{\langle e, x, 1^n \rangle : M \text{ accepts } x \text{ space bounded by } n\},$$

is PSPACE polynomial time m-complete.

Proof. This is more or less identical to Theorem 7.2.1, and is left to the reader. (Exercise 7.4.4). □

It is possible to identify a number of natural problems complete for PSPACE. Typically such problems are inspired by the above characterization of PSPACE in terms of QBF. Quantifier alternations naturally correspond to strategies in games. Player A has a winning strategy, by definition, if there is first move that A can make, such that for every possible second move by Player B, there exists a third move in the game that A can make, such that for every fourth move that B can make ... A wins. For example, for the natural analog of the game GO, determining who has a winning strategy is PSPACE complete. (Lichtenstein and Sipser [LS80])

We will analyse one of the natural combinatorial games for which determining winning strategies is PSPACE complete. This game is called GENERALIZED GEOGRAPHY. The game is played on a directed graph G. The play begins with player I putting a token on a distinguished vertex of G. This now become marked. The players then play alternatively and move the token along an edge incident with the current position of the token to an unmarked vertex. The player who cannot move to an unmarked vertex loses. A strategy for this game is a function making choices:

Definition 7.4.1. Let $\Sigma = V(G)$. A *strategy* in this context is a function $f : \Sigma^* \to V(G)$, such that $f(\sigma) = v$ implies that σv is a legal sequence of moves of the game. In particular, if u is the last vertex in σ (i.e. $\sigma(|\sigma|) = u$) then uv is a directed edge and v is not marked.

GENERALIZED GEOGRAPHY
Input : An instance of generalized geography on a directed graph G.
Question : Does player I have a winning strategy?

Theorem 7.4.1 (Schaefer [Sch78]). GENERALIZED GEOGRAPHY *is PSPACE complete.*

Proof. First GENERALIZED GEOGRAPHY is in PSPACE. Note that after each play of the game, a vertex is marked. Thus the maximum length of a strategy is n where G has n vertices. We can build a machine which cycles through, lexicographically all the possible strategies player I could use. Of course, each play of I could be followed by any vertex incident with the current play of 1 by II, so overall checking strategies would take around n^2 much space as we would need to keep track of what we have done. Certainly polynomial in $|G|$. If I has a winning strategy then we will see it and stop. If we cycle through all possible strategies for I and see no win, II must win.

Now for hardness. Suppose we are given an instance X of QBF with $X = Q_1 v_1 Q_2 v_2 \ldots Q_n v_n F(v_1, \ldots, v_n)$, and, without loss of generality, $Q_1 = \exists$ and $Q_n = \exists$, and $Q_i \neq Q_{i+1}$ for all i. Here B is a formula of propositional logic in conjunctive normal form of the form $c_1 \wedge \ldots c_m$, with c_j clauses. Each boolean variable is represented by a diamond of 4 vertices, $\{v_{i,k} : 0 \leqslant k \leqslant 2\} \cup \{\overline{v}_{i,1}\}$. The edges are $\{v_{i,0}v_{i,1}, v_{i,0}\overline{v}_{i,1}, v_{i,1}v_{i,2}, \overline{v}_{i,1}v_{i,2}\}$. The idea here is that if we hit $v_{i,0}$ we must choose one of the directed edges from it and they will either be $v_{i,1}$ representing a positive choice for v_i or $\overline{v}_{i,1}$ which represents a negative choice.

Additionally, we will have edges $v_{i,2}v_{i+1,0}$, and also one vertex c_j for each clause c_j. We need edges $v_{n,2}c_j$ for $1 \leqslant j \leqslant m$, and a path of length 2 from c_j to $v_{i,1}$ if $v_i \in c_j$, and from c_j to $\overline{v}_{i,1}$ if $\neg v_i \in c_j$. Declare that $v_{1,0}$ is distinguished. This completes the description of G.

Example 7.4.1. Here is an example;

$$X = \exists u \forall b \exists c \forall d [(a \vee \neg b \vee c) \wedge (b \vee \neg d)]$$

See Figure 7.1 below.

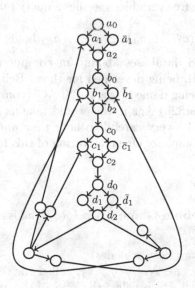

Fig. 7.1: Example of GENERALIZED GEOGRAPHY

We complete the proof. Player I will choose a vertex after $v_{1,0}$. This will correspond to a choice of value for v_1, and player II. Because the path to $v_{2,0}$ is of length 2, it is player II who makes a choice at $v_{2,0}$. Player II wishes to make a choice making X false, etc. Thus we see a direct emulation of the quantifiers. At the end, when we hit $v_{n,2}$, the parity of the number of quantifier alternations means that it is player II who get to choose a clause vertex c_j. Player I will then get to choose one of the paths leading back to the main frame of the graph. That is, it is choosing one of the literals. Player II must choose the vertex to which the path is connected. If player I has chosen this variable $v_{i,1}$ or $\bar{v}_{i,1}$ already, then player I wins, as II has no play. If I has chosen badly, II will be able to choose one of $v_{i,1}$ or $\bar{v}_{i,1}$, and then I would need to choose $v_{i,2}$, which it cannot. If player II gets to choose for this diamond, then they would choose $v_{i,1}$ if $\neg v_i \in c_j$, and choose $\bar{v}_{i,1}$ if $v_i \in c_j$. That is, I can win the game iff X is true. $\square$

The undirected version of GENERALIZED GEOGRAPHY is in P, using the Edmonds Matching algorithm for general graphs. Edmond's Algorithm is somewhat out of the scope of the present book so we refer the reader to Fraenkel, Scheinerman, and Ullman [FSU83].

One area where PSPACE complete problems abound is that of *relational databases*. This is an area of computer science where logic is used to make queries to a database. Formulae of the relevant language are built up from $x_1 = x_2$ together with $R_i(x_1, \ldots, x_n)$ where the x_i are variables or constants over the appropriate domains, and then use boolean connectives and quantifiers. A formula φ with free variables specifies a query relation to the database d,

$$r_\varphi = \{(a_1, \ldots, a_n) : \varphi(a_1, \ldots, a_n) \text{ holds in } d\}.$$

Relational queries and databases abound in computer science with whole languages such as XML being developed for them. Being so close to logic, it is easy to show that seeing if one is true is PSPACE complete in general. (See, for example, Vardi [Var82].) Again, this would take us a bit afar to cover.

The reader should be very careful to note that *not all of these strategy games are* PSPACE complete. One illustration of this fact is provided by the following game:

Definition 7.4.2 (Pebble Game). A *pebble game* is a quadruple (N, R, S, T) consisting of

(i) N, a finite set of vertices (or nodes)
(ii) $R \subseteq \{(x, y, z) : x, y, z \in N, x \neq y \neq z\}$, called the set of *rules*.
(iii) $S \subset N$, the start set. $|S|$ is called the *rank* of the game.
(iv) $T \in N$, called the terminal node of the game.

The pebble game is played by two players I and II, who alternatively move pebbles on the nodes. At the beginning of the game, pebbles are placed on all the start nodes. If $(x, y, z) \in R$ and there are pebbles upon x and y but not on z, then a player can move a pebble from x to z. The winner is the first player to put a pebble on T or can force the other player into a position where he cannot move.

PEBBLE GAME PROBLEM
Input : A pebble game (N, R, S, T), and a positive integer k.
Question : If $|S| = k$ does player I has a winning strategy.

A theorem somewhat beyond the scope of this course is the following:

Theorem 7.4.2 (Adachi, Iwata and Kasai [AIK84]). *Let* $L \in$ DTIME(n^k). *Then* L *is* $O(n \log n)$*-reducible- using space* $O(n \log n)$*-to the* $(2k + 1)$*-pebble game problem.*

In the above, the "$O(n \log n)$-reducible" means that the reduction is not only polynomial but has low degree. Note that the theorem is proven by a generic reduction from multi-tape Turing Machines. The interested reader can find this presented in [DH10] where the construction is used for other purposes. A consequence of this result, proven in [AIK84] is that the following problem is also hard.

The CAT AND MOUSE GAME is a quintuple $G = (X, E, c, M, v)$ with X a set of vertices, E a set of edges, $c \in X$, $M \subseteq X$ (the mice), and $v \in X$ (the cheese). In the game, player I (the cat) begins with his token on c and player II begins with tokens on each member of M. Players play alternatively and can move tokens from vertex x to vertex y provided $xy \in E$. Two tokens of player II cannot occupy the same vertex. Player I wins if he can place his token on a vertex with one of player II's token. Player II wins if she can place one of her tokens on vertex v even if it is occupied by player I's token. Player I plays first. For each k, and $|M| = k$, this game is also hard for DTIME(n^k).

Whilst we are on the subject, the reader might wonder about the higher classes like E and EXP. The usual method (see Theorem 7.2.1) shows that the we have a complete problem.

Theorem 7.4.3 (Meyer and Stockmeyer [MS72]).

(i) The language below is complete for EXP.

$$L = \{\langle e, x, n \rangle : \varphi_e \text{ accepts } x \text{ in } \leqslant 2^{|n|} \text{ many steps}\}.$$

(ii) The language below is complete for NEXP.

$$L = \{\langle e, x, n \rangle : \text{some computation path of } \varphi_e \text{ accepts } x \text{ in } \leqslant 2^{|n|} \text{ many steps}\}.$$

There are natural problems for these classes, but we will not look at them in this first course.

7.4.1 Exercises

Exercise 7.4.4 Prove Theorem 7.4.1.

Exercise 7.4.5 (Schaeffer see [LS80]) PLANAR GENERALIZED GEOGRA-
PHY is the same as GENERALIZED GEOGRAPHY, except that G must be planar. Show that PLANAR GENERALIZED GEOGRAPHY is PSPACE complete.

Exercise 7.4.6 (Book [Boo72]) Show that if $E \subseteq NP$, then $EXP \subseteq NP$.

7.5 Further Variations on the Theme : Advice Classes and Randomized Reductions

7.5.1 Introduction

There has been an explosion of work on computational complexity theory.
There are two reasons for this. First, as we have seen, we really don't understand computational complexity very well. Second, it is very important.

Most of the material in the "basic advanced" text by Kozen [Koz06] is not
found in this course. For example, we do not look at LOG vs P, and problems
complete for P under logspace reductions, interactive proofs, circuit classes
such as AC_0 and NC, Toda's Theorem, counting complexity and #P, the
PCP Theorem, etc.

Our goal is to give a thorough *grounding* in the ideas of computational
complexity, at least from our personal bias. However, we will cover in this
last section of the chapter, two additional topics. One which is not usually
covered in a first course concerns *advice classes*, and the second concerns
randomization. The first material will be useful when we look at *kernelization
lower bounds* in §9.5. The second is but a first foray into randomization but
allows us to see an important theorem which shows that, under randomized
reductions, determining if a formula X of propositional logic has a solution is
as hard as SAT *even if we promise that X has only one satisfying assignment
if it has any.* This result was the key to many of the of the major theorems
of the late 20th century.

7.5.2 Advice and Nonuniform Complexity Classes

Definition 7.5.1 (Karp and Lipton [KL80]).

1. An *advice* function $f : \Sigma^* \to \Sigma^*$ has the property that if $|x| = |y|$ then $f(x) = f(y)$, and hence for a length we can write unambiguously $f(n)$ for $f(x)$ if $|x| = n$. The advice is called *polynomial* if the length of $f(n)$ is $O(n^c)$ for some constant c. The function f does not need to be computable at all.
2. For a language L and complexity class $\mathcal{C}$ we say that $L \in \mathcal{C}/\text{Poly}$ if there is a polynomial advice function f, and a language $\widehat{L} \in \mathcal{C}$ such that for all x,

$$x \in L \text{ if and only if } \langle x, f(|x|) \rangle \in \widehat{L}.$$

In particular, P/Poly is the collection of languages which are in P with nonuniform advice. This may seem a strange concept at first glance, since for any language L, the language $S_L = \{1^n : n \in L\}$ is in P/Poly, as the only thing of length n that could *possibly* be in S_L is a string of the form 1^n; and that is the advice. Hence there will be uncountably many languages in P/Poly.

One of the great results of classical structural complexity theory is that P/Poly is exactly the languages L that can be decided by a (nonuniform) family of circuits C_n, one for each input size n, where $|C_n| \leqslant p(n)$ for some polynomial p (Pippinger [Pip79]). Another important result proven by Karp and Lipton [KL80] is that if SAT $\subseteq$ P/log, then P = NP. These results would take us a bit far to include, and we refer the reader to Balcázar and J. Díaz and J. Gabarró [BDG90] for an excellent account of these results. Instead, we will concentrate on some results which will be important for us later in this book.

In this section, we will define an *advice* hierarchy which is an advice analog of the polynomial hierarchy. We will see that a collapse of an advice hierarchy implies a related collapse in the polynomial time hierarchy.

To begin this analysis, we define the advice hierarchy by appending "/Poly" to the end of everything. We show that a collapse of some level of PH/Poly propagates upwards. For ease of notation, we will drop the superscript "p" from, for instance, Σ_i^p, and write Σ_i, when the context is clear that we are concerning ourselves with the *polynomial time hierarchy* or its analogues.

Lemma 7.5.1 (Yap [Yap83]). *For all $i \geqslant 1$, $\Pi_i \subseteq \Sigma_i/\text{Poly}$ implies $\Sigma_{i+1}/\text{Poly} = \Sigma_i/\text{Poly}$.*

Proof. First it is easy to see that for any complexity class $\mathcal{C}$, (co-$\mathcal{C}$)/Poly is the same as co-($\mathcal{C}$/Poly). It is enough to prove this for $i = 1$ and the rest works entirely analogously. Suppose that $\Pi_1 \subseteq \Sigma_1/\text{Poly}$ and let $L \in \Sigma_2/\text{Poly}$. Since $L \in \Sigma_2/\text{Poly}$, we have some advice function f, and $Q \in \Sigma_2$ with $x \in L$

iff $\langle x, f(x) \rangle \in Q$, and hence for some polynomial time computable R, $x \in L$ if and only if $\exists y[\forall z R(\langle x, f(x) \rangle, y, z)]$. Because $\Pi_1 \subseteq \Sigma_1/\text{Poly}$, we can replace the Π_1 relation in the brackets $[\ ]$ by a Σ_1/Poly one. We see that for each y, there are polynomial time relations R'_y and advice function g such that $x \in L$ if and only if $\exists y \exists y'(R'_y(\langle x, f(x) \rangle, y, \langle y, g(x, f(x), y) \rangle))$. To finish we need to amalgamate all the advice so that the advice will depend only upon x. However, all the advice depends only on x and y and the range of these is $\leqslant (|x| + |y|)^c = |x|^{c'}$ for some c, c'. As it is only the lengths, we can concatenate the advice together in a self-delimiting way, and only have a polynomial amount of advice that depends on $|x|^{c'}$. Thus we can construct a polynomial time relation R'' and an advice function h, such that $x \in L$ if and only if $\exists y R''(\langle x, h(x) \rangle, y)$, as required. $\square$

Corollary 7.5.1 (Yap [Yap83]). *The following are equivalent.*

1. $\Pi_i \subseteq \Sigma_i/\text{Poly}$.
2. $\Sigma_i \subseteq \Pi_i/\text{Poly}$.
3. $\text{PH}/\text{Poly} = \Pi_i/\text{Poly} = \Sigma_i/\text{Poly}$.

   The result we will need in §9.5 is the following. It is a translation of an advice collapse to a PH collapse.

Theorem 7.5.1 (Yap [Yap83]). *For $i \geqslant 1$,*

$$\Sigma_i \subseteq \Pi/\text{Poly} \text{ implies } \Sigma_{i+2} = \text{PH}.$$

Proof. We show that $\Sigma_3 \subseteq \Pi_3$, assuming $\text{NP} \subseteq \text{co-NP}/\text{Poly}$. The general case is analogous. Let $L \in \Sigma_3$ so that $x \in L$ if and only if $\exists y \forall z R(x, y, z) \in \text{SAT}$ for some poly relation R, since SAT is NP-complete. Now the assumption is that $\text{NP} \subseteq \text{co-NP}/\text{Poly}$. This gives us an advice function α and a poly relation C such that

$$\psi \in \text{SAT} \text{ if and only if } \forall v C(\psi, \alpha(|\psi|), v).$$

We see $x \in L$ if and only if $\exists y \forall z \forall v C(R(x, y, z), \alpha(|R(x, y, z)|), v)$.

   Using the fact that all strings, etc., are poly bounded in $|x|$, say by n^c, we can build an advice function β which concatenates all advice strings of length $\leqslant n^c$. So we would have (without loss of generality), $x \in L$ if and only if $\exists y \forall z \forall v C(R(x, y, z), \beta(|R(x, y, z)|), v)$. Notice that this shows that $L \in \Sigma_2/\text{Poly}$. The idea of the proof is to now "guess" the advice and arrange matters so that we can use a self-reduction to verify we have the correct advice. That is, we will modify the relation C to get a relation H so that $x \in L$ if and only if $\exists w \exists y \forall z \forall z \forall v H(R(x, y, z), w, v)$. This self-reduction process is achieved as follows. Let $R(x, y, z) = R$ have variables $x_1, \dots, x_m$. We describe the procedure $\text{Red}(R)$ as follows.

1. Ask $\forall v C(R, w, v)$? If yes, continue; or else no, reject.
2. Set $x_1 = 0$, and ask if $\forall v C(R(x_1 = 0), w, v)$? If yes, continue, else set $x_1 = 1$.

3. Repeat this for $x_2, \ldots, x_m$.
4. If we get to the end then we have an assignment for R, and we can check if that assignment satisfies R. If yes, we return a yes, if not reject.

We claim that $x \in L$ if and only if $\exists w \exists y \forall z \forall v C[(R(x, y, z), w, v) \wedge \text{Red}(R(x, y, z))]$.

For the forward implication, we can set $w = \beta(x)$. For the reverse implication, if some w actually passes $\text{Red}(R(x, y, z))$ then $R(x, y, z) \in$ SAT for all z and some y, and hence $x \in L$. Moreover, if $x \in L$ then some w will work. The claim now follows and hence $L \in \Sigma_2$, giving the result. □

We remark that advice classes became popular for quite a while, as there was a plan to tackle P vs NP using them. The argument was that circuits are non-dynamic objects and perhaps proving lower bounds for them would be easier than dealing with moving heads on Turing Machines and the like. Also the feeling was that P vs NP has little to do with uniformity. Thus if we could show that there were problems in NP requiring exponential sized circuits, we would be done. Unfortunately, the best lower bounds proven so far for languages in NP in inputs of size m is $3m$.

7.5.3 Valiant-Vazirani and BPP

In this last section we will meet another complexity class not routinely encountered, but of significant importance. It concerns *randomized reductions*. We will be using elementary probability theory, which is fundamentally counting. The probability that a string is chosen of length n is 2^{-n}, technically we will we be using the uniform distribution (i.e. unbiased coin-toss met in school).

Definition 7.5.2. We say that a language $L \in$ RP if there is a non-deterministic polynomial time Turing machine M such that for all x,

- $x \in L$ implies $M(x)$ accepts for at least $\frac{3}{4}$ many computation paths on input x.
- $x \notin L$ implies that $M(x)$ has no accepting paths.

We define the class BPP similarly, except we replace the second item by

- $x \notin L$ implies that $M(x)$ has at most $\frac{1}{4}$ paths accepting.

In the above, the numbers $\frac{3}{4}$ and $\frac{1}{4}$ are arbitrary. For a randomly chosen computation path y, the probability that $M(x)$ accepts for certificate y, is

$\mathrm{Pr} M(x,y) \geqslant \frac{3}{4}$, and likewise that M rejects. Using repetition, we can amplify the computational path as we like[8]. In particular, for any $L \in \mathrm{RP}$ or BPP, and any polynomial n^d, if you build a new Turing machine M which replicates the original one n^{c+d} many times, you can make sure that

- $x \in L$ implies that the probability that M accepts is $\geqslant 1 - 2^{-n^d}$ and
- $x \notin L$ implies M does not accept (RP) or the probability that M accepts is $\leqslant 2^{-n^d}$ (BPP).

This process is referred to as (probability) *amplification*.

It is easily seen that $\mathrm{P} \subseteq \mathrm{RP} \subseteq \mathrm{NP}$, and that $\mathrm{RP} \subseteq \mathrm{BPP}$. There is one additional relationship known.

Theorem 7.5.2 (Sipser [Sip83], Lautemann [Lau83]). $\mathrm{BPP} \subseteq \Sigma_2 \cap \Pi_2$.

Proof. Let $L \in \mathrm{BPP}$. Assume that we have a machine M amplified as above, with acceptance probability $\geqslant 1 - 2^{-n}$ and rejection $\leqslant 2^{-n}$. Let

$$A_x = \{y \in \{0,1\}^m \mid M(x,y) \text{ accepts}\}.$$

$$R_x = \{y \in \{0,1\}^m \mid M(x,y) \text{ rejects}\}.$$

Here $m = n^c$ is the length of a computation path of $M(x)$. Notice that for $x \in L$, $|A_x| \geqslant 2^m - 2^{m-n}$ and $|R_x| \leqslant 2^{m-n}$. Conversely, if $x \notin L$, $|R_x| \geqslant 2^m - 2^{m-n}$ and $|A_x| \leqslant 2^{m-n}$.

We need the following technical lemma. Let (for this proof) $\oplus$ denote the *exclusive or* operation on binary vectors, that is, bitwise sum modulo 2.

Lemma 7.5.2. $x \in L$ *if and only if* $\exists z_1 \dots, z_m [\forall i (|z_i| = m) \land |\{y \oplus z_j \mid 1 \leqslant j \leqslant m \land y \in A_x\}| = 2^m]$.

Proof. First suppose $x \notin L$. Then $|A_x| \leqslant 2^{m-n}$. For any $z_1, \dots, z_m$, $\{y \oplus z_j \mid 1 \leqslant j \leqslant m \land y \in A_x\} = \cup_{j=1}^m \{y \oplus z_j \mid y \in A_x\}$. Note that for sufficiently large n, $\sum_{j=1}^m |\{y \oplus z_j \mid y \in A_x\}| = \sum_{j=1}^m |A_x| \leqslant m2^{m-n} < 2^m$. Therefore, $|\{y \oplus z_j \mid 1 \leqslant j \leqslant m \land y \in A_x\}| \neq 2^m$. Conversely, if $x \in L$, $|R_x| \leqslant 2^{m-n}$. We need $z_1, \dots, z_m$ such that for all d, $\{d \oplus z_j \mid 1 \leqslant j \leqslant m\} \not\subseteq R_x$. Each $z_1, \dots, z_m$ for which there is a d with $\{d \oplus z_j \mid 1 \leqslant j \leqslant m\} \subseteq R_x$ is determined by a subset of R_x of size m, and a string d of length m. There are $\leqslant (2^{m-n})^m$. of the former, and $\leqslant 2^m$ of the latter, giving a total of $\leqslant (2^{m-n})^m 2^m = 2^{m(m-n+1)} < 2^{m^2}$. But there are 2^{m^2} many choices for $z_1, \dots, z_m$ of length m, and hence there is some choice $z_1, \dots, z_m$, with $\{d \oplus z_j \mid 1 \leqslant j \leqslant m\} \not\subseteq R_x$. $\square$

Now we can complete the proof of the Sipser-Lautemann Theorem. Since BPP is closed under complementation, it is enough to show $L \in \Sigma_2$. To decide if $x \in L$,

---

[8] Assuming that we have a supply of independent random coins. Each repetition is independent of the previous one, and if the probability of non-acceptance is $\frac{1}{4}$ then doing this twice yields $\frac{1}{4}^2 = \frac{1}{16}$, and so forth.

- Guess $z_1, \ldots, z_m$.
- Generate all potential d of length m.
- See, in polynomial time if $d \in \{y \oplus z_j \mid 1 \leqslant j \leqslant m \wedge y \in A_x\}$, which is the same as $\{d \oplus z_j \mid 1 \leqslant j \leqslant m\} \cap A_x \neq \emptyset$; by checking $M(x, d \oplus z_j)$ for each j.

This matrix defines a Σ_2 predicate. Hence $L \in \Sigma_2$ by the lemma. $\square$

We next explore means for showing that problems are likely to be intractable, using randomized polynomial-time reductions. In some sense, this result and the one above were the key to many results in modern complexity like Toda's Theorem[9] and the PCP Theorem[10]. The problem of interest is whether there is unique satisfiability for a CNF formula.

UNIQUE SATISFIABILITY (USAT)
Instance : A CNF formula φ.
Question : Does φ have exactly one satisfying assignment?

Theorem 7.5.3 (Valiant and Vazirani [VV86]). $\mathrm{NP} \subseteq \mathrm{RP}^{\mathrm{USAT}}$.

Proof. The method is based around a certain algebraic construction. Let V be an n dimensional vector space over $\mathrm{GF}[2]$ the two element field. For this field we remind the reader that for any subspace Q of V, $Q^{\perp} = \{z \mid z \cdot w = 0 \text{ for all } w \in Q\}$, is the *orthogonal complement* of Q.

Lemma 7.5.3 (Valiant and Vazirani [VV86]). *Let S be a nonempty subset of V. Let $\{b_1, \ldots, b_n\}$ be a randomly chosen basis of V and let $V_j = \{b_1, \ldots, b_{n-j}\}^{\perp}$, so that $\{0\} = V_0 \subset V_1 \subset \cdots \subset V_n = V$ is a randomly chosen tower of subspaces of V, and V_j has dimension j. Then the probability*

$$Pr(\exists j(|V_j \cap S| = 1)) \geqslant \frac{3}{4}.$$

Proof. First note that if either $|S| = 1$ or $0 \in S$ then we are done. Hence we assume $|S| \geqslant 2$ and $S \cap V_0 = \emptyset$. It is easy to see that there is some least j with $|S \cap V_j| \geqslant 2$. Since we are working in $\mathrm{GF}[2]$, this means that the dimension of the span $\mathrm{sp}(V \cap V_j)$ is also at least 2.

Consider the hyperplane $H = \{b_{n-k+1}\}^{\perp}$. ($H$ is the collection of vectors $\{v \in V : v \cdot b_{n-k+1} = 0\}$.) Let I be a maximal (linearly) independent subset of $S \cap V_k$. Notice that $V_{k-1} = V_k \cap H$, and $I \cap H \subseteq S \cap V_k \cap H = S \cap V_{k-1}$.

---

[9] This theorem states that if we consider polynomial time algorithms that have access to an oracle giving the number of accepting paths on a nondeterministic polynomial time Turing machine ($P^{\#P}$) then we can compute any level of the polynomial time hierarchy in polynomial time; that is, $\mathrm{PH} \subseteq P^{\#P}$.

[10] This is a characterization of NP in terms of "polynomially checkable proofs," which is beyond the scope of this book. We discussed this earlier. Kozen [Koz06] has a nice and accessible discussion of the PCP Theorem.

Now to establish the result, it is enough to show that $\Pr(S \cap V_{k-1}) = \emptyset \leqslant \frac{1}{4}$. Hence it is enough to show that $\Pr(I \cap H) = \emptyset \leqslant \frac{1}{4}$. Notice that $\Pr(I \cap H) = \emptyset \leqslant \max_{m \geqslant 2}\{\Pr(I \cap H) = \emptyset \mid \dim(\mathrm{sp}(I)) = m)\}$. To estimate $\Pr((I \cap H) = \emptyset \mid \dim(\mathrm{sp}(I)) = m)$, notice that since H is a hyperplane in a vector space, if $\dim(\mathrm{sp}(I)) = m$, then $\dim(\mathrm{sp}(I \cap H)) \in \{m, m-1\}$. However, since if $\dim(\mathrm{sp}(I) = m$ and $\dim(\mathrm{sp}(I) = m)$ then $I \subseteq H$ and hence $I \cap H \neq \emptyset$. This means that $\Pr(I \cap H) = \emptyset \mid \dim(\mathrm{sp}(I) = m \wedge \dim(\mathrm{sp}(I) = m)$ will be bounded by $\Pr((I \cap H) = \emptyset \mid \dim(\mathrm{sp}(I) = m \wedge \dim(\mathrm{sp}(I \cap H) = m - 1)$. Since we are using random bases, the probability we are after, therefore, is the probability that we have a set of m independent vectors $\widehat{I}$ and a random hyperplane $\widehat{H} = \{x\}^{\perp}$ and $\widehat{I} \cap \widehat{H} = \emptyset$. That is, for all $y \in I$, $y \cdot x = 0$. Now we calculate using inclusion/exclusion: $\Pr(y \cdot x = 0$ for all $y \in \widehat{I}) = 1 - \Pr(\exists y \in \widehat{I} \mid y \cdot x = 0) = 1 - (\sum_{i=1}^{|\widehat{I}|}(-1)^{i+1} \sum_{J \subseteq \widehat{I} \wedge |J|=i} \Pr(x \in J^{\perp}))$. This quantity equals $1 - (\sum_{i=1}^{m}(-1)^{i+1}\binom{m}{i}2^{-i}) = \sum_{i=0}^{m}\binom{m}{i}(-1)^{i}2^{-i} = (1 - \frac{1}{2})^{m}$. This quantity is maximized for $m = 2$, giving the bound $\frac{1}{4}$. $\square$

Now we turn to the proof of the Valiant-Vazirani Theorem. Let φ be an instance of SATISFIABILITY. Let n be the number of variables $\{x_1, \ldots, x_n\}$ in φ. We use the random basis $\{b_i : 1 \leqslant i \leqslant n\}$ to construct the tower $V_0 \subset \cdots \subset V_n = V$ of the lemma. This requires $O(n^2)$ random bits. The set S will be the set of *satisfying* assignment of $\{x_1, \ldots, x_n\}$. The key idea is that we can regard $(x_1, \ldots, x_n)$ as n-bit vectors in $\mathrm{GF}[2]^n$. All we need to do is to construct a formula ξ_i stating that a particular assignment lies in V_i. Then the relevant formula is $\varphi \wedge \xi_i$. The reduction, regarded as a Turing reduction, is essentially: for each $i = 0, \ldots, n$, see if $\varphi \wedge \xi_i$ is in USAT. By the lemma above, this succeeds with probability $\frac{3}{4}$, if there is some satisfying assignment for φ (i.e. $S \neq \emptyset$), and never returns "yes" if φ is not satisfiable. This can also be regarded as a randomized polynomial time m-reduction by guessing i, and making the error as small as desired using amplification. $\square$

The reader will note that this proof also gives a reduction for SAT *with a uniqueness promise*, that is, for formulae ρ such that if if ρ has a satisfying assignment it will have exactly 1. Note also that the reduction is a polynomial time Turing reduction as we need to ask many questions. More on this in Chapter 8.

Chapter 8
Some Structural Complexity

Abstract We develop some general results on computational complexity. We show that normal methods which relativize are insufficient to decide many natural questions about complexity class separations, including P vs NP. We prove Ladner's Theorem showing that the polynomial time degrees are dense.

Key words: Oracles, polynomial time degrees, Ladner's theorem

8.1 Introduction

In this chapter, we will briefly look at a part of computational complexity theory akin to the chapter on deeper computability theory, where we *abstract* the problems, and concentrate on the inner workings of the reductions. Whilst this might seem artificial, nevertheless we shall see that there are a number of interesting results. Some showing that there are a plethora of complexities of languages, and some point at limitations of approaches.

8.1.1 Oracles

In our analysis in classical computability we have used the heuristic principle that "everything relativizes". For example, the jump is defined as the relativization of the halting problem

$$X' = \{e : \Phi_e^X(e) \downarrow\}.$$

Then we see that for any X, $X <_T X'$. For the proof we simply say "use the same proof as we did for the proof of the undecidability of halting problem, but putting X's everywhere". (Chapter 5.)

© The Author(s), under exclusive license to Springer Nature Switzerland AG 2024   217
R. Downey, *Computability and Complexity*, Undergraduate Topics
in Computer Science, https://doi.org/10.1007/978-3-031-53744-8_8

It is not true that "everything relativizes" in classical computability theory, as shown by Shore [Sho79], but "everything normal" does. It is therefore reasonably surprising that facts about P vs NP *do not* relativize. We will soon prove the following result.

Techniques which relativize will not suffice to show $P = NP$, nor show $P \neq NP$.

This is the upshot of the next theorem. What techniques relativize? Any *normal* diagonalization relativizes, and any *normal* simulation relativizes. In the below, we use the relativization of P^X, interpreted to be the collection of languages $\{L : L \leqslant_T^P X\}$, that is, those for which there is a deterministic oracle Turing machine running in polynomial time and accepting L with oracle X. Similarly, we take NP^X to be the set of languages for which there is a nondeterministic oracle Turing machine M running in polynomial time and with oracle X, such that $x \in L$ iff some computation path of M^X accepts x.

Theorem 8.1.1 (Baker, Gill and Solovay [BGS75]). [1]

1. There is a computable oracle A such that $P^A \neq NP^A$.
2. There is a computable oracle B such that $P^B = NP^B = PSPACE^B$.

Proof. 1.
We construct a computable oracle A and a language $L \in NP^A$ to meet the requirements

$$R_{\langle e,k \rangle} : L \text{ is not accepted by } \Phi_e^A \text{ in } \leqslant n^k \text{ steps.}$$

We define $L = \{1^n : \exists x(|x| = n \wedge x \in A)\}$. Clearly $L \in NP^A$.

Construction The construction is essentially a finite extension argument. At stage s we will decide the fate of all elements x of length s. Here by fate we mean whether $x \in A$. The construction works by dealing with the requirements R_j in order of j. Suppose that we are currently dealing with $j = \langle e, k \rangle$. We will have inductively defined thresholds for R_q and $q < \langle e, k \rangle$. These represent lengths that $R_{\langle e,k \rangle}$ needs to respect for higher priority requirements. We will also have an absolute constant d which is more or less 2 allowing for simulations and decisions.

At stage s, if s has not reached the threshold of the $R_{\langle e',k \rangle}$ for $\langle e', k \rangle < j$, or $d \cdot s^k > 2^s$, put nothing of length s into A. That is, $A_{s+1} = A_s$. If s has

---

[1] It is worth mentioning that in [BGS75], the authors mention 2. was also discovered, independently, by Albert Meyer with Michael Fischer and by H. B. Hunt III, and 1. was obtained independently by Richard Ladner.

reached the threshold of the $R_{\langle e',k \rangle}$ for $\langle e',k \rangle < j$, and $2^s > d \cdot s^k$, compute $\Phi_e^{A_s^*}(1^s)$. Here A_s^* is the same as A_s and also has no elements of length s put into A_{s+1}. Note that we can do this action in more or less s^k many steps.

Case 1 $\Phi_e^{A_s^*}(1^s) = 1$, then we can diagonalize against Φ_e by

(i) Adding nothing to L by asking that nothing of length s enters $A \setminus A_s$.
(ii) Making sure that $A_s^* \prec A$ by making sure that the threshold of $R_{\langle e,k \rangle}$ exceeds the use of the computation $\Phi_e^{A_s^*}(1^s)$.

This makes

$$L(1^s) = 0 \neq 1 = \Phi_e^A(1^s).$$

This is a "win by luck", as we don't need to add things to A.

Case 2 $\Phi_e^{A_s^*}(1^s) = 0$. Now we are not winning by luck, and need to put 1^s into L. *This is the key point of the proof.* The computation of $\Phi_e^{A_s^*}(1^s) = 0$ runs for s^k many steps. Thus, *since $2^s > s^k$ and the computation $\Phi_e^{A_s^*}(1^s) = 0$ can only ask s^k many questions, there must be some string x with $|x| = s$* **not** *queried in the computation.* By the Use Principle (Proposition 5.5.1), we can put x into A_{s+1} (putting $1^s \in L$) but the computation $\Phi_e^{A_{s+1}}(1^s) = 0$ *remains identical to $\Phi_e^{A_s^*}(1^s) = 0$ as x is not addressed in the computation.* We will call this the *invisible point* and this idea is also used in Theorem 8.1.2 below. It follows that

$$L(1^s) = 1 \neq 0 = \Phi_e^{A_s^*}(1^s).$$

We set the next threshold to exceed the use of the computation, and move to $j+1$.

End of Construction

It is clear that A is computable and we diagonalize Φ_e and k.

Now we turn to 2.

This is easy. Let B be any PSPACE complete language. These are evidently closed under polynomial time Turing reductions. Thus $P^B = \text{PSPACE}^B$.   □

There have been many oracle separation and collapse results. It is argued that these results that techniques which relativize don't suffice to solve questions regarding the relationships between complexity classes. One notable one, also from the original paper [BGS75], is the following. Its significance is that in Proposition 5.1.1, we observed that Δ_1^0 is the same as computable. That is, being computable is the same as being c.e. and co-c.e.. The natural complexity analog is that "if language L is in NP $\cap$ co-NP then L $\in$ P". That is, co-NP $\cap$ NP = P. This analog *might* be true, but proving it will require a proof that does not relativize, as we see in the next result. We remark that most experts think that co-NP $\cap$ NP $\neq$ P, but this is yet another longstanding conjecture.

Theorem 8.1.2 (Ladner, Fischer and Meyer in [BGS75]). *There is a computable oracle B with*

$$P^B \neq NP^B \text{ and } NP^B = co\text{-}NP^B,$$

and hence

$$P^B \neq co\text{-}NP^B \cap NP^B.$$

Proof. This construction is significantly more complex than the proof of Theorem 8.1.1. Again, we will have $L \in NP^B$ to meet the requirements

$$R_{\langle e,k \rangle} : L \text{ is not accepted by } \Phi_e^B \text{ in } \leqslant n^k \text{ steps}.$$

Again, we define $L = \{1^n : \exists x(|x| = n \wedge x \in A)\}$ so $L \in NP^A$. In this construction, the diagonalization requirements are met only using *odd* length n's. That is, as with the proof of Theorem 8.1.1, we will diagonalize L by, if necessary, putting some 1^n into L by putting something into B which the computation $\Phi_e^B(1^n) = 0$ cannot see.

To make $NP^B \subseteq co\text{-}NP^B$, we will work with the following NP^B complete set.

$$K^B = \{\langle e, x, 1^n \rangle : \text{ some computation path of } \Phi_e^B \text{ of length } \leqslant n \text{ accepts } x\}.$$

The usual argument shows that K^B is NP^B-complete. The reader is invited to prove that this language is NP^B complete in Exercise 8.1.3. It is important that the reader notices that to decide if $z \in K^B$, we only look at elements of length smaller than $|z|$. We will build B so that

$$x \notin K^B \text{ iff } \exists y(|y| = |x| \wedge xy \in B)\}.$$

Notice that we are using elements of length $2 \cdot |x|$ to decide if $x \in K^B$.

By symmetry, this also makes $co\text{-}NP^B \subseteq NP^B$, and so $NP^B = co\text{-}NP^B$.

Even length elements are used for coding and odd length ones for diagonalization. The argument is again a priority argument and has the notion of elements being cancelled. B is built by finite extensions. At stage 0, $B_0 = \emptyset$.

Construction Stage $s + 1$

The induction hypothesis is that we have defined B_s for all elements of length $\leqslant s$, and whilst it might be defined on elements z with $|z| > s$, $B_s(z) = 0$ for such z.

Case 1. $s = 2m$. We work towards diagonalization. Let $x = 2^s$, so that $|x| = s + 1$ is odd. Look for the least uncancelled $\langle e, n \rangle < s$ such that

(i) B_s is not yet defined on any element of length $s + 1$.
(ii) $(s + 1)^k < 2^m$.

In this case we will cancel $\langle e, k \rangle$ diagonalize as in Theorem 8.1.1. That is we will be running Φ_e upon input x with possible incarnations of B_{s+1} as oracles.

As in Theorem 8.1.1, we will define B_{s+1} to either be the empty extension B_s^* of length $(s+1)^k$ (in the case that $\Phi_e^{B_s^*}(x) = 1$), or B_{s+1} includes an invisible element of length s, and is otherwise empty.

Case 2 $s = 2m + 1$. For each x of length $m + 1 = \frac{s+1}{2}$, we know already that

$$x \in K^B \text{ iff } x \in K^{B_s},$$

since B_s is already defined on all elements of length $\leqslant |x| - 1 \leqslant s$.

For each x of length $m+1$, we will need a $y = y_x$ of length $m+1$ such that B_s is as yet undefined on xy. Then we can define B_{s+1} to be the extension of B_s with

$$B_{s+1}(xy_x) = \begin{cases} 1 \text{ if } x \notin K^{B_s} \\ 0 \text{ otherwise.} \end{cases}$$

This gives $xy_x \in B$ iff $x \notin K^B$.
End of Construction

Verification The key is a small combinatorial argument that y exists for each x, in Case 2 of the construction. The only stages which could have involved elements of length $s + 1$ are those of the form $2n + 1$, with $n \leqslant m$. Each of these could restrain at most 2^n many elements from future entry into B to preserve some computation at the stage we dealt with them, and cancelled the relevant $\langle e, k \rangle$. There are 2^{m+1} strings of length $m + 1$. The number of y's restrained is at most

$$\sum_{n \leqslant m} 2^n < 2^{m+1}.$$

thus we can choose one y for each x where necessary. $\square$

We remark that it is known that if P = co-NP∩NP, then factoring is in P, and it would follow that cryptosystems like RSA would be insecure. So presumably bankers out there hope that P $\neq$ co-NP∩NP.

8.1.2 Exercises

Exercise 8.1.3 Show that $L = \{\langle e, x, 1^n \rangle : \text{ some computation path of } \Phi_e^B \text{ length } \leqslant n \text{ accepts } x\}$ is always NP^B-complete.

Exercise 8.1.4 Define $B = \{\langle e, x, 1^n \rangle : \text{ some computation path of } \Phi_e^B \text{ length } \leqslant n \text{ accepts } x\}$. Show that NP^B =P^B. You will need to show that B's definition is not circular.

Exercise 8.1.5 (Baker, Gill and Solovay [BGS75]) Show that there is a computable oracle A with

$$P^A = NP^A \cap \text{co-}NP^A \neq NP^A.$$

8.1.3 The Role of Oracles and Other Indications that We Don't Know How to Approach P vs NP

While the computability-theorists love relativization, it is fair to say that the complexity theory community has a strange relationship with oracles. In the 10 years after the Baker, Gill and Solovay [BGS75] paper there were many, many oracle results showing complexity class separation/collapse results could not be settled by argument that relativize in the way that they do in computability theory. The key question is:

> What do oracle results tell us about the questions we are actually interested in, the non-relativized questions?

One result which generated a lot of hope was by Bennett and Gill [BG81]. Bennet and Gill showed that for a *random oracle* X, $P^X \neq NP^X$, with probability 1. Bennett and Gill argued that

> "random oracles, by their very structurelessness, appear more benign and less likely to distort the relations among complexity classes than the other oracles used in complexity theory and recursive function theory, which are usually designed expressly to help or frustrate some class of computations."

However, this hope was yet another forlorn one. Within two years Stuart Kurtz [Kur83] had disproved this "random oracle hypothesis" (that is, the hypothesis that if something was true relative to a random oracle, then it was actually true unrelativized). Also see Chang et. al. [CCG+94] for another solution, but more below. Kurtz's counterexample was a bit artificial as it involved relativizing something which was already relativized. So perhaps something might be salvaged for "natural problems".

Anyway, the only real way we seemed to know how to separate classes was diagonalization, and the only way to achieve coincidence was to use simulation. What methods might be used if not these?

The next major advance came in Shamir [Sha92]. It built on earlier work of Lund, Fortnow, Karloff and Nisan [LFKN92]. Shamir's paper was concerned with a class called IP for *interactive protocol*. An interactive proof system (protocol) for a language L is a generalization of being in NP where we use probabilistic notions in an interactive game between a *prover* and a *verifier*. The prover is trying to convince the verifier whether they should believe $x \in L$. After a polynomial number of rounds the verifier will accept if it believes $x \in L$ and reject if it believes $x \notin L$. We need that the probability that the protocol accepts if $x \in L$ is at least $\frac{2}{3}$ (which can be amplified) and the probability that the verifier rejects is at least $\frac{2}{3}$ if $x \notin L$. The verifier has a supply of random coins for this protocol. Here is an example of this idea: GRAPH NON-ISOMORPHISM is in IP.

The protocol begins with two graphs G, H with n vertices. We want the verifier to accept if $G \not\cong H$ and reject otherwise. The following protocol is repeated a constant number of times.

(i) The verifier uses its random coin to compute graphs $G_1 \cong G$ and $H_1 \cong H$. (These are random permutations of the vertices.)
(ii) The verifier flips a coin and sends one of the pair (G, H_1) or (G, G_1) to the prover.
(iii) The prover must return to the verifier which of (G, H_1) or (G, G_1) was chosen by the verifier.

The only way that the prover can determine which of (G, H_1) or (G, G_1) was chosen by the verifier is if it can distinguish between G_1 and H_1. If $G \cong H$, then $H_1 \cong G$. Thus, the prover will always be able to determine which it was sent by simply testing if the second graph is isomorphic to G. Therefore, if the two graphs are not isomorphic and the prover and the verifier both follow the protocol, the prover can always determine the result of the coin flip. However, if $H \cong G$, then G_1 and H_1 are indistinguishable. Hence the prover cannot distinguish between (G, H_1) or (G, G_1) any better than random guessing.

What's the computational power of this strange model? It was great shock when Shamir proved IP = PSPACE. Introduced by Lund, Fortnow, Karloff and Nisan in [LFKN92], the method Shamir used was quite new. It is called *arithmetization*. What we do is take a QBF formula X and turn it into a number by replacing $\vee$ by $+$, $\wedge$ by $\cdot$, $\neg x$ by $(1 - x)$, $\exists x$ by $\sum_{x=0}^{1}$, and $\forall x$ by $\prod_{x=0}^{1}$. Then X is true iff the number evaluates to $n \neq 0$. So why not just compute this and solve X, showing that P = PSPACE? The problem is that the number is likely too big to be evaluated in polynomial time.

So, what Shamir did was to move the whole evaluation problem into arithmetic over prime field (i.e. replacing the sums and products by evaluations in the field), but the prime of polynomial size relative to the input. This alleviates the problems of numbers being too large since modular arithmetic can be done in polynomial time (relative to the prime). In the protocol, first we ask the oracle "What is the valuation for X is in the prime field?". Then the protocol, roughly stripped one quantifier at each step, treating the formula without the sum (product) as a sum (product) of polynomials and ask the oracle to supply the value of the evaluation. You can verify that the oracle does not lie by checking random points of the field to evaluate the "polynomial" and see if it evaluates correctly. For example if we had $f(z) = h(z) + q(z)$ as the claimed value modulo p, then this could be checked by substituting random $q \in GF(P)$, and see if evaluated correctly. The protocol uses the fact that modular arithmetic is fast. The point, roughly, is that the only way that we can be fooled is if the q chosen randomly happens to be the point of intersection of two curves over $GF(p)$, for a large prime p. This is highly unlikely. Moreover lies propagate through the rest of the proof; the first lie by the oracle is never forgiven, at least with high probability. An amusing presentation of this argument and an interesting discussion about the effects of worldwide communication and research can be found in Babai [Bab90].

Don't be too concerned if the above seems a bit vague. The point is that the proof used a new method from Lund. et. al. [LFKN92], *arithmetization*. What is especially interesting was that when Shamir gave his proof, we knew already that there was *an oracle separating the classes*; in fact *relative to a random oracle*, $\text{IP}^A \neq \text{PSPACE}^A$, so arithmetization *does not relativize*. Furthermore IP = PSPACE gave a natural counterexample to the random oracle hypothesis.

Arithmetization has had a number of successes, notably the famous PCP Theorem by Arora et. al. [ALM+98]. Sadly, in 2009, Scott Aaronson and Avi Wigderson [AW09] introduced yet another new method called "algebraization". Using this new method, Aaronson and Wigderson proved that arithmetization, at least as we have used it, cannot settle P vs NP.

One last approach which had seen some minor successes used *circuit complexity*. Here we imagine languages being accepted by boolean circuits. So we have a circuit C such that $x \in L$ iff the circuit accepts x. Maybe that we could show that there were languages in NP without small (polynomial sized) circuits. (In fact this would prove that NP $\not\subseteq$ P / Poly.) The intuitive argument was that that circuits did not have moving parts like Turing Machines. So maybe we might be able to analyse them better than, for example, Turing Machines with all those heads and tapes and the like. One early success was that we could show that certain problems could not be done by constant depth circuits (Hastad [Has87]), CLIQUE could not be solved by polynomial sized (in fact subexponential) *monotone* circuits (i.e. no inverters (negation gates)) by Razborov [Raz85]. Both of these celebrated papers used clever combinatorial arguments, and involved probabilistic arguments. Hastad's involved a certain gate "switching" lemma, and Razborov's involved randomly setting the values of certain gates. So maybe there was hope as we could use apparently more powerful methods if we tried to prove something stronger involving P / Poly, and nonuniformly separate classes.

However, the hope of separating things using circuits-at least as used so far-was also dashed by Razborov and Rudich [RR97], who analysed the method used in these kind of arguments, calling them "natural proofs" (which has a definition). They showed that no natural proof (in their sense) will suffice for settling P vs NP.

The upshot seems to be that *all known methods seem to face genuine barriers*, and so new ideas will be needed. One important question is "suppose that P $\neq$ NP, or even P = NP but not through a feasible algorithm". What should we do? In the next chapters we will address this question. As we also mention there, another question is: "Does NP $\neq$ P matter for a class of questions which are easily solved well in practice, in spite of the fact we know them to be NP-complete?" And another related open question in "Why is this, anyway; that is, what is it about certain problems which we know have hard instances, and yet seem easily solvable for instances occurring in natural problems?".

Remark 8.1.1. **Personal Musings**. I finish this section with some personal musings. We have to be very careful with what we mean by these limitation results. IP = PSPACE is proven by a simulation, only one that filters through a kind of reduction to arithmetic formulas. So what do we mean by "normal simulation?". Also proving lower bounds for monotone circuits certainly used diagonalization, only filtering through a randomization process, so what counts as "normal diagonalization?" Finally, part of the problem, the mixing of oracles and bounded reductions, can be indicated by classical computability. A truth table reduction $A \leqslant_{tt} B$ is a given by a computable function f, such that, on input x, $f(x)$ computes a boolean expression $\sigma(x)$, and $x \in A$ iff B models $\sigma(x)$ where $x_i \in \sigma(x)$ is interpreted as i. For example, $f(4) = \sigma(4)$ might say $((x_1 \vee \neg x_3) \wedge x_5) \vee x_{17}$, so that $4 \in A$ iff either $(5 \in B$ and at least one of $(1 \in B)$ or $(3 \notin B))$ or $(17 \in B)$ holds. It is known that there is a "minimal" c.e truth table degree: namely $\mathbf{a} > \mathbf{0}$, such that there is no $\mathbf{c}$ with $\mathbf{a} > \mathbf{c} > \mathbf{0}$. (Kobzev [Kob78]) On the other hand, Mohrherr [Moh84] showed that the truth table degrees above $\mathbf{0}'$ are dense, meaning, in particular, that if $\mathbf{a} > \mathbf{0}'$, there is always a $\mathbf{c}$ with $\mathbf{a} > \mathbf{c} > \mathbf{0}'$. Does this mean that the existence of minimal truth table degrees does not relativize? Well no, the point is that Mohrherr's Theorem concerns the truth table degrees above $\mathbf{0}'$, and in those degrees we have not relativized the $x \mapsto \sigma(x)$ function. The reason I mention Mohrherr's Theorem is that truth table reduction is much more like polynomial time Turing reduction in that giving access to an oracle does not change the nature of the access mechanism, it is still a computable computation of $\sigma(x)$, and similarly for $A \leqslant_T^P B$, we still only have polynomial access to the oracle. They are only *partial relativizations*. In classical relativization in computability theory, everything is relativized including the access mechanism, and it is not altogether clear what that would mean for a polynomial. In the case of $\leqslant_{tt}$ then we would need to relativize this to $\emptyset'$, and then we do find that there is a degree c.e. relative to $\emptyset'$ which a minimal $\leqslant_{tt}^{\emptyset'}$ "truth table" degree, so Kobzev's theorem is resurrected.

Certainly there have been some attempts to re-define the concept of an oracle in computational complexity theory, but the situation remains murky at best.

8.2 Polynomial Time Degrees and Ladner's Theorem

The use of $\leqslant_T$ and $\leqslant_m$ are ubiquitous in classical computability theory; particularly in undecidability proofs. The analogous fact is true in computational complexity theory where we replace the reductions by $\leqslant_T^P$ and $\leqslant_m^P$. In classical computability theory, we studied the Turing degrees and solved Post's Problem for the c.e. degrees. In our narrative about the resource bounded computations, so far we have only used these reductions in hardness and completeness proofs, and in the last section proved oracle results.

In this short section we will look briefly at the structural behaviour of the induced degree structures. Here, if A is a language, we will let $[A]_R = \{B : B \equiv_R A\}$ where R is one of $\leqslant_m^P$ or $\leqslant_T^P$, the polynomial time m-degree and polynomial time T-degree of A respectively. As usual the degrees of either class is an upper-semilattice with join induced by

$$A \oplus B = \{0\sigma, 1\tau : \sigma \in A, \tau \in B\}.$$

The following is one of the classical results from the early days of computational complexity theory.

Theorem 8.2.1 (Ladner [Lad75]).
The polynomial time Turing degrees of computable languages form a dense partial ordering. That is, if $A <_T^P B$ with A, B computable, then there exists computable X with $A <_T^P X <_T^P B$.

Proof. Whilst this is not the original Ladner proof, it is the one most used. We will be using a diagonalization argument which works roughly by "blowing holes in B". We will construct $X = A \oplus C$, and hence will be concerned with C. We will build a linear time computable relation R with R defined on strings of the form 1^s. We will define

$$x \in C \text{ iff } |x| = s \wedge R(1^s) = 1 \wedge x \in B.$$

Clearly $C \leqslant_m^P B$. We need to meet the requirements

$$R_{2\langle e,k \rangle} : \Phi_e^X \neq B \text{ in time } |x|^k.$$

$$R_{2\langle e,k \rangle+1} : \Phi_e^A \neq X \text{ in time } |x|^k.$$

We meet the requirements in the order $R_0, R_1, \ldots$.

Certification As A and B are computable we know that their characteristic functions are computable. We can slow down the computations running the computable functions computing χ_A and χ_B one step at a time, so that at stage s, we would compute s steps in the computation of $\{\chi_A(n), \chi_B(n) : |n| \leqslant s\}$, not letting strings of length bigger than n halt before they do for strings of length n. Notice that A and B might be very complicated computable sets and take a very long time to compute. So, under this process at stage s we might only have determined membership of A or B for strings of length $n << s$. Once we have reached a stage s where we know A (or B) for strings of length n we will say that $A \restriction n$ (or $B \restriction n$)) is *certified* at stage t for $t \geqslant s$.

Construction The construction works by letting some R_i assert control of the construction from some stage $s = s_i$ until it is met at some stage t_i. At that point, we would move on to R_{i+1}.

At stage s, suppose that we are working on requirement $R_{2\langle e,k\rangle}$. We will have set $R(1^u) = 0$ for all of the stages $u \leqslant s$ where we have been working towards satisfying this requirement. This forces $C(y) = 0$ for all y with $|y| = u$. The idea is the following. Suppose that $R_{2\langle e,k\rangle}$ held control for all stages. Then X would be of the form $A \oplus C$ where there is some n such that for all strings x of length $> n$, $C(x) = 0$. This follows since the only things we put into C would have been the finitely many things entering C before $R_{2\langle e,k\rangle}$ asserted control. Hence $X \equiv_m^P A$, as X is A plus a finite piece. But then

$$\Phi_e^X = B \text{ in time } |x|^k \text{ implies } B \leqslant_T^P A,$$

a contradiction. So we *know* that if we make $C = \emptyset$ long enough, we must have created a disagreement somewhere.

The idea is to *look back* and see where we did this. That is, at stage $s+1$, whilst $R_{2\langle e,k\rangle}$ is asserting control, we examine initial segments of A and B for s many steps. We do s steps of the following procedure:

Look at strings x of length $\leqslant s$, and A and B on strings of length $\leqslant s$. See if there is a computation of $\Phi_e^{X_s}(x) \neq B_s(x)$ via computations that are certified at stage s (and in particular, X_s is certified on the use of the computation).

Now, notice that we might have made a decision long ago at stage t which would have caused a disagreement $\Phi_e^X(x) \neq B(x)$, but it might take us a very long time to actually find that disagreement at some stage s with $t << s$. But eventually we will observe this previously created disagreement.

In the construction, if we have not observed a disagreement, then keep $R_{2\langle e,k\rangle}$ as the requirement asserting control and keep $R(1^{s+1}) = 0$.

If we have observed a disagreement looking back, then we pass control to $R_{2\langle e,k\rangle+1}$ setting $R(1^{s+1}) = 1$.

The case that $R_{2\langle e,k\rangle+1}$ has control at stage s is similar. We will have $R(1^s) = 1$ and hence C is copying B. If C successfully copied B and we failed to meet $R_{2\langle e,k\rangle+1}$, then $\Phi_e^A = X$, and C almost equals B. Thus $B \leqslant_T^P A$, a contradiction. So again we will keep looking back till we find a disagreement and then release control, switch to $R(1^{s+1}) = 0$ and let $R_{2\langle e',k'\rangle}$ assert control, where $\langle e',k'\rangle = \langle e,k\rangle + 1$. $\square$

The method of proof above is called *delayed diagonalization*. The reader should note that $C \leqslant_m^P B$ in the construction above. This observation gives three nice corollaries.

Corollary 8.2.1 (Ladner [Lad75]). *The polynomial time m-degrees of computable sets is a dense partial ordering.*

Proof. Run the proof above with the assumption $A <_m^P B$. $\square$

Corollary 8.2.2 (First observed by Shinoda and Slaman-unpublished). *The polynomial time Turing degrees of all languages are dense.*

Proof. Now we run the same proof using an oracle for B. In linear time relative to an oracle for B we can compute a slow enumeration for B, and moreover as $A \leqslant_T^P B$, we have a polynomial time procedure $\Gamma^B = A$ in time $|x|^d$ for some d on inputs of length x. Thus we can run the whole construction using B as an oracle. $\square$

Corollary 8.2.3 (Ladner [Lad75]).

If $P \neq NP$, then there are languages in $NP \setminus P$ which are not NP-complete.

Proof. By Exercise 7.2.3, if $C \leqslant_m^P B$ and $B \in NP$, then $C \in NP$. We can apply Ladner's Theorem with $A = \emptyset$ and B coding SAT, giving an NP language X with X not NP-complete. $\square$

Corollary 8.2.3 yields the following question:

Question 8.2.1. Assuming that $P \neq NP$, are there any *natural* languages $L \in$ NP which are not NP-complete?

We have no idea. We have some possible candidates, two noteworthy ones are below.

GRAPH ISOMORPHISM
 Input : Two graphs G_1, G_2.
Question : Is $G_1 \cong G_2$?
The next one is not a decision problem but an NP-search problem.

FACTORIZATION
 Input : An integer n.
Question : Find the prime factors of n.

FACTORIZATION is a very important problem, even with the version where we promise that n has two prime factors. The assumed hardness of this problem is the basis of the RSA cryptosystem. The *only* reason we think it is hard is that we don't know how to do it.

GRAPH ISOMORPHISM is a very interesting case. Many workers believe that it is in P. In 2016, Lázló Babai [Bab16] announced that GRAPH ISOMORPHISM is in DTIME($n^{\log n}$), quasi-polynomial time. There was an error in the proof, but this seems to have been patched according to a 109 page online version (in 2018) on the University of Chicago website. All of the evidence seems to suggest that GRAPH ISOMORPHISM is easy. It is known to be in polynomial time for wide classes of graphs such as bounded degree (Luks [Luk82]). Assuming Babai's claim, which is currently unpublished in a peer reviewed journal, GRAPH ISOMORPHISM being NP-complete would imply that all NP-complete problems, such as SAT, would lie in the class DTIME($n^{\log n}$), which is thought to be highly unlikely. From the published literature, we know that if GRAPH ISOMORPHISM is NP-complete then the polynomial time hierarchy collapses to two or fewer levels. The proof of the result below is a wee bit too complex to include in this course.

Theorem 8.2.2 (Schöning [Sch88]). *If* GRAPH ISOMORPHISM *is NP-complete, then* $\Sigma_2^P = \Sigma_k^P$ *for all* $k \geqslant 2$.

For more fine-grained results we refer the reader to the book [KST93].

8.2.1 Exercises

Exercise 8.2.3 Suppose that A and B are computable languages. We say that A and B form a *minimal pair*, if $A \not\leqslant_T^P B$ and $B \not\leqslant_T^P A$, and for all X, if $X \leqslant_T^P A, B$, then $X \in$ P. Use delayed diagonalization to construct a minimal pair. (Hint: Making $\Phi_e^B \neq A$ in time $|x|^k$ and similarly for $\Phi_e^A \neq A$ is relatively easy since you can make one side empty for a long time and the other can be diagonalized using a witness. To make the minimal pair happen meet requirements

$$R_{e,i,k,d} : \Phi_e^A = \Phi_i^B \text{ running in times } |x|^k \text{ and } |x|^d, \text{ respectively} \rightarrow \Phi_e^A \in \text{P}.$$

The easiest way to do this is to make one side or the other empty for a long time making the computation in P.)

Exercise 8.2.4 The original proof of Ladner used stretching. To the language C we need for $X = A \oplus C$ is constructed by building a function f. Now we will put $z = x01^{f(|x|)} \in C$ iff $x \in B$. That is z is a "stretched form" of x. The idea is that in a computation $\Phi_e^{A \oplus C}(x)$ running in time $|x|^k$, we can only ask questions of length $\leqslant |x|^k$. In the construction, we would be building f and make sure that f is built in polynomial time. When we compute $\Phi_e^{A \oplus C}(x)$, once $f(x)$ is bigger than $|x|^k$, the value of B on x will not effect the value of $\Phi_e^{A \oplus C}(x)$. Once $R_{\langle e,k \rangle}$ asserts control, we will define f accordingly to have $f(|x|) > |x|^k$ for $|x| > s$. Now, if $\Phi_e^{A \oplus C} = B$, then you can argue that the very computation $\Phi_e^{A \oplus C}(x)$ allows us to compute $B(x)$ which then allows us to compute C on longer z which then allows us to compute B on longer x's.

Using these ideas give a proof of the density of the polynomial time Turing degrees of computable sets using stretching.

Chapter 9
Parameterized Complexity

Abstract We look at the basics of parameterized complexity. This is a
method which seeks to find tractability by limiting some parameter in the
input. We analyse methods for proving parameterized tractability and also
give some basic results the completeness and hardness theory. We also look at
limitations of the methods and XP-optimality. The latter gives methods for
proving various algorithms are more or less optimal, subject to complexity
considerations.

Key words: Parameterized complexity, kernelization, bounded search trees,
W-hierarchy, XP, fixed-parameter tractability, iterative compression

9.1 Introduction

In this section we will look at a coping strategy which seems to have turned
out to be important for a wide class of practical problems. It is called *parameterized complexity*, or more recently *multivariant complexity* or *fine-grained
complexity*. The idea is to look more deeply into the *way* that a problem is
intractable; what is *causing* the intractability. The area was introduced by
Downey and Fellows [DF92, DF93, DF95a, DF95b].

In this book we have seen the story of classical complexity. It begins with
some problem we wish to find an efficient algorithm for. Now, what do we
mean by efficient? We have idealized the notion of being efficient by being in
polynomial time. Having done this, we discover that the only algorithm we
have for the given problem is to try all possibilities and this takes $\Omega(2^n)$ for
instances of size n. What we would like is to *prove* that there is no algorithm
running in feasible time. Using our idealization that feasible is the same as
polynomial, this equates to showing that there is no algorithm running in
polynomial time.

© The Author(s), under exclusive license to Springer Nature Switzerland AG 2024   231
R. Downey, *Computability and Complexity*, Undergraduate Topics
in Computer Science, https://doi.org/10.1007/978-3-031-53744-8_9

Suppose that we succeed in showing that there is no polynomial time algorithm. This would mean to us is that we would (i) need to try some other method to solve the problem such as some kind of approximate solution because (ii) we could give up on showing that there was a polynomial time algorithm.

In this course, we have seen the story continues with the following rhetoric. In spite of the efforts of a number of researchers, for many problems whose best solution known is brute force complete search. However, there is no *proof* that the problem is not in polynomial time. These problems are are NP-complete or worse. This means that we have a *practical* "proof" of hardness in that if any of the problems were in polynomial time, all would be; and secondly showing them to be in polynomial time would show that acceptance for a polynomial time nondeterministic Turing machine would be also. The philosophical argument is that a nondeterministic Turing machine is such an opaque object, without any obvious algebraic structure, that it seems impossible to see if it has an accepting path without trying all of them.

The methodology above seems fine as a *first foray* into *feasible* computation. However, for practical computation, it seems that we ought to refine the analysis to make it more *fine-grained*. Firstly, when we show that something is NP-complete or worse, what we are focusing on is the worst case behaviour. Second, the analysis takes the input as being measured by its *size* alone. You can ask yourself the question:

When in real life do we know nothing else about a problem than its *size*?

The answer is *never*; or at least only in cryptography, where this is done by design.

For instance, the problem is planar, tree-like, has many parameters bounded, etc. The idea behind parameterized complexity is to try to exploit the *structure* of the input to get some practical tractability. That is, we try to understand what aspect of the problem is to blame for the combinatorial explosion which occurs. If this parameter can be controlled then we would have achieved practical tractability.

Anybody working in software engineering will know that it is important to design tools specific to the type of problem at hand. Suppose that you are concerned with relational databases. Typically the database is huge, and the queries are relatively small. Moreover, "real life" queries are queries *people* actually ask. Hence, such queries tend to be also of low logical complexity. Furthermore, in areas like computational biology, the number 4 is typical, and structure of things like DNA is *far* from random. The *main idea* of parameterized complexity is to design a paradigm that will address complexity issues in the situation where we know in advance that certain parameters will be likely bounded and this might significantly affect the complexity. Thus in the database example, an algorithm that works very efficiently for small formulas with low logical depth might well be perfectly acceptable in practice.

Thus, parameterized complexity is a refined complexity analysis, driven by the idea that in real life data is often given to us naturally with an underlying structure which we might profitably exploit. The idea is not to replace polynomial time as the *underlying* paradigm of feasibility, but to provide a set of tools that refine this concept, allowing some exponential aspect in the running times by allowing us either to use the given structure of the input to arrive at feasibility, or develop some relevant hardness theory to show that the kind of structure is not useful for this approach.

This simple idea is pretty obvious once you think about it. For example, in Chapter 2.2, we showed regular language acceptance is in linear time. But this is really not quite true: it is only true *if* the language is presented to us as, say, a regular expression, whereas it could be a language presented as the output of a Turing machine, then by Rice's Theorem, regular language acceptance case acceptance is *undecidable*. The *point* is that we only really care about regular languages when they are given to us in a structured way, namely via regular expressions.

By way of motivation, we recall three basic combinatorial decision problems VERTEX COVER, DOMINATING SET, INDEPENDENT SET.

The reader should recall that in a graph G a *vertex cover* is where vertices cover edges: that is $C = \{v_1, \ldots, v_k\}$ is a vertex cover iff for each $e \in E(G)$, there is a $v_i \in C$ such that $v_i \in e$. The reader should recall that a *dominating set* is the situation where vertices cover vertices: $D = \{v_1, \ldots, v_k\}$ is a dominating set iff for all $v \in V(G)$, either $v \in D$ or there is an $e \in E(G)$ such that $e = \langle v_i, v \rangle$ for some $v_i \in D$. Finally an *independent set* is a collection of vertices no pair of which are connected. In Chapter 7, we proved all of the related decision problems, VERTEX COVER, INDEPENDENT SET, DOMINATING SET are NP-complete.

These kind of problems can occur as mathematical models of data occurring in computational biology. Suppose we had a conflict graph of some data from this area. Because of the nature of the data we know that it is likely the conflicts are at most about 50 or so, but the data set is large, maybe 10^{12} points. We wish to eliminate the conflicts, by identifying those 50 or fewer points. Let's examine the problem depending on whether the identification turns out to be a dominating set problem or a vertex cover problem.

- DOMINATING SET Essentially the only known algorithm for this problem is to *try all possibilities*. (Does this sound familiar?) Since we are looking at subsets of size 50 or less then we will need to examine all $(10^{12})^{50}$ many possibilities. Of course, this is completely impossible.

- VERTEX COVER There is now an algorithm running in time $O(1.2738^k + kn)$ (Chen et. al. [CKX10]) for determining if an G has a vertex cover of size k. This and and structurally similar algorithms has been implemented and is practical for n of unlimited practical size and k large. The relevant k has been increasing all the time, evolving from about 400. Such algorithms have been have been successfully implemented (e.g. Langston ey. al. [LPS$^+$08]) on instances on graphs with millions of nodes and vertex covers the thousands. Moreover, this last work is on actual biological data.

The following table from Downey-Fellows [DF98] exhibits the difference between the parameter k being part of the exponent like DOMINATING SET or as part of the constant like VERTEX COVER. This table compares of a run time of $\Omega(n^k)$ vs $2^k n$.

	$n = 50$	$n = 100$	$n = 150$
$k = 2$	625	2,500	5,625
$k = 3$	15,625	125,000	421,875
$k = 5$	390,625	6,250,000	31,640,625
$k = 10$	1.9×10^{12}	9.8×10^{14}	3.7×10^{16}
$k = 20$	1.8×10^{26}	9.5×10^{31}	2.1×10^{35}

Table 9.1: The Ratio $\frac{n^{k+1}}{2^k n}$ for Various Values of n and k.

In classical complexity a decision problem is specified by two items of information:
(1) The input to the problem.
(2) The question to be answered.

In parameterized complexity there are three parts of a problem specification:
(1) The input to the problem.
(2) The aspects of the input that constitute the parameter.
(3) The question to be answered.

Thus *one* parameterized version of VERTEX COVER is the following:

VERTEX COVER
Instance: A graph $G = (V, E)$.
Parameter: A positive integer k.
Question: Does G have a vertex cover of size $\leqslant k$?

We could, for instance, parameterize the problem in other ways. For example, we could parameterize by some width metric, some other shape of the graph, planarity etc. Any of these would enable us to seek hidden tractability in the problem at hand.

9.2 Formal Definitions

There are several varieties of parametric tractability, but for this course, I will stick to the basic "strongly uniform" variety most commonly used in practice.

Definition 9.2.1 (Downey and Fellows [DF92]). A *parameterized language* is $L \subseteq \Sigma^* \times \Sigma^*$ where we refer to the second coordinate as the *parameter*.

It does no harm to think of $L \subseteq \Sigma^* \times \mathbb{N}^1$.

Here is the main definition of what computational tractability is:

Definition 9.2.2 (Downey and Fellows [DF92]). A parameterized language L is *(strongly) fixed parameter tractable* (FPT), iff there is a computable function f, a constant c, and a (deterministic) algorithm M such that for all x, k,

$$\langle x, k \rangle \in L \text{ iff } M(x, k) \text{ accepts,}$$

and the running time of $M(x, k)$ is $\leqslant f(k) \cdot |x|^c$.

It is not difficult to show that the multiplicative constant in the definition can be replaced by an additive one, so that $L \in FPT$ iff L can be accepted by a machine in time $O(|x|^c) + f(k)$ for some computable f (Exercise 9.3.8.). In the case of VERTEX COVER we have $f(k) = 1.2738^k$, and the O is 2. One nice notation useful here is the O^* notation which ignores the polynomial part be it additive or multiplicative and is only concerned with the exponential part. Using this notation, the algorithm for (parameterized) VERTEX COVER is said to be $O^*(2^k)$. The table on the web site

http://fpt.wikidot.com/fpt-races

lists 35 (at the time of writing) basic problems which are fixed parameter tractable with (mostly) practical algorithms, and for which there are current "races" for algorithms with the best run times.

---

[1] Flum and Grohe [FG06] have an alternative formulation where the second coordinate is a function $\kappa : \Sigma^* \to \Sigma^*$, but I prefer to keep the second parameter as a string or number.

9.2.1 Discussion

Now you might (in some cases validly) complain about the presence of an
arbitrarily bad computable function f. Could this not be like, for example
Ackermann's function? This is a true enough complaint, *but* the argument
also applies to *polynomial time*. Could not polynomial time allow for running
times like $n^{30,000,000}$? As noted by Edmonds [Edm65], the practical algorithm
builder's answer tends to be that "in real life situations, polynomial time
algorithms tend to have small exponents and small constants." That certainly
was true in 1965, but as we will see in §9.3.4, this is no longer true. The same
heuristic applies here. By and large, for most practical problems, at least until
recently, the $f(k)$'s tended to be manageable and the exponents reasonable.

In fact, an important offshoot of parameterized complexity theory is that
it does (sometimes) provide tools to show that *bad constants* or *bad exponents* for problems with algorithms *running in polynomial time cannot* be
eliminated, modulo some reasonable complexity assumption. More on this in
§9.3, especially in §9.3.4.

One of the key features of the theory is a wide variety of associated techniques for proving parametric tractability. We will discuss them in §9.4, but
before we do so, let's examine the associated hardness theory.

9.3 Parametric Intractability

Since we are woefully bad at *proving* problems to not be in polynomial time,
we replace provable intractability, and instead resort to a completeness theory.
That is, we have a hardness theory, NP-completeness, based on the assumption that certain canonical problems like SAT are not in polynomial time.
The two key ingredients of a hardness theory are

  (i) a notion of hardness and
  (ii) a notion of "problem A could be solved efficiently if we could solve
       problem B"; that is a notion of reducibility.

In Chapter 7 we have seen the classic theory of NP completeness (i) is
achieved by the following:

NONDETERMINISTIC TURING MACHINE ACCEPTANCE
Input: A nondeterministic Turing Machine M and a number e.
Question: Does M have an accepting computation in $\leqslant |M|^e$ steps?

The Cook-Levin argument is that a Turing machine is such an opaque
object that it seems that there would be no way to decide if M accepts,
without essentially trying the paths. If we accept this thesis, then we probably
should accept that the following problem is not $O(|M|^c)$ for any fixed c and

is probably $\Omega(|M|^k)$ since again our intuition would be that all paths would need to be tried:

SHORT NONDETERMINISTIC TURING MACHINE ACCEPTANCE
Input: A nondeterministic Turing Machine M.
Parameter: A number k.
Question: Does M have an accepting computation in $\leqslant k$ steps?

So here is a notion of hardness. I would find it difficult to believe that NP $\neq$ P, but that SHORT NONDETERMINISTIC TURING MACHINE ACCEPTANCE could be in FPT, for example, solved in $O(|M|^3)$ for any path length k. In fact, as we will soon see, SHORT NONDETERMINISTIC TURING MACHINE ACCEPTANCE not in FPT is closely related to the statement n-variable 3SAT not being solvable in subexponential time, DTIME($2^{o(n)}$).

Thus to show DOMINATING SET is likely not FPT, it would be enough *if we could solve* DOMINATING SET *in time* $O(n^c)$ *by for each fixed* k, *then we could have an* $O(n^c)$ *algorithm for* SHORT NONDETERMINISTIC TURING MACHINE ACCEPTANCE. To formalize this programme, we will need a notion of a reduction. Our principal working definition for parameterized reductions is the following which is the m-reduction form.

Definition 9.3.1 (Downey and Fellows [DF92]). Let L, L' be two parameterized languages. We say that $L \leqslant_{fpt} L'$ iff there is an algorithm M, are computable functions g and f and a constant c, such that

$$M : \langle G, k \rangle \mapsto \langle G', k' \rangle,$$

so that

(i) $M(\langle G, k \rangle)$ runs in time $\leqslant g(k)|G|^c$.
(ii) $k' \leqslant f(k)$.
(iii) $\langle G, k \rangle \in L$ iff $\langle G', k' \rangle \in L'$.

Note that by taking the maximum, we could have assumed that $f = g$. However, in many practical reductions $f(k)$ will be much smaller that $g(k)$.

Example 9.3.1. A simple example of a parametric reduction is from k-CLIQUE to k-INDEPENDENT SET, where the standard reduction is parametric. The reader should recall that this takes an instance (G, k) of k-CLIQUE and transforms it into (H, k) where $V(G) = V(H)$ and $xy \in E(G)$ iff $xy \notin E(H)$. We remark in passing that it is somewhat unusual that classical NP-completeness reductions turn out to be parameterized reductions, as we discuss below.

The following is a consequence of Cai, Chen, Downey and Fellows [CCDF97], and Downey and Fellows [DF95b]. We will soon work up to the proofs.

Theorem 9.3.1. *The following are hard for* SHORT NONDETERMINISTIC TURING MACHINE ACCEPTANCE: INDEPENDENT SET, DOMINATING SET.

Following Karp [Kar73], and then four decades of work, we know that thousands of problems are all NP-complete. They are all reducible to one another and hence seem to have the same classical complexity. On the other hand, with parameterized complexity, as we now see, we have theory which separates VERTEX COVER from DOMINATING SET and INDEPENDENT SET. With such refined reducibilties, it seems highly unlikely that the hardness classes would coalesce into a single class like NP-complete. And indeed we think that there are *many* hardness class. We have seen in the theorem above that SHORT NONDETERMINISTIC TURING MACHINE ACCEPTANCE $\equiv_{fpt}$ INDEPENDENT SET. However, we *do not* think that DOMINATING SET $\leqslant_{fpt}$ INDEPENDENT SET.

9.3.1 A basic hardness class: $W[1]$

A standard parameterized version of the satisfiability problem of Cook-Levin is the following.

WEIGHTED (CNF) SAT
Input: A CNF formula X.
Parameter: A number k.
Question: Does X have a true assignment of (Hamming) weight k (here the weight is the number of variables set to true)?

Similarly, we can define WEIGHTED s-CNF SAT where the clauses have only s many literals. Classically, using a padding argument, we know that CNF SAT $\equiv_m^P$ 3-SAT, as we saw in Corollary 7.2.1. Recall that to do this for a clause of the form $\{q_1, \ldots, q_k\}$ we add extra variables z_j and turn the clause into several as per: $\{q_1, q_2, z_1\}$, $\{\overline{z_1}, q_3, z_2\}$, etc. Now this is definitely *not* a parametric reduction from WEIGHTED CNF SAT to WEIGHTED 3 CNF SAT because a weight k assignment could go to any other weight assignment for the corresponding instance of 3-CNF SAT.

We define infinitely many parameterized complexity classes as

$$W[1, s] = \{L \subseteq \Sigma^* \times \Sigma^* : L \leqslant_{fpt} s\text{-CNF SAT}\}.$$

The letter "W" stands for *weft*[2] which comes from the original papers. In [DF92, DF93, DF95a, DF95b], the classes were introduced as *circuits* with gates representing $\wedge$, $\vee$ and inverters representing $\neg$. They had n many inputs for boolean variables and a single output at the bottom. These circuits were

---

[2] A term from weaving.

classified via depth and weft, where depth is the the length of the longest path to the output and weft is the depth when only large gates are considered. Then, for example, a CNF formula would have a single large $\wedge$ gate at the bottom with a layer of small or large $\vee$ gates (representing clauses) feeding into it, perhaps with inverters between the inputs and the $\vee$ gates. In the case of CNF the gates would be large as there is no bound beyond n on their size. In the case of s-CNF, they are small. The difference between s-CNF and s'-CNF is that we bound the size of the small gates by s and s' respectively.

In the original model, the circuits did not have to be be boolean circuits, only have a single output and be of polynomial depth. For if the circuits had reasonably high depth, then something at level k could have an output into something of level $k' << k$. In boolean circuits this only happens with $k' = k - 1$ (modulo inverters). In Figure 9.1, we have an example of a weft 2 depth 5 decision circuit.

However, in [DF92, DF95a, DF95b], a normalization theorem is proven to say that we can regard these circuits as being boolean, and representing formulae. This proof is a combinatorial argument and we have decided to omit it, and take as a fact that we can *define* the complexity classes using boolean circuits (formulae).

Convention 9.3.2 For this Chapter we will also use the notation $+$ for $\vee$ and $\cdot$ for $\wedge$, as this makes the material more consistent with usage in the area. So a CNF formula is a product of sums.

This then allows us to reduce circuits of polynomial depth and weft 1 back to one of the classes $W[1, s]$ We are about to prove a basic result in parameterized complexity. The result says that weighted (antimonotone) 2-CNF SAT is complete for $W[1]$. It is important that the reader see the difference between the combinatorics needed for a NP-completeness proof and one for parameterized complexity. Do not be daunted by this proof, for whilst it is reasonably long, it is combinatorial, and does not need high powered mathematics.

We are defining the basic class as

$$W[1] = \bigcup_{s=1}^{\infty} W[1, s].$$

Theorem 9.3.3 (Downey and Fellows [DF95b]).

$$W[1, s] = W[1, 2] \text{ for all } s \geqslant 2,$$

so that $W[1] = W[1, 2]$.

Recall that a circuit (boolean formula) C is termed *monotone* if it does not have any inverters. Equivalently, C corresponds to boolean expressions having only positive literals. If we restrict the definition of $W[1, s]$ to fam-

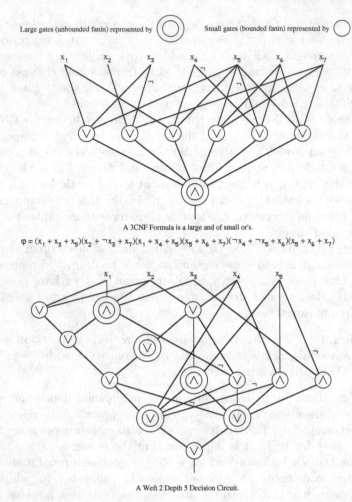

Fig. 9.1: Examples of Circuits

ilies of monotone circuits, we obtain the family of parameterized problems
MONOTONE $W[1, s]$.

Similarly, we define a formula to be ANTIMONOTONE if all the literals
in the clauses are negated variables. Thus, in an antimonotone circuit, each
fanout line from an input node goes to a *not* gate (and in the remainder of the
circuit, there are *no* other *not* gates). The restriction to families of antimonotone circuits yields the classes of parameterized problems ANTIMONOTONE
ANTIMONOTONE $W[1, s]$.

We begin the proof of Theorem 9.3.3 with the the following preliminary
result. This has an interesting parametric reduction.

Theorem 9.3.4 (Downey and Fellows [DF95b]).
  $W[1,s] = $ ANTIMONOTONE $W[1,s]$ *for all* $s \geqslant 2$.

Proof. [3] The plan of our argument is to identify a problem (s-Red/BLUE
NONBLOCKER) that belongs to antimonotone $W[1,s]$, and then show that
the problem is hard for $W[1,s]$. RED/BLUE NONBLOCKER is the following
parameterized problem.

RED/BLUE NONBLOCKER
Input: A graph $G = (V, E)$ where V is partitioned into two colour classes
$V = V_{\text{red}} \cup V_{\text{blue}}$.
Parameter: A positive integer k.
Question: Is there is a set of red vertices $V' \subseteq V_{\text{red}}$ of cardinality k such
that every blue vertex has at least one neighbor that does not belong to V'.

The *closed neighborhood* of a vertex $u \in V$ is the set of vertices $N[u] = \{x : x \in V \text{ and } x = u \text{ or } xu \in E\}$.

It is easy to see that the restriction of RED/BLUE NONBLOCKER to graphs
G with blue vertices of maximum degree s belongs to antimonotone $W[1,s]$
since the product-of-sums (Remember: and of ors) boolean expression

$$\prod_{u \in V_{\text{blue}}} \sum_{x_i \in N[u] \cap V_{\text{red}}} \neg x_i$$

has a weight k truth assignment if and only if G has size k nonblocking set.

Such an expression corresponds directly to a formula meeting the defining conditions for antimonotone $W[1,s]$. We will refer to the restriction of
RED/BLUE NONBLOCKER to graphs with blue vertices of maximum degree bounded by s as s-RED/BLUE NONBLOCKER. We next argue that s-
Red/BLUE NONBLOCKER is complete for $W[1,s]$.

By our definition of $W[1,s]$, we can assume that we are given a normalized
s-CNF boolean expression. Thus, let X be a boolean expression in conjunctive
normal form with clauses of size bounded by s. Suppose X consists of m
clauses $C_1, \ldots, C_m$ over the set of n variables $x_0, \ldots, x_{n-1}$. We show how to
produce in polynomial time by local replacement, a graph $G = (V_{\text{red}}, V_{\text{blue}}, E)$
that has a nonblocking set of size $2k$ if and only if X is satisfied by a truth
assignment of weight k.

Before we give any details, we give a brief overview of the construction,
whose component design is outlined in Figure 9.2. There are $2k$ "red" components arranged in a line. These are alternatively grouped as blocks from
$V_{\text{red}} = V_1 \cup V_2$ ($V_1 \cap V_2 = \emptyset$), with a block of vertices from V_1 and then V_2
to be precisely described below. The idea is that V_1 blocks should represent
a positive choice (corresponding to a literal being true) and the V_2 blocks
corresponding to the "gap" until the next positive choice. We think of the

---

[3] The reduction for RED/BLUE NONBLOCKER from Downey-Fellows [DF95b] contained a
flaw that was spotted by Alexander Vardy. The reduction used here is from [DF13].

V_1 blocks as $A(0), \ldots, A(k-1)$. [With a V_2 block between successive $A(i)$, $A(i+1)$ blocks, the last group following $A(k)$.] We will ensure that for each pair in a block, there will be a blue vertex connected to the pair and nowhere else (these are the sets V_3 and V_5). This device ensures that at most one red vertex from each block can be chosen, and since we must choose $2k$, this ensures that we choose *exactly* one red vertex from each block. The reader should think of the V_2 blocks as arranged in k columns. Now, if i is chosen from a V_1 block, we will ensure that column i gets to select the next gap. To ensure this we connect a blue degree 2 vertex to i and each vertex not in the i-th column of the next V_2 block. Of course, this means that if i is chosen, since these blue vertices must have an unchosen red neighbor, we must choose from the i-th column. The final part of the component design is to enforce consistency in the next V_1 block; that is, if we choose i and have a gap choice in the next V_2 block, column i, of j, then the next chosen variable should be $i+j+1$. Again, we can enforce this by using many degree 2 blue vertices to block any other choice. (These are the V_6 vertices.) Also, we keep the k choices for V_1 blocks in ascending order $1 \leqslant c_1 \leqslant \cdots \leqslant c_k \leqslant n$ with c_i in $A(i)$ by the use of extra blue enforcers, the blue vertices V_8. The idea here is that we ensure that for each j in $A(i)$ and each $j' \leqslant j$ in $A(q)$ with $(q > i)$, there is a blue vertex v in V_8 adjacent to both j and j'. This ensures that if j is chosen in block $A(i)$, then we cannot choose any $j' \leqslant j$ in any subsequent V_1 block. The last part of the construction is to force consistency with the clauses. We do this as follows. For each way a nonblocking set can correspond to making a clause false, we make a blue vertex and join it up to the s relevant vertices. This ensures that they cannot *all* be chosen. (This is the point of the V_7 vertices.) We now turn to the formal details.

The red vertex set V_{red} of G is the union of the following sets of vertices:
$V_1 = \{a[r_1, r_2] : 0 \leqslant r_1 \leqslant k-1, 0 \leqslant r_2 \leqslant n-1\}$,
$V_2 = \{b[r_1, r_2, r_3] : 0 \leqslant r_1 \leqslant k-1, 0 \leqslant r_2 \leqslant n-1, 1 \leqslant r_3 \leqslant n-k+1\}$.

The blue vertex set V_{blue} of G is the union of the following sets of vertices:
$V_3 = \{c[r_1, r_2, r_2'] : 0 \leqslant r_1 \leqslant k-1, 0 \leqslant r_2 < r_2' \leqslant n-1\}$,
$V_4 = \{d[r_1, r_2, r_2', r_3, r_3'] : 0 \leqslant r_1 \leqslant k-1, 0 \leqslant r_2, r_2' \leqslant n-1, 0 \leqslant r_3,$
$\qquad r_3' \leqslant n-1$ and either $r_2 \neq r_2'$ or $r_3 \neq r_3'\}$,
$V_5 = \{e[r_1, r_2, r_2', r_3] : 0 \leqslant r_1 \leqslant k-1, 0 \leqslant r_2, r_2' \leqslant n-1, r_2 \neq r_2',$
$\qquad 1 \leqslant r_3 \leqslant n-k+1\}$,
$V_6 = \{f[r_1, r_1', r_2, r_3] : 0 \leqslant r_1, r_1' \leqslant k-1, 0 \leqslant r_2 \leqslant n-1,$
$\qquad 1 \leqslant r_3 \leqslant n-k+1, r_1' \neq r_2+r_3\}$,
$V_7 = \{g[j, j'] : 1 \leqslant j \leqslant m, 1 \leqslant j' \leqslant m_j\}$,
$V_8 = \{h[r_1, r_1', j, j'] : 0 \leqslant r_1 < r_1' \leqslant k-1 \text{ and } j \geqslant j'\}$.

In the description of V_7, the integers m_j are bounded by a polynomial in n and k whose degree is a function of s, which will be described below. Note that since s is a fixed constant independent of k, this is allowed by our definition of reduction for parameterized problems.

For convenience, we distinguish the following sets of vertices:
$A(r_1) = \{a[r_1, r_2] : 0 \leqslant r_2 \leqslant n-1\}$,

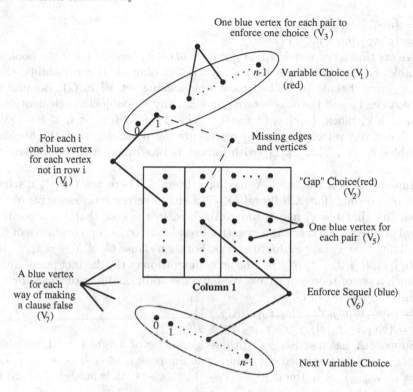

Fig. 9.2: Gadget for RED/BLUE NONBLOCKER

$B(r_1) = \{b[r_1, r_2, r_3] : 0 \leqslant r_2 \leqslant n - 1, 1 \leqslant r_3 \leqslant n - k + 1\}$,
$B(r_1, r_2) = \{b[r_1, r_2, r_3] : 1 \leqslant r_3 \leqslant n - k + 1\}$.

The edge set E of G is the union of the following sets of edges. In these descriptions, we implicitly quantify over all possible indices for the vertex sets $V_1, \ldots, V_8$.

$E_1 = \{a[r_1, q]c[r_1, r_2, r_2'] : q = r_2 \text{ or } q = r_2'\}$,
$E_2 = \{b[r_1, q_2, q_3]d[r_1, r_2, r_2', r_3, r_3'] : \text{either } (q_2 = r_2 \text{ and } q_3 = r_3)$
$\qquad \text{or } (q_2 = r_2' \text{ and } q_3 = r_3')\}$,
$E_3 = \{a[r_1, r_2]e[r_1, r_2, q, q']\}$,
$E_4 = \{b[r_1, q, q']e[r_1, r_2, q, q']\}$,
$E_5 = \{b[r_1, r_2, r_3]f[r_1, r_1', r_2, r_3]\}$,
$E_6 = \{a[r_1 + 1 \bmod n, r_1']f[r_1, r_1', r_2, r_3]\}$,

$E_7 = \{a[r_1, j]h[r_1, r_1', j, j']\},$
$E_8 = \{a[r_1', j']h[r_1, r_1', j, j']\}.$

We say that a red vertex $a[r_1, r_2]$ *represents the possibility* that the boolean variable x_{r_2} may evaluate to *true* (corresponding to the possibility that $a[r_1, r_2]$ may belong to a $2k$-element nonblocking set V' in G). Because of the vertices V_8 and the edge sets E_7 and E_8, any k nonblocking elements one from each V_1 block, $\{a[r_j, r_{i_j}] : j = 0, \ldots, k-1\}$, $i_0 < i_1 \ldots < i_k \leqslant k - 1$. We say that a red vertex $b[r_1, r_2, r_3]$ *represents the possibility* that the boolean variables $x_{r_2+1}, \ldots, x_{r_2+r_3-1}$ (with indices reduced mod n) may evaluate to *false*.

Suppose C is a clause of X having s literals. There are $O(n^{2s})$ distinct ways of choosing, for each literal $l \in C$, a single vertex representative of the possibility that $l = x_i$ may evaluate to *false*, in the case that l is a positive literal or in the case that l is a negative literal $l = \neg x_i$, a representative of the possibility that x_i may evaluate to *true*. For each clause C_j of X, $j = 1, \ldots, m$, let $R(j, 1), R(j, 2), \ldots, R(j, m_j)$ be an enumeration of the distinct possibilities for such a set of representatives. We have the additional sets of edges for the clause components of G:

$E_9 = \{a[r_1, r_2]g[j, j'] : a[r_1, r_2] \in R(j, j')\},$
$E_{10} = \{b[r_1, r_2, r_3]g[j, j'] : b[r_1, r_2, r_3] \in R(j, j')\}.$

Suppose X has a satisfying truth assignment τ of weight k, with variables $x_{i_0}, x_{i_1}, \ldots, x_{i_{k-1}}$ assigned the value *true*. Suppose $i_0 < i_1 < \cdots < i_{k-1}$. Let $d_r = i_{r+1 \,(\mathrm{mod}\, k)} - i_r \;(\mathrm{mod}\, n)$ for $r = 0, \ldots, k - 1$. It is straightforward to verify that the set of $2k$ vertices

$$N = \{a[r, i_r] : 0 \leqslant r \leqslant k - 1\} \cup \{b[r, i_r, d_r] : 0 \leqslant r \leqslant k - 1\}$$

is a nonblocking set in G.

Conversely, suppose N is a $2k$-element nonblocking set in G. It is straightforward to check that a truth assignment for X of weight k is described by setting those variables *true* for which a vertex representative of this possibility belongs to N, and by setting all other variables to *false*.

Note that the edges of the sets E_1 (E_2) which connect pairs of distinct vertices of $A(r_1)$ $[B(r_1)]$ to blue vertices of degree 2 enforce that any $2k$-element nonblocking set must contain exactly one vertex from each of the sets $A(0), B(0), A(1), B(1), \ldots, A(k-1), B(k-1)$. The edges of E_3 and E_4 enforce (again, by connections to blue vertices of degree 2) that if a representative of the possibility that x_i evaluates to *true* is selected for a nonblocking set from $A(r_1)$, then a vertex in the i-th row of $B(r_1)$ must be selected as well, representing (consistently) the interval of variables set false (by increasing index because of the E_7 and E_8 edges) until the "next" variable selected to be *true*. The edges of E_5 and E_6 ensure consistency between the selection in $A(r_1)$ and the selection in $A(r_1 + 1 \bmod n)$. The edges of E_9 and E_{10} ensure that a consistent selection can be nonblocking if and only if it does not happen that there is a set of representatives for a clause witnessing that every literal in the clause evaluates to *false*. (There is a blue vertex for every such possible set of representatives.)   □

Theorem 9.3.4 provides the starting point for demonstrating the following dramatic collapse of the $W[1]$ stratification.

Theorem 9.3.5. $W[1] = W[1,2]$.

Proof. It suffices to argue that for all $s \geqslant 2$, ANTIMONOTONE $W[1,s] = W[1,2]$. The argument here consists of another change of variables. Let C be an antimonotone $W[1,s]$ circuit for which we wish to determine whether a weight k input vector is accepted. We show how to produce a circuit C' corresponding to an expression in conjunctive normal form with clause size 2, that accepts an input vector of weight

$$k' = k2^k + \sum_{i=2}^{s} \binom{k}{i}$$

if and only if C accepts an input vector of weight k. (The circuit C' will, in general, not be antimonotone, but this is immaterial by Theorem 9.3.4. Actually, in [DF95b] Downey and Fellows used another reduction that only needs $k' = k^{s+1} + \sum_{i=2}^{s} \binom{k}{i}$ and is hence polynomial in k for a fixed s.)

Let $x[j]$ for $j = 1, \ldots, n$ be the input variables to C. The idea is to create new variables representing all possible sets of at most s and at least two of the variables $x[j]$. Let $A_1, \ldots, A_p$ be an enumeration of all such subsets of the input variables $x[j]$ to C. The inputs to each *or* gate g in C (all negated, since C is antimonotone) are precisely the elements of some A_i. The new input corresponding to A_i represents that all of the variables whose negations are inputs to the gate g have the value *true*. Thus, in the construction of C', the *or* gate g is replaced by the negation of the corresponding new "collective" input variable.

We introduce new input variables of the following kinds:

1. One new input variable $v[i]$ for each set A_i for $i = 1, \ldots, p$ to be used as above.
2. For each $x[j]$, we introduce 2^k copies $x[j,0], x[j,1], x[j,2], \ldots, x[j, 2^k - 1]$.

In addition to replacing the *or* gates of C as described above, we add to the circuit additional *or* gates of fanin 2 that provide an enforcement mechanism for the change of variables. The necessary requirements can be easily expressed in conjunctive normal form with clause size 2, and thus can be incorporated into a $W[1,2]$ circuit.

We require the following implications concerning the new variables:

1. The $n \cdot 2^k$ implications, for $j = 1, \ldots, n$ and $r = 0, \ldots, 2^k - 1$,

$$x[j,r] \Rightarrow x[j, r+1 \pmod{2^k}].$$

2. For each containment $A_i \subseteq A_{i'}$, the implication

$$v[i'] \Rightarrow v[i].$$

3. For each membership $x[j] \in A_i$, the implication

$$v[i] \Rightarrow x[j, 0].$$

It may be seen that this transformation may increase the size of the circuit by a linear factor exponential in k. We make the following argument for the correctness of the transformation.

If C accepts a weight k input vector, then setting the corresponding copies $x[i, j]$ among the new input variables accordingly, together with appropriate settings for the new "collective" variables $v[i]$, yields a vector of weight k' that is accepted by C'.

For the other direction, suppose C' accepts a vector of weight k'. Because of the implications in (1) above, exactly k sets of copies of inputs to C must have value 1 in the accepted input vector (since there are 2^k copies in each set). Because of the implications described in (2) and (3) above, the variables $v[i]$ must have values in the accepted input vector compatible with the values of the sets of copies. By the construction of C', this implies there is a weight k input vector accepted by C. $\square$

We have now done most of the work required to show that the following parameterizations of well-known problems are complete for $W[1]$:

CLIQUE
Instance : A graph $G = (V, E)$.
Parameter : A positive integer k.
Question : Is there a set of k vertices $V' \subseteq V$ that forms a complete subgraph of G (that is, a clique of size k)?

Theorem 9.3.6 (Downey and Fellows [DF95b]).

(i) INDEPENDENT SET *is complete for* $W[1]$.
(ii) CLIQUE *is complete for* $W[1]$.

Proof. It is easy to observe that INDEPENDENT SET belongs to $W[1]$. By Theorems 9.3.4 and 9.3.5, it is enough to argue that INDEPENDENT SET is hard for ANTIMONOTONE $W[1, 2]$. Given a boolean expression X in conjunctive normal form (product of sums) with clause size 2 and all literals negated, we may form a graph G_X with one vertex for each variable of X and having an edge between each pair of vertices corresponding to variables in a clause. The graph G_X has an independent set of size k if and only if X has a weight k truth assignment.

(ii) This follows immediately by considering the complement of a given graph. The complement has an independent set of size k if and only if the graph has a clique of size k. $\square$

Finally, we are in a position to establish our analog of the Cook-Levin Theorem, which we restate for the reader's convenience. In fact, we can now state an extended form.

Theorem 9.3.7 (Parameterized Cook-Levin Theorem). *The following are are complete for* $W[1]$:

(i) **(Cai, Chen, Downey and Fellows [CCDF97])** SHORT TURING MACHINE ACCEPTANCE.
(ii) WEIGHTED 2-SATISFIABILITY.
(iii) s-RED/BLUE NONBLOCKER *for* $s \geqslant 2$.
(iv) INDEPENDENT SET.
(v) CLIQUE.

Proof. We need only prove that SHORT TURING MACHINE ACCEPTANCE is $W[1]$-complete to finish the result. To show hardness for $W[1]$, we reduce from CLIQUE. $W[1]$-hardness will then follow by Theorem 9.3.6(ii). Let $G = (V, E)$ be a graph for which we wish to determine whether it contains a k-clique. We have shown how to construct a nondeterministic Turing machine M that can reach an accept state in $k' = f(k)$ moves if and only if G contains a k-clique. The Turing machine M is designed so that any accepting computation consists of two phases. In the first phase, M nondeterministically writes k symbols representing vertices of G in the first k tape squares. (There are enough symbols so that each vertex of G is represented by a symbol.) The second phase consists of making $\binom{k}{2}$ scans of the k tape squares, each scan devoted to checking, for a pair of positions i and j, that the vertices represented by the symbols in these positions are adjacent in G. Each such pass can be accomplished by employing $O(|V|)$ states in M dedicated to the ij-th scan.

In order to show membership in $W[1]$, it suffices to show how the SHORT TURING MACHINE ACCEPTANCE problem for a Turing machine $M = (\Sigma, Q, q_0, \delta, F)$ and positive integer k can be translated into one about whether a circuit C accepts a weight k' input vector, where C has depth bounded by some t (independent of k and the Turing machine M) and has only a single large (output) *and* gate, with all other gates small. We arrange the circuit so that the k' inputs to be chosen to be set to 1 in a weight k' input vector represent the various data: (1) the i-th transition of M, for $i = 1, \ldots, k$, (2) the head position at time i, (3) the state of M at time i, and (4) the symbol in square j at time i for $1 \leqslant i, j \leqslant k$. Thus, we may take $k' = k^2 + 3k$. In order to force exactly one input to be set equal to 1 among a pool of input variables (for representing one of the above choices), we can add to the circuit for each such pool of input variables, and for each pair of variables x and y in the pool, a small "not both" circuit representing $(\neg x \vee \neg y)$. It might seem that we must also enforce (e.g., with a large *or* gate) the condition "at least one variable in each such pool is set true"—but this is actually unnecessary, since in the presence of the "not both" conditions on each pair of input variables in each pool, an accepted weight k' input vector *must have* exactly one variable set true in each of the k' pools. Let n denote the total number of input variables in this construction. We have $n = O(k\delta + k^2 + k|Q| + k^2|\Sigma|)$ in any case.

The remainder of the circuit encodes various checks on the consistency of the above choices. These consistency checks conjunctively determine whether the choices represent an accepting k-step computation by M, much as in the proof of Cook-Levin Theorem, Theorem 7.2.6. These consistency checks can be implemented so that each involves only a bounded number b of the input variables. For example, we will want to enforce that if five variables are set true indicating particular values of the tape head position at time $i + 1$ and the head position, state, scanned symbol, and machine transition at time i, then the values are consistent with δ. Thus, $O(n^5)$ small "checking" circuits of bounded depth are sufficient to make these consistency checks; in general, we will have $O(n^b)$ "checking" circuits for consistency checks involving b values. All of the small "not both" and "checking" circuits feed into the single large output *and* gate of C. The formal description of all this is laborious but straightforward. (Exercise 9.3.15.) □

Phew! The reader should note that henceforth we will only need the the statement of the parameterized version the Cook-Levin Theorem. Thus, Theorem 9.3.7 can be used as a *Black Box*; that is as a starting point for hardness proofs.

We remark that Theorem 9.3.7 depends crucially on there being no bound on the size of the Turing machine alphabets in the definition of the problem. If we restrict SHORT TURING MACHINE ACCEPTANCE to Turing machines with $|\Sigma|$ bounded by some constant b, then the number of configurations is bounded by $b^k|Q|k$ and the problem becomes fixed-parameter tractable (see Exercise 9.3.12).

9.3.2 Exercises

Exercise 9.3.8 Show that L is in FPT iff there is a machine M and a computable functions g, and constant d, accepting $\langle x, k \rangle$ in time $O(|x|^d + g(k))$. (Hint: If you take the c from Definition 9.2.2, then think of $d = c + 1$, and note that $|x|^d$ will eventually dominate $f(k) \cdot |x|^c$, whatever f is.)

Exercise 9.3.9 Show that the parameterized problem asking whether a graph has k disjoint *edges* is $W[1]$-complete.

Exercise 9.3.10 Suppose that $r \geqslant 1$. We say that a graph G is *r-regular* if all vertices of g have degree $\leqslant r$. Show that for all $r \geqslant 1$, the versions of CLIQUE and INDEPENDENT SET for r-regular graphs are $W[1]$-complete. (Hint: The problem is only interesting when the target parameter $k \leqslant r$, because if $k > r$

the answer is no. So figure out a way to lift the degrees of vertices. Think about taking r copies of the graph G.)

Exercise 9.3.11 Show that if we consider RED/BLUE NONBLOCKER where all degrees are bounded by s, then it is FPT.

Exercise 9.3.12 Prove that the problem RESTRICTED SHORT TURING MACHINE ACCEPTANCE is FPT. This is the restriction of SHORT TURING MACHINE ACCEPTANCE to machines with Σ (the number of states) bounded by some fixed constant b.

Exercise 9.3.13 Prove that the INDEPENDENT SET is FPT when restricted to:

1. graphs with fixed maximum degree m,
2. planar graphs. (This is a bit harder.)

Exercise 9.3.14 Prove that if determining if a monotone s-CNF formula has a satisfying assignment of weight k is FPT.

Exercise 9.3.15 Using the classical proof of the Cook-Levin Theorem, Theorem 7.2.6, as a model carefully complete the details of the proof of Theorem 9.3.7 (i).

Exercise 9.3.16 Prove that the following problem IRREDUNDANT SET is in $W[1]$.

IRREDUNDANT SET
Instance : A graph $G = (V, E)$, a positive integer k.
Parameter : A positive integer k.
Question : Is there a set $V' \subseteq V$ of cardinality k having the property that each vertex $u \in V'$ has a private neighbour? (A *private neighbour* of a vertex $u \in V'$ is a vertex u' (possibly $u' = u$) with the property that for every vertex $v \in V'$, $u \neq v$, $u' \notin N[v]$.)
(Hint: Let $G = (V, E)$ be a graph for which we wish to determine if G has a k-element irredundant set. We construct the circuit C corresponding to the following boolean expression E. The variables of E are:

$$\{p[i, x, y] : 1 \leqslant i \leqslant k, x, y \in V\} \cup \{c[i, x] : 1 \leqslant i \leqslant k, x \in V\}$$

The expression E has the clauses:
(1) $(\neg p[i, x, y] + \neg p[i, x', y'])$ for $1 \leqslant i \leqslant k$ and either $x \neq x'$ or $y \neq y'$, $x, x', y, y' \in V$.
(2) $(\neg c[i, x] + \neg c[i, x'])$ for $1 \leqslant i \leqslant k$ and $x \neq x'$, $x, x' \in V$.
(3) $(c[i, x] + \neg p[i, x, y])$ for $1 \leqslant i \leqslant k$ and $x, y \in V$.
(4) $(\neg c[j, u] + p[i, x, y])$ for $1 \leqslant i, j \leqslant k$, $i \neq j$ and $u \in N[y]$.

Now argue that the circuit C accepts a weight $2k$ input vector if and only if G has a k-element irredundant set.)

Exercise 9.3.17 (Fellows, Hermelin, Rosamond, Vialette [FHRV09])
Using a reduction from CLIQUE, *show that the following problem is complete for* $W[1]$.

k-Multicoloured Clique
Input : *A graph G with a vertex colouring $\chi : V(G) \to \{1, \ldots, k\}$.*
Parameter : k
Question : *Does G have a colourful clique of size k? Here the clique being colourful means that if x, y are in the clique, then $\chi(x) \neq \chi(y)$.*

9.3.3 The W-hierarchy

The current belief of computer scientists is that there is *no parametric reduction at all* from Weighted Cnf Sat to Weighted 3 Cnf Sat. Naturally, we cannot prove that this is the case, as it would imply P$\neq$NP, but what we can do is develop a more refined classification of combinatorial problems based on the belief that no reduction is possible.

Extending this reasoning further, we view Weighted Cnf Sat as a formula that is a product of sums. We can similarly define Weighted t-Pos Sat as the weighted satisfiability problem for a formula X in product of sums of product of sums... with t alternations.

We then define[4] the following:

Definition 9.3.2 (Downey and Fellows [DF92, DF95a]).

$$W[t] = \{L \subset \Sigma^* \times \Sigma^* : L \leqslant_{fpt} \text{Weighted } t - \text{Pos Sat}\}.$$

What about general circuits, or boolean expressions with no bound on the depth. Weighted Sat if we have no restriction on the formula. Downey and Fellows [DF92, DF95a] defined three classes on the top of the hierarchy:

Definition 9.3.3.

(i) $W[P]$ denotes the collection of parameterized languages fpt-reducible to the following problem.

Weighted Circuit Sat
Input : A decision circuit of X size polynomial in inputs $\{x_1, \ldots, x_n\}$.
Parameter : An integer k.
Question : Does X have a weight k satisfying assignment?

---

[4] Originally these classes were defined using bounded weft circuits, but for our purposes boolean formulae suffice, although the fact that the two definitions are equivalent requires a combinatorial proof.

(ii) $W[SAT]$ is the same as $W[P]$ except that we ask that the circuit is a boolean one corresponding to formula of propositional logic.
(iii) XP is the class of all languages L fpt-reducible to a language $\widehat{L}$ with

$$\widehat{L}^{(k)} \in \mathrm{DTIME}(n^k),$$

for all k, where $\widehat{L}^{(k)} =_{\mathrm{def}} \{\langle \sigma, k \rangle \in \widehat{L}\}$. $\widehat{L}^{(k)}$ is called the k-th *slice* of $\widehat{L}$.

These definitions collectively give the *W-hierarchy* below

$$FPT \subseteq W[1] \subseteq W[2] \subseteq W[3] \ldots W[SAT] \subseteq W[P] \subseteq XP.$$

Direct diagonalization (i.e. essentially the Time Hierarchy Theorem) proves the following:

Proposition 9.3.1. $FPT \neq XP$.

Proof. Suppose that FPT⊆XP. Define a language L, such that, for each k, $L^{(k)}$ is in $\mathrm{DTIME}(n^{k+1}) \setminus \mathrm{DTIME}(n^k)$, given by the Hierarchy Theorem. Then $L \in$ XP. Suppose that $L \in$ FPT. Thus, by definition, there is a p, and M with

$$\langle x, k \rangle \in L \text{ iff } M \text{ accepts } (\langle x, k \rangle) \text{ in time } |x|^p.$$

Thus, for all k, $L^{(k)} \in \mathrm{DTIME}(n^p)$, a contradiction. $\square$

No other containments are known to be proper. It would take too much space to properly analyse these classes. We will state without proof some completeness results, and refer the reader to Downey and Fellows [DF13] or Flum and Grohe [FG06] for more details.

Theorem 9.3.18 (Downey and Fellows [DF92, DF95a]). DOMINATING SET *is $W[2]$-complete*.

Theorem 9.3.19 (Cesati and Di Ianni [CI97]). *The problem below is complete for $W[2]$*

SHORT MULTI-TAPE NTM
Input : *A multi-tape nondeterministic Turing Machine M with m tapes and a word w.*
Parameter : *Positive integers k and m. (That is $\langle k, m \rangle$)*
Question : *Does M have an accepting computation of w in $\leqslant k$ many steps?*

Of course, for NP-completeness multi-tape and single tape make no difference. We mention one relatively simple $W[P]$ completeness result.

Theorem 9.3.20 (Abrahamson, Downey and Fellows [ADF93]). *The following problem is complete for $W[P]$.*

SHORT CIRCUIT SATISFIABILITY
Instance : *A decision circuit C with at most n gates and $\leqslant k \log n$ inputs.*
Parameter : *A positive integer k.*
Question: *Is there any setting of the inputs that causes C to output 1?*

Proof. To see that it is $W[P]$-hard, take an instance C of WEIGHTED CIRCUIT SATISFIABILITY, with parameter k and inputs $x_1, \ldots, x_n$. Let $z_1, \ldots, z_{k \log n}$ be new variables. Using lexicographic order in polynomial time (independent of k) we can have a surjection from $Z = \{z_1, \ldots, z_{k \log n}\}$ to the (characteristic function of the) k-element subsets of $X = \{x_1, \ldots, x_n\}$. Representing this as a circuit with inputs Z and outputs X we can put this circuit on top of C to for C' so that C accepts a weight k input iff C' accepts some input.

That $W[P]$ contains SHORT CIRCUIT SATISFIABILITY is equally easy, just use the polynomial embedding of the $k \log n$ into the $k+1$-element subsets of n for n sufficiently large. $\square$

There are many problems complete for $W[P]$, and we confine ourselves to stating two from [ADF93].

DEGREE 3 SUBGRAPH ANNIHILATOR
Instance : A graph G.
Parameter : A positive integer k.
Question : Is there a set V' of at most k vertices of G, such that the subgraph induced by $V(G) \setminus V'$ has no minimum degree 3 subgraph?

k-LINEAR INEQUALITIES
Instance : A system of (rational) linear inequalities.
Parameter : A positive integer k.
Question : Can the system be made consistent over the rationals by deleting at most k of the inequalities?

There is even a version of parameterized space, based on QBFSAT, called $AW[*]$. As we have seen in §7.4, PSPACE complete problems we met are mostly based on games. The parameterized version is based on the question of the existence of winning strategies in k moves. For example, the following is shown in [ADF93] to be $AW[*]$ complete.

SHORT GENERALIZED GEOGRAPHY
Input : An instance of GENERALIZED GEOGRAPHY
Parameter : An integer k.
Question : Can Player I win in at most k moves?

Details can be found in [DF13], Chapter 26. Rather than dwelling on hardness we turn to using this parameterized technology to analyse some classical computational complexity.

9.3.4 Showing Known Algorithms are Likely Very Bad Indeed; Especially in PTAS's

Parameterized intractability has been used to show that certain algorithms within P are very unlikely to have reasonable running times; that is assuming $W[1] \neq$ FPT.

A lot of effort has gone into trying to combat intractability. As per Garey and Johnson [GJ79], polynomial time approximation schemes (PTAS's) are one of the main traditional methods. Whilst we discuss them in greater detail in §10.2, we will see how they relate to parameterized complexity in this section.

PTAS's concern (classical) optimization problem. For concreteness, we look at a *minimization* problem but the *maximization* is dual. Thus, out goal is to minimize some relation R on the instance. For example, MINIMAL VERTEX COVER is the problem of finding the smallest vertex cover in a graph G.

Definition 9.3.4. For a minimization problem of minimizing R on a class of instances $\mathcal{C}$, a *polynomial time approximation scheme* (PTAS) is an algorithm $A(\cdot, \cdot)$ which, for each fixed $\varepsilon > 0$, runs in polynomial time, and for $G \in \mathcal{C}$, will compute a solution S, with

$$(1 + \varepsilon)|Q| \leqslant |S|.$$

Here Q is a minimal sized solution for the instance G.

Many ingenious polynomial time approximation schemes have been invented. The traditional argument for PTAS's is that: even though we can't solve the problem, we can get an approximate solution in polynomial time, so that is probably good enough. We will discuss this approach more fully in §10.2. As we discuss there, often the wonderful PCP theorem of Arora *et al.* [ALM+98] shows that no such PTAS exists. But what to do if we have a PTAS and it has very poor running time?

Let's look at some relatively recent examples, taken from some recent major conferences such as STOC, FOCS and SODA, etc. Some of these algorithms seem not that practical. We follow [Dow03].

- Arora [Aro96] gave a $\mathcal{O}(n^{\frac{3000}{\varepsilon}})$ PTAS for EUCLIDEAN TSP
- Chekuri and Khanna [CK00] gave a $\mathcal{O}(n^{12(\log(1/\varepsilon)/\varepsilon^8)})$ PTAS for MULTIPLE KNAPSACK
- Shamir and Tsur [ST98] gave a $\mathcal{O}(n^{2^{2^{\frac{1}{\varepsilon}}}-1})$ PTAS for MAXIMUM SUBFOREST
- Chen and Miranda [CM99] gave a $\mathcal{O}(n^{(3mm!)^{\frac{m}{\varepsilon}}+1})$ PTAS for GENERAL

MULTIPROCESSOR JOB SCHEDULING
- Erlebach *et al.* [EJS01] gave a $\mathcal{O}(n^{\frac{4}{\pi}(\frac{1}{\varepsilon^2}+1)^2(\frac{1}{\varepsilon^2}+2)^2})$ PTAS for MAXIMUM INDEPENDENT SET for geometric graphs.

For the sake of this section, it is not important what these problems are, save to say that they were PTAS's for problems of arising from natural optimization problems, and seen as important enough to appear in conferences widely held to publish cutting edge results.

Table 9.2 below calculates some running times for these PTAS's with a 20% error; $\varepsilon = \cdot 2$.

Reference	Running Time for a 20% Error
Arora [Aro96]	$\mathcal{O}(n^{15000})$
Chekuri and Khanna [CK00]	$\mathcal{O}(n^{9,375,000})$
Shamir and Tsur [ST98]	$\mathcal{O}(n^{958,267,391})$
Chen and Miranda [CM99]	$> \mathcal{O}(n^{10^{60}})$ (4 Processors)
Erlebach *et al.* [EJS01]	$\mathcal{O}(n^{523,804})$

Table 9.2: The Running Times for Some Relatively Recent PTAS's with 20% Error.

Now, by anyone's measure, a running time of $n^{500,000}$ is bad and $n^{9,000,000}$ is even worse[5]! The optimist would argue that these examples are important in that they prove that PTAS's exist, and are but a first foray. The optimist would also argue that with more effort and better combinatorics, we will be able to come up with some $n \log n$ PTAS for the problems. For example, Arora [Aro97] also came up with another PTAS for EUCLIDEAN TSP, but this time it was nearly linear and practical.

But this situation is akin to P vs NP. Why not argue that some exponential algorithm is but a first foray into the problem. With more effort and better combinatorics we will find a feasible algorithm for SATISFIABILITY. What if a lot of effort is spent in trying to find a practical PTAS's without success? As with P vs NP what is desired, is either an *efficient* PTAS (EPTAS), or a proof that no such PTAS exists. A primary use of NP completeness is to give compelling evidence that many problems are unlikely to have better than exponential algorithms generated by complete search. The trouble is, these examples are in polynomial time. Lower bounds are hard to come by there. Here's where you might make parameterized complexity your friend.

If the reader studies the examples above, they will realize that a source of the appalling running times is the $\frac{1}{\varepsilon}$ in the exponent. One method that has

---

[5] Funny story; when I spoke about this material in an invited lecture reported in [Dow03], someone from the audience (of very good computer scientists) said to me "Oh, I guess that's why I could not get the algorithm [CM99] to run!".

worked in such examples is to parameterize the problem by taking $k = \frac{1}{\varepsilon}$ as the relevant parameter.

The reader attuned now to parameterized complexity will see that the following is the correct notion.

Definition 9.3.5. An optimization problem Π has an *efficient P-time approximation scheme* (EPTAS) if it can be approximated to a goodness of $(1+\varepsilon)$ of optimal in time $f(k)n^c$ where c is a constant t and $k = 1/\varepsilon$.

Arora gave an EPTAS for the EUCLIDEAN TSP in [Aro97], but for all of the other PTAS's mentioned above, the possibility of such an improvement remains open.

But we have a strategy: Prove the problem is $W[1]$-hard parameterized by $k = \frac{1}{\varepsilon}$ and you prove that the problem has no EPTAS, assuming that $W[1] \neq FPT$. Parameterizing this way, the following is not hard to see. Historically, it was first observed by Bazgan [Baz95] and also Cai and Chen [CC97].

Suppose that Π_{opt} is an optimization problem, and that Π_{param} is the corresponding parameterized problem, where the parameter is the value of an optimal solution. Then Π_{param} is fixed-parameter tractable if Π_{opt} has an EPTAS.

We will confine ourselves to quoting one application of this approach:

In a well-known paper, Khanna and Motwani [KM96] introduced three planar logic problems towards an explanation of PTAS-approximability. Their suggestion is that "hidden planar structure" in the logic of an optimization problem is what allows PTASs to be developed. One of their core problems was the following.

PLANAR TMIN
Input : A collection of Boolean formulas in DNF (sum-of-products) form, with all literals positive, where the associated bipartite graph is planar (this graph has a vertex for each formula and a vertex for each variable, and an edge between two such vertices if the variable occurs in the formula).
Output : A truth assignment of minimum weight (i.e., a minimum number of variables set to *true*) that satisfies all the formulas.

For this optimization problem we will parameterize by the solution weight:

GAP PLANAR TMIN
Input : An instance of TMIN, and a rational $\varepsilon > 0$ such that either the optimal solution is $\leqslant k$, or is bigger than $(1 + \varepsilon) \cdot k$.
Parameter : $\lceil \frac{1}{\varepsilon} \rceil$.
Goal : Decide if the optimal solution is $\leqslant k$.

Theorem 9.3.21 (Fellows, Cai, Juedes and Rosamond in [Fel02]).
PLANAR TMIN *parameterized by solution weight and* $\lceil \frac{1}{\varepsilon} \rceil$ (*i.e.* GAP PLANAR TMIN *is hard for* $W[1]$ *and therefore does not have an EPTAS unless FPT* $= W[1]$).

Proof. We follow the simplified proof in Marx [Mar08]. We reduce from (MAXIMUM) CLIQUE, Given G and t we construct an instance of TMIN with $k = 2t(2t+1)$ many variables such that G has a clique of cardinality t implies there is a solution instance of weight $\leqslant k$, and if G has no clique of size t then the instance has no solution of weight at most $\varepsilon = \frac{1}{2k} \cdot k$. Marx calls this formulation GAP PLANAR TMIN.

Suppose that $V(G) = \{1, \ldots, n\}$. We construct the instance with $2t(t+1)$ many variables. We arrange them into blocks $A_{i,j}$ and $B_{i,j}$ for $1 \leqslant i, j \leqslant t$, with each block containing n variables. We use $a_{i,j,s}$ for variables in block $A_{i,j}$ and similarly $b_{i,j,s}$ for $B_{i,j}$, for $s = 1, \ldots, n$.

We will construct t^2 DNF's such that $\psi_{i,j}$ contains only variables from blocks $A_{i,j-1}, B_{i-1,j}$. The formulae are defined as

$$\psi_{i,j} = \begin{cases} \bigvee_{s=1}^{n}(a_{i,j-1} \wedge a_{i,j,s} \wedge b_{i-1,j,s} \wedge b_{i,j,s}) \text{ if } i = j, \\ \bigvee_{xy \in E(G)}(a_{i,j-1,x} \wedge a_{i,j,x} \wedge b_{i-1,j,y} \wedge b_{i,j,y}), \text{ otherwise.} \end{cases}$$

The following diagram might be useful:

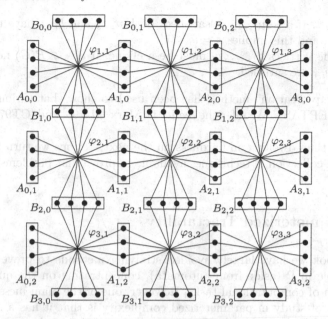

Fig. 9.3: The Blocks

Then $\Psi = \vee_{1 \leqslant i,j \leqslant n} \psi_{i,j}$. The formula is clearly planar by construction[6]. The optimum solution is either k or at least $k+1 > (1+\varepsilon) \cdot k - (1+\frac{1}{2k}) \cdot k = k + \frac{1}{2}$.

If G has a clique $\{v_1, \ldots, v_t\}$, then for every i, j we set $a_{i,j,v_i} \in A_{i,j}$ and $b_{i,j,v_j} \in B_{i,j}$ to *true*. This is a weight $2t(t_1)$ satisfying assignment for the formula Ψ. To see this there are two cases. If $i = j = r$, then $\psi_{i,j}$ is satisfied by $a_{i,j-1,v_r} \wedge a_{i,j,v_r} \wedge b_{i-1,j,v_r} \wedge b_{i,j,v_r}$. If $i \neq j$, then $\psi_{i,j}$ is satisfied by $a_{i,j-1,v_i} \wedge a_{i,j,v_i} \wedge b_{i,j-1,v_j} \wedge b_{i,j,v_j}$, this being a term since $v_i v_j \in E(G)$.

If G has not clique of size t, and there is a satisfying assignment of weight at most $(1+\varepsilon) \cdot k$, then first note that $(1+\varepsilon) \cdot k < k+1$. Therefore the weight of the assignment is at most k. Therefore each block has at most one true variable, otherwise there are are not enough true assignments to go round. Thus the weight is precisely k and there is exactly one true variable in each block.

For each $\psi_{i,j}$ there is an s with the term containing both $a_{i,j,s}$ and $a_{i,j-1,s}$. This means that if $\psi_{i,j}$ is satisfied, then the true variables in blocks $A_{i,j-1}$ and $A_{i,j}$ have the same index. This is true for all blocks with first variable i. Therefore there is a value v_j such that a_{i,j,v_i} is true for all $1 \leqslant j \leqslant t$. Similarly for each j there is a v_j with b_{i,j,b_j} true. We claim that $v_i = v_i'$ for

---

[6] The original construction from, for example, Fellows [Fel02] has a more complicated construction using crossing widgets to enforce planarity.

all $i \leqslant i \leqslant t$. If a term $\psi_{i,j}$ is satisfied, both $a_{i,i,s}$ and $b_{i,i,s}$ have to be true for some s, and this implies $s = v_i = v'_i$.

Then the vertices $a_1, \ldots, v_t$ must be a clique. If $v_i v_j \notin E(G)$ none of the terms in $\psi_{i,j}$ are satisfied. $\square$

This "gap creation" method has been used to show that various PTAS's cannot be EPTAS's. We refer the reader to Marx [Mar08] and [CT97, Mar05a, Mar05b].

Rather than dwelling on hardness, in the next sections we turn to looking at the rich collection of techniques available to prove that problems are FPT.

9.4 Parameterized Tractability

In this book we have largely avoided studying methods to prove problems are solvable in P (aside from automata), concentrating on the questions of calibration of computational tasks. However, one of the main messages from 30+ years of study in parameterized complexity is that it has a distinctive positive toolkit, and we will take time to look as a few of the methods.

The underlying idea is that to try to deal with a combinatorial problem, we should have a discourse with the problem. We should seek to understand which parameters cause the combinatorial explosion. We then attack the problem via parameters to find hidden tractability. **When in doubt, parameterize**.

In this section we will confine ourselves to three practical techniques which have emerged. Later, we vaguely point at further techniques relying on deeper mathematics at the end of the section.

9.4.1 Bounded Search Trees

A fundamental source of high running times is *branching* in algorithms. A very crude idea to limit the running time is to keep this branching small and a function of the parameter.

Example 9.4.1. VERTEX COVER
For instance, given a parameter k to see if a graph G has a size k vertex cover, we can work as follows. Take any edge $e = vw$, and note that one of v or w must be in the cover. So we begin a search tree, by having leaves labeled by v and w, and for each leaf recursively do this for the graphs gotten by deleting any edge covered by v and w respectively. The depth of this process

for a k-vertex cover is k and then we can decide if G has a vertex cover in time $O(2^k|G|)$ by checking the 2^k paths of the tree.

Whilst this is a simple idea, to use it in genuine algorithms, you need to appeal to some kind of combinatorics to improve performance. An illustration of this idea is that if we can make the search tree smaller than the complete binary tree of length k, then the performance will improve. Notice that, if G has no vertex of degree three or more, then G consists of a collection of cycles, and this is very quick to check. Thus we can assume we have vertices of higher degree than 2. For vertex cover of G we must have either v or *all of its neighbours*, so we create children of the root node corresponding to these two possibilities. The first child is labeled with $\{v\}$ and $G - v$, the second with $\{w_1, w_2, \ldots, w_p\}$, the neighbours of v, and $G - \{w_1, w_2, \ldots, w_p\}$. In the case of the first child, we are still looking for a size $k - 1$ vertex cover, but in the case of the second child we need only look for a vertex cover of size $k - p$, where p is at least 3. Thus, the bound on the size of the search tree is now somewhat smaller than 2^k. It can be shown that this algorithm runs in time $O([5^{\frac{1}{4}}]^k \cdot n)$, and note that $5^{\frac{1}{4}} \approx 1.446$. In typical graphs, there are many vertices of higher degree than 3, and hence this works even faster.

The best algorithms along these lines use more complex branching rules. For example, Niedermeier [Nie06] uses the following branching rules.

BRANCHING RULE VC1:
If there is a degree one vertex v in G, with single neighbour u, then there is a minimum size cover that contains u. Thus, we create a single child node labeled with $\{u\}$ and $G - u$.

BRANCHING RULE VC2:
If there is a degree two vertex v in G, with neighbours w_1 and w_2, then either both w_1 and w_2 are in a minimum size cover, or v together with *all other neighbours* of w_1 and w_2 are in a minimum size cover.

BRANCHING RULE VC3:
If there is a degree three vertex v in G, then either v or all of its neighbours are in.

We remark that using these three rules, (using a recurrence relation) it can be shown that if there is a solution of size at most k then the size of the corresponding search tree has size bounded above by $O(1.47^k)$. More involved rules of similar ilk exploring the *local structure* of neighbourhoods in graphs, and *iterative compression* discussed below, result in the algorithm of Chen et. al. [CKX10]) with running time $O(1.2738^k)$ the current fastest algorithm.

There are a number of problems for which the method of bounded search trees is the only method which establishes the problem in FPT, or at least the is the fastest parameterized method. Bounded search trees have been particularly successful in computational biology with problems like the CLOSEST

STRING problem [GNR01] and MAXIMUM AGREEMENT FOREST problem [HM07].

In passing I remark that this method is inherently parallelizable and as we see is often used in conjunction with other techniques. The method for VERTEX COVER can be found discussed in [AKLSS06].

Bounded search tree techniques are discussed in great depth in Cygan et. al. [CFK+16].

9.4.2 Exercises

Exercise 9.4.1 Prove that the problem called HITTING SET, parameterized by k and hyperedge size d, is FPT. A hypergraph is like a graph except that an edge is simply a subset of $V(G)$. In this problem, we are given a hypergraph where the maximum number of vertices in a hyperedge is bounded by $d \geqslant 2$. Note that for $p = 2$ this is simply VERTEX COVER.

Exercise 9.4.2 Prove that DOMINATING SET restricted to graphs of maximum degree t is solvable in time $O((t + 1)^k |G|)$.

Exercise 9.4.3 (Downey and Fellows [DF95c]) Let q be fixed. Prove that the following problems are solvable in time $O(q^k |E|)$.

WEIGHTED MONOTONE q-CNF SATISFIABILITY
Input : A boolean formula E in conjunctive normal form, with no negated variables and minimum clause size q.
Parameter : A positive integer k.
Question: Does E have a weight k satisfying assignment?

WEIGHTED $\leqslant q$-CNF SATISFIABILITY
Input : A boolean formula E in conjunctive normal form, with maximum clause size q.
Parameter : A positive integer k.
Question : Does E have a weight k' satisfying assignment for some $k' \leqslant k$?
(Hint: For the first one, simply modify the method used for VERTEX COVER. For the other, first see if there is a clause with only unnegated variables. If not, then we are done. If there is one, fork on the clause variables.)

Exercise 9.4.4 *[Kaplan, Shamir and Tarjan [KST94]] A graph G is called *chordal* or *triangulated* if every cycle of length $\geqslant 4$ contains a chord. If $G = \langle V, E \rangle$ is a graph, then a *fill-in* of G is a graph $G' = \langle V, E \cup F \rangle$ such that G' is chordal. The MINIMUM FILL-IN problem, which is also called CHORDAL GRAPH COMPLETION is the following:

MINIMUM FILL-IN
Input : A graph G.

Parameter : A positive integer k.

Question: Does G have a fill-in G' with $|F| \leqslant k$?

This is an important problem in computational biology. If k is allowed to vary, Yannakakis [Yan81] has proven the problem to be NP-complete. Use the method of search trees to show that the parameterized version above is FPT[7].

(Hint: This is not easy and comes from a research level paper. Let c_n denote $\binom{2n}{n} \cdot \frac{1}{1+n}$, the n-th Catalan number. First, prove that the number of minimal triangulations of a cycle with n vertices is c_{n-2}. Build a search tree as follows. The root of the tree $\mathcal{T}$ is the graph G. To generate the children of a node G' of $\mathcal{T}$, first find a chordless cycle C of G' (i.e., of length $\geqslant 4$) and let the children of the node labeled with G' correspond to the minimal triangulations of C. This only adds at most $c_{|C|-2}$ many children. Note that if we construct the complete tree, then each minimal triangulation of G will correspond to at least one leaf. Restrict the search to traverse paths in $\mathcal{T}$ that involve only k additional edges. you need to prove that a chordless cycle can be found in a graph in time $O(|E|)$, to prove that all the minimal triangulations of a cycle C can be generated in $O(|C|)$ time and that the total number of nodes of $\mathcal{T}$ we need to visit is bounded by $2 \cdot 4^{2k}$. These observations give the running time of the algorithm to be $O(2^{4k}|E|)$.)

9.4.3 Kernelization

This is again a simple basic idea. If we can make the problem *smaller* then the search will be *quicker*. This idea of shrinking is a *data reduction* or *preprocessing* idea, and is the heart of many heuristics.

Whilst there are variations of the idea below, the simplest version of kernelization is the following.

Definition 9.4.1 (Kernelization).

Let $L \subseteq \Sigma^* \times \Sigma^*$ be a parameterized language. A reduction to a problem kernel, or kernelization, comprises replacing an instance (I, k) by a reduced instance (I', k'), called a problem kernel, such that

(i) $k' \leqslant k$,

(ii) $|I'| \leqslant g(k)$, for some function g depending only on k, and

(iii) $(I, k) \in L$ if and only if $(I', k') \in L$.

---

[7] Kaplan, Shamir and Tarjan [KST94] have used (essentially) the Problem Kernel Method of the next section to give a $O(k^5|V||E| + f(k))$ for suitably chosen f. Finally, Leizhen Cai [Cai96] also used the problem kernel method another way to give a $O(4^k((k+1)^{-3/2})[|E(G)||V(G)| + |V(G)|^2]$ time algorithm. The current champion for this problem is Fomin and Villanger [FV11] and runs in time $0^*(2^{\sqrt{k} \log k})$.

The reduction from (I, k) to (I', k') must be computable in time polynomial in $|I| + |k|$.

There are other notions, where the kernel may be another problem (often "annotated") or the parameter might increase, but, crucially, the size of the kernel depends *only on* k.

Here are some natural reduction rules for a kernel for VERTEX COVER.

REDUCTION RULE VC1:
Remove all isolated vertices.

REDUCTION RULE VC2:
For any degree one vertex v, add its single neighbour u to the solution set and remove u and all of its incident edges from the graph.

These rules are obvious. Sam Buss (see [DF98]) originally observed that, for a simple graph G, any vertex of degree greater than k must belong to every k-element vertex cover of G (otherwise all the neighbours of the vertex must be included, and there are more than k of these).
This leads to our last reduction rule.

REDUCTION RULE VC3:
If there is a vertex v of degree at least $k + 1$, add v to the solution set and remove v and all of its neighbours.

After exhaustively applying these rules, we get to a graph (G', k'), where no vertex in the reduced graph has degree greater than $k' \leqslant k$, or less than two. Then simple combinatorics shows that if such a reduced graph has a size k vertex cover, its must have size $\leqslant k^2$. This is the size k^2 kernelization.

Now we can apply the bounded depth search tree rule to this reduced graph, and get an algorithm for vertex cover running in time $O(g(k^2))$, where $g(n)$ is the running time of the best algorithm, obtained by bounded depth search trees on a graph of size n, but for for trees of depth k, of the previous section. Specifically, this is $O(n)$ (for the kernelization) plus $2^k \cdot k^2$ using the dumbest of the branching algorithms.

As observed by Langston and his team for problems in sequence analysis (e.g. for DNA pattern matching, [AKLSS06]), and as articulated by Niedermeier and Rossmanith [NR00], better running times can be obtained by *interleaving* depth-bounded search trees and kernelization. That is, first kernelize, then either begin or perform a bounded search tree, and the *re-kernelize* the children, and repeat. This really does make a difference. In [Nie06] the 3-HITTING SET problem (See Exercise 9.4.1) is given as an example. An instance (I, k) of this problem can be reduced to a kernel of size k^3

in time $O(|I|)$, and the problem can be solved by employing a search tree of size 2.27^k. Compare a running time of $O(2.27^k \cdot k^3 + |I|)$ (without interleaving) with a running time of $O(2.27^k + |I|)$ (with interleaving).

We also remark that there are many strategies of reduction rules to *shrink* the kernel. These include things like *crown reductions* (Abu-Khzam et. al. [AKCF+04, AKLSS06]), and other crown structures which generalize the notion of a degree 1 vertex having its neighbours in the vertex cover, to more complicated structures which resemble "crowns" attached to the graph[8].

Clearly, another game is to seek the smallest kernel. For instance, we know by Nemhauser and Trotter [NT75] a size $2k$ kernel (i.e. the number of vertices) is possible for VERTEX COVER. Note that with this kernel size, we get an algorithm running in time $O(n) + g(2k, k) \cdot 2k$, where $g(2k, k)$ is the best we can do for finding a vertex cover in a graph with 2k many vertices, and seeking a vertex cover of size k. A natural question is "can we do better?". As we see in §9.5, modulo some complexity considerations, sometimes we can show lower bounds on kernels. (Clearly, if $P = NP$ then all have constant size kernels, so some assumption is needed.) Using a classical assumption called the *unique games conjecture* (something apparently stronger than P$\neq$NP) we can prove that there is an $\varepsilon > 0$ such that no kernel of size $(1 + \varepsilon) \cdot k$ can be achieved. We mention this in passing, and urge the reader to follow up this line.

The state of the art in the theory of kernelization can be found in Fomin, Lokshtanov, Saurabh, and Zehavi [FLSZ19]. This is a 529 page book devoted to kernelization so the reader can see that the area of kernelization is technically very deep and we have only addressed the very basic technique.

Remark 9.4.1. If L is *strongly* FPT, meaning that there is a machine accepting $L^{(k)}$ in time $f(k) \cdot |x|^k$, with f a computable function, then it is kernelizable (Exercise 9.4.5.). The kernel obtained may not be of polynomial size in k. It is not true that FPT=*polynomial size kernel* unless the complexity world is different than we think it is. We discuss this issue in §9.5.

---

[8] Specifically, a *crown* in a graph $G = (V, E)$ consists of an independent set $I \subseteq V$ (no two vertices in I are connected by an edge) and a set H containing all vertices in V adjacent to I. A crown in G is formed by $I \cup H$ iff there exists a size $|H|$ maximum matching in the bipartite graph induced by the edges between I and H, that is, every vertex of H is matched. It is clear that degree-1 vertices in V, coupled with their sole neighbours, can be viewed as the most simple crowns in G. If we find a crown $I \cup H$ in G, then we need at least $|H|$ vertices to cover all edges in the crown. Since all edges in the crown can be covered by admitting at most $|H|$ vertices into the vertex cover, there is a minimum size vertex cover that contains all vertices in H and no vertices in I. These observations lead to the reduction rules based on deleting crowns.

9.4.4 Exercises

Exercise 9.4.5 Show that a language $L \subseteq \Sigma^* \times \Sigma^*$ is strongly FPT iff it is kernelizable. (Hint: Suppose that M accepts $L^{(k)}$ in running time $f(k) \cdot |x|^k$ with f computable. Notice that $|x|^{k+1}$ eventually dominates $c \cdot |x|^k$ and hence $L^{(k)}$ can "eventually" be accepted in time $|x|^k$, and the kernel will be the exceptions.)

Exercise 9.4.6 Show that the following problem is FPT by the method of reduction to a problem kernel:

SET BASIS (Garey and Johnson, [GJ79])
Instance : A collection C of subsets of a finite set S.
Parameter : A positive integer k.
Question : Is there a collection B of subsets of S with $|B| = k$ such that, for every set $A \in C$, there is a subcollection of B whose union is exactly A?

Exercise 9.4.7 (Bodlaender [Bod11]) Construct in polynomial time a kernel of size $O(k \log k)$ for the following problem.

WEIGHTED MARBLES
Input : A sequence of marbles, each with an integer weight and a colour.
Parameter : A positive integer k.
Question : Can we remove marbles of a total cost $\leqslant k$, such that for each colour, all marbles of that colour are consecutive?

(Hint: Consider the following reduction rules: Rule 1: If we have two consecutive marbles of the same colour, replace it them by one with the sum of the weights. Call a colour *good* if there is only one marble with this colour. Rule 2: Suppose two successive marbles both have a good colour. Give the second the colour of the first. Apply these exhaustively. No two successive marbles will be of the same colour, and no two successive marbles will have a good colour. The number of marbles is at most twice $(+1)$ the number with a bad colour. This gives Rule 3: If there are at least $2k + 1$ bad coloured marbles, say No. Finally for a kernelization we also apply exhaustively Rule 4: If a marble has weight $> k + 1$, give it weight $k + 1$.)

Exercise 9.4.8 (Downey and Fellows [DF13]) Prove that the following problem is FPT.

GROUPING BY SWAPPING (Garey and Johnson, [GJ79])
Instance : A finite alphabet Σ, a string $x \in \Sigma^*$, and a positive integer k.
Parameter : A positive integer k.
Question : Is there a sequence of k or fewer adjacent symbol interchanges that transforms x into a string x' in which all occurrences of each symbol $a \in \Sigma$ are in a single block?

Exercise 9.4.9 (Niedermeier [Nie06]) Give a problem kernel for the following problem.

MATRIX DOMINATION
Instance : An $n \times n$ matrix with entries in $\{0, 1\}$. X.
Parameter: A positive integer k.
Question : Find a set C of at most k 1's such that all the other nonzero entries are in the same row or column as at least one member of C.

9.4.5 Iterative Compression

This technique was first introduced in a paper by Reed, Smith and Vetta in 2004 [RSV04] and more or less re-discovered by Karp [Kar11]. Although currently only a small number of results are known, it seems to be applicable to a range of parameterized minimization problems, where the parameter is the size of the solution set. Most of the currently known iterative compression algorithms solve *feedback set problems* in graphs, problems where the task is to destroy certain cycles in the graph by deleting at most k vertices or edges. In particular, the K-GRAPH BIPARTISATION problem, where the task is to find a set of at most k vertices whose deletion destroys all odd-length cycles, has been shown to be FPT by means of iterative compression [RSV04]. This had been a long-standing open problem in parameterized complexity theory.

Definition 9.4.2 (Compression Routine).
A *compression routine* is an algorithm that, given a problem instance I and a solution of size $k + 1$, either calculates a smaller solution of size k or proves that the given solution is of minimum size.

Here is the iterative compression routine for our much attacked friend VERTEX COVER. It is slightly modified in that we begin with an empty cover, but this is of no consequence. More importantly, in the algorithm below, if there is *no* size k solution at some stage, this certified "no" is a hereditary property. Therefore, as we only *add* vertices and edges, the instance cannot become a "yes"; so we are safe to abort.

The routine is to add the vertices of the graph, one at a time; if the relevant problem is of the compressible type, we will have a running solution for the induced subgraph of size k or at some stage get a certificate that there is no such solution for G. The algorithm modifies the current solution at each compression stage to make another one for the next stage.

Algorithm 9.4.10 (Iterative Compression for Vertex Cover)

1. Initially, set $C_0 = \emptyset$, and $V_0 = \emptyset$.

2. Until $V = V_s$ or the algorithm has returned "no", set $V_{s+1} = V_s \cup \{v\}$, where v is some vertex in $V - V_s$. The inductive hypothesis is that C_s is a vertex cover of $G[V_s]$ of size $\leqslant k$. Consider $G[V_{s+1}]$.

3. Clearly $C_s \cup \{v\}$ is a size $\leqslant k+1$ vertex cover of $G[V_{s+1}]$. If it has size $\leqslant k$, go to the next step. If not, we will seek to compress it. Let $\widehat{C} = C \cup \{v\}$. Consider all possible partitions of $\widehat{C}$ into two sets, $D \sqcup Q$. The idea is that the vertices of D will be *discarded*, and the vertices of Q *kept* in the search for the "next" vertex cover (of size k). We attempt to construct a suitable replacement H for the discarded part D of the vertex cover $C_s \cup \{v\}$.

   For each such partition, and for each edge $e = uv$ not covered (i.e. $u, v \notin Q \cup H$), if $u, v \in D$, then there is no possible vertex cover of $G[V_{s+1}]$ that contains no vertices of D, the vertices to be discarded from $C_s \cup \{v\}$.

   Otherwise, H must cover all the edges that have one endpoint in D and the other outside of Q. Thus, if uv is not covered by $H \cup Q$, find a vertex w, one of u or v not in D, and set $H := H \cup \{w\}$.

4. At the end of this if $H \cup Q$ has size $\leqslant k$, set $C_{s+1} = H \cup Q$ and go to step $s + 2$. Whenever $|H \cup Q|$ exceeds k then stop, and move to a new partition.

5. If we try all partitions, and fail to move to step $s + 2$, then stop and return "no."

There are at most 2^{k+1} different partitions to consider in each step, so the algorithm requires time at most $O(2^k |G|)$.

The parametric tractability of the method stems from the fact that each intermediate solution considered has size bounded by some $k' = f(k)$, where k is the parameter value for the original problem. It works very well with *monotone* problems, where if we get an intermediate no then the answer is definitely *NO*. Note that many minimization problems are not monotone in this sense. For example, a NO instance $(G = (V, E), k)$ for k-DOMINATING SET can be changed to a YES instance by means of the addition of a single vertex that is adjacent to all vertices in V.

Niedermeier [Nie06] has an excellent discussion of this technique, which would seem to have a lot of applications.

9.4.6 Exercises

Exercise 9.4.11 (Reed, Smith, Vetta [RSV04]) * (This is definitely not easy. But I have given solutions.)

(i) Prove the following Lemma.

Lemma 9.4.1. *Suppose that X is a minimum edge bipartization of $G =$ (V, E). Suppose that Y is any set of edges disjoint from X. Then the following are equivalent.*

a. *Y is an edge bipartization set for G.*

b. *There is a valid two-colouring Φ of V such that Y is an edge cut in $G - X$ between $\Phi^{-1}(B)$ and $\Phi^{-1}(W)$. That is, all paths between the black and the white vertices inc include an edge in Y.*

(ii) Using this lemma, show that the following problem is FPT using iterative compression. In fact, it is solvable in time $O(2^k km^2)$ for a graph with m edges. (The reader will see a full proof for this problem in the solutions. The method is akin to that used for VERTEX COVER.)

EDGE BIPARTIZATION

Instance : A graph $G = (V, E)$.

Parameter : An integer k.

Question : Can I delete k or fewer edges and turn G into a bipartite graph?

9.4.7 Not-Quite-Practical FPT Algorithms*

There are a number of distinctive techniques used in parameterized complexity which are "not-quite-practical" FPT algorithms, in the sense that the running times are not feasible in general, but can be in certain circumstances. Additionally, some can be randomized and ones using logical metatheorems can later admit considerable refinement in practice for a *specific* problem. These techniques include colour-coding and dynamic programming on bounded width graph decompositions. Since this survey is meant to be brief, I will only allude to these techniques. The reader should either read Downey and Fellows [DF13], or Cygan et. al. [CFK+16]. I will mention only one in some detail.

It is called *Colour Coding.*

Colour Coding is useful for problems that involve finding small subgraphs in a graph, such as paths and cycles. Introduced by Alon et. al. [AYZ94], it can be used to derive seemingly efficient randomized FPT algorithms for several subgraph isomorphism problems.

It remains in the "not quite practical" basket due to the large numbers needed to implement it. Here is a brief description of how the method works. We will apply the problem to k-PATH which seeks to find a (simple) path of k vertices in G. What we do is to *randomly* colour the vertices of the graph with k colours, and look for a *colourful* solution, namely one with k vertices of one of each colour.

In more detail, to find a colourful path of length k, k-COLOURFUL PATH can be solved by the following dynamic programming procedure. First we fix a vertex $s \in V(G)$ and then for any $v \in V(G)$, and $i \in \{1, \ldots, k\}$, we define

$$\mathcal{C}_s(i, v) = \{R \subseteq \{1, \ldots, k\} \mid G \text{ has a colourful } i\text{-path } P \text{ from}$$
$$s \text{ to } v \text{ such that } \mathbf{col}(P) = R\}.$$

The key to the efficiency is that $\mathcal{C}_s(i, v)$ stores sets of colours in paths of length $i - 1$ between s and v, instead of the paths themselves.

Clearly, G has a colourful k-path starting from s iff $\exists v \in V(G) : \mathcal{C}_s(k, v) \neq \emptyset$. The dynamic programming is based on the following relation:

$$\mathcal{C}_s(i, v) = \bigcup_{v' \in N_G(v)} \{R \mid R \setminus \{\chi(v)\} \in \mathcal{C}_s(i - 1, v')\}$$

Here $\chi(v)$ denotes the colour of v. Observe that $|\mathcal{C}_s(i, v)| \leqslant \binom{k}{i}$ and, for all $v \in V(G)$, $\mathcal{C}_s(i, v)$ can be computed in $O(m \cdot \binom{k}{i} \cdot i)$ steps (here, m is the number of edges in G). For all $v \in V(G)$, one can compute $\mathcal{C}_s(k, v)$ in $O(\sum_{i=1,\ldots,k} m \cdot \binom{k}{i} \cdot i) = O(2^k \cdot k \cdot m)$ steps. Thus, checking whether G coloured by χ has a colourful path of length k in $O(2^k \cdot k \cdot m \cdot n)$ steps (just apply the above dynamic programming for each possible starting vertex $s \in V(G)$). Now applying this method to randomly chosen colourings gives the desired randomized algorithm.

The two keys to this idea are

(i) we can check for colourful paths quickly. We only need to look at the relationship of the next colour class to a subset of the current one, rather than keeping track of many paths, and,
(ii) if there is a simple path then the probability that it will have k colours for a random colouring is $\frac{k!}{k^k}$ which is bounded below by e^{-k}.

Then, given (i) and (ii), we only need repeat process enough to fast probabilistic algorithm.

Theorem 9.4.12 (Alon, Yuster and Zwick [AYZ94]). k-PATH *can be solved in expected time* $2^{O(k)}|E|$.

Alon, Yuster and Zwick demonstrated that the colour coding technique could be applied to a number of problems of the form asking "is G' a subgraph of G?"

What has this to do with parameterized complexity? It turns out that this method can be used as an engine for FPT algorithms via a de-randomization process. A k-*perfect family of hash functions* is a family $\mathcal{F}$ of functions (colourings) taking $[n] = \{1, \ldots n\}$ onto $[k]$, such that for all $S \subseteq [n]$ of size k there is a $f \in \mathcal{F}$ whose restriction to is bijective (colourful). It is known that

k-perfect families of $2^{O(k)} \log n$ linear time hash functions exist. Using the hash functions allows us to deterministically simulate the probabilistic algorithm by running through the hashings, and hence gives a deterministic $2^{O(k)} |E| \log |V|$ algorithm for k-PATH. More such applications can be found in Downey and Fellows [DF13], and Niedermeier [Nie06], and Cygan et. al. [CFK+16].

The $O(k)$ in the exponent hides evil, and the de-randomization method at present seems far from practical, since the hashing families only begin at large numbers.

Note that the method does not work when applied to things like k-CLIQUE to be shown randomized FPT because (i) above *fails*. The important part of the dynamic programming method was that a path was represented by its beginning v_0 and some vertex v_i, and to extend the path only needed *local knowledge*; namely the colours used so far and v_i. This fails for CLIQUE, and would need $\binom{n}{s}$ at step s in the clique case.

There are many, many more methods used for proving problems FPT. They involve methods using automata on graphs which are treelike, or pathlike, methods based on logic, methods based on "well-quasi-ordering" theory, which can give proofs that certain problems are FPT and hence in P for a fixed k, where we have *no idea* what the algorithm actually is (an idea going back to Fellows and Langston [FL88]). We cannot do justice to these methods in a short chapter like this, and refer the reader to Downey and Fellows [DF13], Flum and Grohe [FG06], Gygan et. al. [CFK+16].

9.4.8 Exercises

Exercise 9.4.13 (Downey, Fellows and Koblitz [DF98]) Use colour-coding to show that the following problem is in P using a randomized algorithm, and hence in FPT using k-perfect hashing.

MULTIDIMENSIONAL MATCHING
Input : A set $M \subseteq X_1 \times \cdots \times X_r$, where the X_i are disjoint sets.
Parameter : The positive integer r and a positive integer k.
Question : Is there a subset $M' \subseteq M$ with $|M'| = k$, such that no two elements of M' agree in any coordinate?

Exercise 9.4.14 Show that the same is true for the following variations of k-PATH.

(i) k-CYCLE which asks for a cycle of length at least k,
(ii) k-EXACT CYCLE which asks for a cycle of length exactly k, and
(iii) k-CHEAP TOUR which takes a weighted graph and asks for a tour of length at least k and cost $\leqslant S$. The parameters here are S and k.

9.5 Kernelization Lower Bounds

We have met the natural and powerful technique of kernelization. Indeed if L is (strongly) FPT then it has a kernel. We have also seen that smaller kernels give faster algorithms. The natural question to ask is the following:

> Which problems have polynomial-size kernels?

Of course, if NP=P, then all parameterized versions of problems in NP have constant size kernels. So we will need some complexity hypothesis in order to establish any lower bounds. There were a number of *ad hoc* approaches to this question in the literature, but relatively recently, some general machinery was introduced to solve such questions. The machinery also allows us to prove results about the density of hard instances of NP-hard questions assuming some complexity assumptions.

In this section we develop the main engine for proving Theorem 9.5.3. We confine ourselves to the the the original framework of Bodlaender, Downey, Fellows and Hermelin [BDFH09]. There have been contributions over many years extending these techniques and we refer the reader to [DF13] and [FLSZ19].

We begin with the notion *distillation algorithms* for NP-complete problems, and then use an argument of Fortnow and Santhanam [FS11] to show that a distillation algorithm for any NP-complete problem implies the collapse of the polynomial hierarchy to at least three levels. Then we define a parametric-analog of a distillation algorithm which we call a composition algorithm. Following this, we show that if a compositional parameterized problem has a polynomial kernel, then its classical counterpart has a distillation algorithm.

Definition 9.5.1 (Harnik Naor [HN06], Bodlaender, Downey, Fellows and Hermelin [BDFH09]). An OR-*distillation algorithm* for a classical problem $L \subseteq \Sigma^*$ is an algorithm that

- receives as input a sequence $(x_1, \ldots, x_t)$, with $x_i \in \Sigma^*$ for each $1 \leqslant i \leqslant t$,
- uses time polynomial in $\sum_{i=1}^{t} |x_i|$,
- and outputs a string $y \in \Sigma^*$ with
  1. $y \in L$ if and only if $x_i \in L$ for some $1 \leqslant i \leqslant t$.
  2. $|y|$ is polynomial in $\max_{1 \leqslant i \leqslant t} |x_i|$.

We can similarly define AND-distillation by replacing the third item by

- $y \in L$ if and only if $x_i \in L$ for *all* $1 \leqslant i \leqslant t$.

That is, given a sequence of t instances of L, an OR-distillation algorithm gives an output that is equivalent to the sequence of instances, in the sense that a collection with at least one yes-instance (*i.e.* instance belonging to L) is mapped to a yes-instance, and a collection with only no-instances is mapped to a no-instance. Consider the case of satisfiability. In that case, it would see that this is easy as we could simply make $y = \vee x_i$. However, doing this would need an algorithm an algorithm using polynomial-time in the *(collective) total size of all instances*. The crux of distillation is that its output must be bounded by a polynomial in *the size of the largest of the instances from the sequence*, rather than in the total length of the instances in the sequence.

It seems highly implausible that NP-complete problems have distillation algorithms. If they do, it collapses polynomial time hierarchy.

Lemma 9.5.1 (Fortnow and Santhanam [FS11]). *If any* NP-*complete problem has an OR-distillation algorithm then* co-NP $\subseteq$ NP/*Poly and hence* $\Sigma_n^P = \Sigma_0^P$, *for all* $n > 3$.

Proof. Let L be an NP-complete problem with a OR-distillation algorithm $\mathcal{A}$, and let $\overline{L}$ denote the complement of L. We show that using $\mathcal{A}$, we can design a non-deterministic Turing-machine (NDTM) that, with the help of polynomial advice, can decide $\overline{L}$ in polynomial-time. This will that show co-NP$\subseteq$NP/*Poly*, and combined with Yap's theorem, Theorem 7.5.1, we get the result.

Let $n \in \mathbb{N}$ be a sufficiently large integer. Denote by $\overline{L}_n$ the subset of strings of length at most n in the complement of L, *i.e.* $\overline{L}_n = \{x \notin L : |x| \leqslant n\}$. By its definition, given any $x_1, \ldots, x_t \in \overline{L}_n$, the distillation algorithm $\mathcal{A}$ maps $(x_1, \ldots, x_t)$ to some $y \in \overline{L}_{n^c}$, where c is some constant independent of t. Any sequence containing a string $x_i \notin \overline{L}_n$ is mapped to a string $y \notin \overline{L}_{n^c}$. The main part of the proof consists in showing that there exists a set $S_n \subseteq \overline{L}_{n^c}$, with $|S_n|$ polynomially bounded in n, such that for any $x \in \Sigma^{\leqslant n}$ we have the following:

- If $x \in \overline{L}_n$, then there exist strings $x_1, \ldots, x_t \in \Sigma^{\leqslant n}$ with $x_i = x$ for some i, $1 \leqslant i \leqslant t$, such that $\mathcal{A}(x_1, \ldots, x_t) \in S_n$.
- If $x \notin \overline{L}_n$, then for all strings $x_1, \ldots, x_t \in \Sigma^{\leqslant n}$ with $x_i = x$ for some i, $1 \leqslant i \leqslant t$, we have $\mathcal{A}(x_1, \ldots, x_t) \notin S_n$.

Given such a set $S_n \subseteq \overline{L}_{n^c}$ as advice, a NDTM M can decide whether a given $x \in \Sigma^{\leqslant n}$ is in $\overline{L}$ as follows: It first guesses t strings $x_1, \ldots, x_t \in \Sigma^{\leqslant n}$, and checks whether one of them is x. If not, it immediately rejects. Otherwise, it computes $\mathcal{A}(x_1, \ldots, x_t)$, and accepts iff the output is in S_n. It is immediate to verify that M correctly determines (in the non-deterministic sense) whether

$x \in \overline{L}_n$. In the remaining part of the proof, we show that there exists such an advice $S \subseteq \overline{L}_{n^c}$ as required above.

We view $\mathcal{A}$ as a function mapping strings from $(\overline{L}_n)^t$ to $\overline{L}_{n^c}$, and say a string $y \in \overline{L}_{n^c}$ *covers* a string $x \in \overline{L}_n$ if there exist $x_1, \ldots, x_t \in \Sigma^{\leqslant n}$ with $x_i = x$ for some i, $1 \leqslant i \leqslant t$, and with $\mathcal{A}(x_1, \ldots, x_t) = y$. Clearly, our goal is to find polynomial-size subset of $\overline{L}_{n^c}$ which covers all strings in $\overline{L}_n$. By the pigeonhole principle, there is a string $y \in Y$ for which $\mathcal{A}$ maps at least $|(\overline{L}_n)^t|/|\overline{L}_{n^c}| = |\overline{L}_n|^t/|\overline{L}_{n^c}|$ tuples in $(\overline{L}_n)^t$ to. Taking the t'th square root, this gives us $|\overline{L}_n|/|\overline{L}_{n^c}|^{1/t}$ distinct strings in $\overline{L}_n$ which are covered by y. Hence, by letting $t = \lg|\overline{L}_{n^c}| = O(n^c)$, this gives us a constant fraction of the strings in $\overline{L}_n$. It follows that we can repeat this process recursively in order to cover all strings in $\overline{L}_n$ with a polynomial (in fact, logarithmic) number of strings in $\overline{L}_{n^c}$. □

We remark that the Fortnow-Santhanam argument has a number of strong consequences for the assumption co-NP $\not\subseteq$ NP/$Poly$. For example we have the following.

Theorem 9.5.1 (Buhrman and Hitchcock [BH08]). *If S is a set NP-hard under $\leqslant_m^P$-reductions, then S must have exponential density unless co-NP$\subseteq$ NP/$Poly$. That is, there is no $\varepsilon > 0$ such that $S_{=n} =_{\text{def}} \{x \in S \mid |x| = n\}$ has less than 2^{n^ε} for almost all n. Furthermore, if $\overline{\text{SAT}} \leqslant_m^P$ any set of of subexponential density, then co-NP$\subseteq$ NP/$Poly$.*

Proof. Let $g(x_1, \ldots, x_n)$ be an *and*-function (i.e. $g(x_1, \ldots, x_n) \in A$ iff $x_i \in A$ for all i) for some set A like $\overline{\text{SAT}}$. That is, as in the Fortnow-Santhanam proof, we can assume that we have a polynomial many one reduction f taking A to S. Then say that $z \in S$ is an NP-proof for $x \in A$ with $|x| = n$ iff there exist $x_1, \ldots, x_n$ such that for all $i, |x_i| = n$ and there exists an i with $x = x_i$ and $f(g(x_1, \ldots, x_n)) = z$.

Again, as the Fortnow-Santhanam proof we argue that there is a string z_1 that is an NP-proof for at least half the strings of $A_{=n}$, and then use recursion, with the advice being the sequence of z_i's. Note that if z is a NP-proof for precisely t strings $x \in A$, then

$$|\{(x_1, \ldots, x_n) \mid |x_i| = n \wedge f(g(x_1, i \ldots, x_n)) = z\}| \leqslant t^n.$$

Assuming that both f and g run in time n^c if we let $m_n = n^{2c^2}$ we see $|f(g(x_1, \ldots, x_n)) = z| \leqslant m_n$. Since S has subexponential density, for large enough n, $|S_{\leqslant m_n}| < 2^n$. Now let t be the largest integer such that some z_1 is a NP-proof for t elements of length n in A. Again as in the Fortnow-Santhanam proof, we observe that for each tuple $(x_1, \ldots, x_n)$, with $x_i \in A$ for all i, $f(g(x_1, \ldots, x_n))$ maps to some fixed z in $S_{\leqslant m_n}$, we see $t^n|S_{\leqslant m_n}| \geqslant |A_{=n}|^n$ which implies $t^n 2^n \geqslant |A_{=n}|$. Therefore $t \geqslant \frac{|A_{=n}|}{2}$. Now recursing on this process again gives the desired sequence $z_1, \ldots, z_p$ of advice and the argument follows as in the Fortnow-Santhanam proof. □

We remark that in [BH08], Buhrman and Hitchcock prove a stronger version of the result above.

Theorem 9.5.2 (Buhrman and Hitchcock [BH08]). *Sets S* NP-*hard under polynomial time Turing reductions making at most* $n^{1-\varepsilon}$ *queries must have exponential density unless* CO-NP $\subseteq$ NP/*Poly*.

We invite the reader to prove this in Exercise 9.5.5. The parametric version of distillation is the following.

Definition 9.5.2 (Bodlaender, Downey, Fellows and Hermelin [BDFH09]). An OR-*composition algorithm* for a parameterized problem $L \subseteq \Sigma^* \times \mathbb{N}$ is an algorithm that

- receives as input a sequence $((x_1, k), \dots, (x_t, k))$, with $(x_i, k) \in \Sigma^* \times \mathbb{N}^+$ for each $1 \leqslant i \leqslant t$,
- uses time polynomial in $\sum_{i=1}^{t} |x_i| + k$,
- and outputs $(y, k') \in \Sigma^* \times \mathbb{N}^+$ with
  1. $(y, k') \subset L$ if and only if $(x_i, k) \in L$ for some $1 \leqslant i \leqslant t$.
  2. k' is polynomial in k.

We can similarly define AND-composition.

A composition-algorithm outputs an equivalent instance to an input sequence in same sense of a distillation algorithm, except that now the parameter of the instance is required to be polynomially-bounded by the maximum of all parameters in the sequence, rather than the size of the instance bounded by the maximum size of of all instances.

We call classical problems with distillation algorithms OR, AND-*distillable problems*, and parameterized problems with composition algorithms OR, AND-*compositional problems*. There is a deep connection between distillation and composition, obtained via polynomial kernelization.

Lemma 9.5.2 (Bodlaender, Downey, Fellows and Hermelin [BDFH09]). *Let L be a compositional parameterized problem whose derived classical problem L_c is* NP-*complete. If L has a polynomial kernel, then L_c is distillable.*

Proof. Let $x_1^c, \dots, x_t^c \in \Sigma^*$ be instances of L_c, and let $(x_i, k_i) \in \Sigma^* \times \mathbb{N}^+$ denote the instance of L from which x_i^c is derived, for all $1 \leqslant i \leqslant t$. Since L_c is NP-complete, there exist two polynomial-time transformations $\Phi : L_c \to$ SAT and $\Psi :$ SAT $\to L_c$. We use the composition and polynomial kernelization algorithms of L, along with Φ and Ψ, to obtain a distillation algorithm for L_c. The distillation algorithm proceeds in three steps.

Set $k = \max_{1 \leqslant i \leqslant t} k_i$. In the first step, we take the subsequence in $((x_1, k_1), \dots, (x_t, k_t))$ of instances whose parameter equals ℓ, for each $1 \leqslant$

$\ell \leqslant k$. We apply the composition algorithm on each one of these subsequences separately, and obtain a new sequence $((y_1, k_1'), \ldots, (y_r, k_r'))$, where (y_i, k_i'), $1 \leqslant i \leqslant r$, is the instance obtained by composing all instances with parameters equaling the i'th parameter value in $\{k_1, \ldots, k_t\}$. In the second step, we apply the polynomial kernel on each instance of the sequence $((y_1, k_1'), \ldots, (y_r, k_r'))$, to obtain a new sequence $((z_1, k_1''), \ldots, (z_r, k_r''))$, with (z_i, k_i'') the instance obtained from (y_i, k_i'), for each $1 \leqslant i \leqslant r$. Finally, in the last step, we transform each z_i^c, the instance of L_c derived from (z_i, k_i''), to a Boolean formula $\Phi(z_i^c)$. We output the instance of L_c for which Ψ maps the disjunction of these formulas to, i.e. $\Psi(\bigvee_{1 \leqslant i \leqslant r} \Phi(z_i^c))$.

We argue that this algorithm distills the sequence $(x_1^c, \ldots, x_t^c)$ in polynomial time, and therefore is a distillation algorithm for L_c. First, by the correctness of the composition and kernelization algorithms of L, and by the correctness of Φ and Ψ, we have

$$
\begin{aligned}
\Psi(\textstyle\bigvee_{1 \leqslant i \leqslant r} \Phi(z_i^c)) \in L_c \ & \text{if and only if } \textstyle\bigvee_{1 \leqslant i \leqslant r} \Phi(z_i^c) \in \text{SAT} \\
& \text{if and only if } \exists i, 1 \leqslant i \leqslant r : \Phi(z_i^c) \in \text{SAT} \\
& \text{if and only if } \exists i, 1 \leqslant i \leqslant r : z_i^c \in L_c \\
& \text{if and only if } \exists i, 1 \leqslant i \leqslant r : (z_i, k_i'') \in L \\
& \text{if and only if } \exists i, 1 \leqslant i \leqslant r : (y_i, k_i') \in L \\
& \text{if and only if } \exists i, 1 \leqslant i \leqslant t : (x_i, k_i) \in L \\
& \text{if and only if } \exists i, 1 \leqslant i \leqslant t : x_i^c \in L_c.
\end{aligned}
$$

Furthermore, as each step in the algorithm runs in polynomial-time in the total size of its input, and since the output of each step is the input of the next step, the total running-time of our algorithm is polynomial in $\sum_{i=1}^{t} |x_i^c|$. To complete the proof, we show that the final output returned by our algorithm is polynomially bounded in $n = \max_{1 \leqslant i \leqslant t} |x_i^c|$.

The first observation is that since each x_i^c is derived from the instance (x_i, k_i), $1 \leqslant i \leqslant t$, we have $r \leqslant k = \max_{1 \leqslant i \leqslant t} k_i \leqslant \max_{1 \leqslant i \leqslant t} |x_i^c| = n$. Therefore, there are at most n instances in the sequence $((y_1, k_1'), \ldots, (y_r, k_r'))$ obtained in the first step of the algorithm. Furthermore, as each (y_i, k_i'), $1 \leqslant i \leqslant r$, is obtained via composition, we know that k_i' is bounded by some polynomial in $\ell \leqslant k \leqslant n$. Hence, since for each $1 \leqslant i \leqslant r$, the instance (z_i, k_i'') is the output of a polynomial kernelization on (y_i, k_i'), we also know that (z_i, k_i'') and z_i^c have size polynomially-bounded in n. It follows that $\sum_{i=1}^{r} |z_i^c|$ is polynomial in n, and since both Φ and Ψ are polynomial-time, so is $\Psi(\bigvee_{1 \leqslant i \leqslant r} \Phi(z_i^c))$. $\square$

Putting it all together we have the following.

Theorem 9.5.3 (Bodlaender, Downey, Fellows and Hermelin [BDFH09])
Unless CO-NP $\subseteq$ NP/*Poly, none of the following* FPT *problems have polynomial kernels:*

- k-PATH, k-CYCLE, k-EXACT CYCLE *(the last two ask whether there is a cycle of length $\geqslant k$ or length exactly k, respectively in a graph G).*

- k-PLANAR SUBGRAPH TEST *(does G have a planar subgraph with at least k vertices) and k-PLANAR INDUCED SUBGRAPH TEST.*
- k, σ-SHORT NONDETERMINISTIC TURING MACHINE COMPUTATION *(this is defined below).*

There are a number of other applications, but we would need to define new concepts, so will stick to these few. Here is the almost trivial lemma we need.

Lemma 9.5.3 (Bodlaender, Downey, Fellows and Hermelin [BDFH09]). *Let L be a parameterized graph problem such that for any pair of graphs G_1 and G_2, and any integer $k \in \mathbb{N}$, we have $(G_1, k) \in L \vee (G_2, k) \in L$ if and only if $(G_1 \cup G_2, k) \in L$, where $G_1 \cup G_2$ is the disjoint union of G_1 and G_2. Then L is compositional.*

Proof. Given $(G_1, k), \ldots, (G_t, k)$, take G to be the disjoint union $G_1 \cup \cdots \cup G_t$. The instance (G, k) satisfies all requirements of Definition 9.5.2.

As a direct corollary of the lemma above, we get that all of the following FPT problems are compositional: k-PATH, k-CYCLE, k-CHEAP TOUR, k-EXACT CYCLE. We remark that in §9.4.7, we showed that k-PATH was FPT by colour coding, and the others were exercises. Other examples include k-CONNECTED MINOR ORDER TEST, in FPT by Robertson and Seymour's Graph Minor Theorem, k-PLANAR CONNECTED INDUCED SUBGRAPH TEST, in FPT due to Eppstein [Epp95][9], and many others.

As an example of a non graph-theoretic problem which is distillable, consider the parameterized variant of the Levin-Cook generic NP-complete problem – the k, σ-SHORT NONDETERMINISTIC TURING MACHINE COMPUTATION problem. In this problem, we receive as input a non-deterministic Turing machine M with alphabet-size σ, and an integer k, and the goal is to determine in FPT-time, with respect to both k and σ, whether M has a computation path halting on the empty input in at most k steps. This problem can be shown to be in FPT by applying the algorithm which exhaustively checks all global configurations of M as per Cesati and Di Ianni [CI97].

Lemma 9.5.4 (Bodlaender, Downey, Fellows and Hermelin [BDFH09]). k, σ-SHORT NONDETERMINISTIC TURING MACHINE COMPUTATION *is compositional.*

Proof. Given $(M_1, (k, \sigma)), \ldots, (M_t, (k, \sigma))$, we can assume that the alphabet of each M_i, $1 \leq i \leq t$, is $\{1, \ldots, \sigma\}$. We create a new NDTM M, which is the disjoint union of all M_i's, in addition to a new unique initial state which is connected the initial states of all M_i by an ε-edge. (That is, by a non-deterministic transition that does not write anything on the tape, nor moves the head.) Note that M has alphabet size σ. Letting $k' = 1 + k$, the instance $(M, (k', \sigma))$ satisfies all requirements of Definition 9.5.2. $\square$

---

[9] We won't define these as they are only examples

What about AND composition and distillation? For a long time, the situation was as above. There was no "classical" evidence that AND-distillation implied any collapse until a remarkable solution to this conundrum was announced by a graduate student from MIT[10].

Theorem 9.5.4 (Drucker [Dru12]). *If* NP-*complete problems have* AND-*distillation algorithms, then* CO-NP $\subseteq$ NP/*Poly.*

Even though there are now shorter proofs, all proofs of Theorem 9.5.4 are complex and need concepts beyond the scope of the present book. AND-composition can be used to show that many important problems such as determining if a graph has "treewidth k" (meaning it has a certain structure widely applied in algorithmic graph theory) cannot have polynomial sized kernels unless collapse occurs, in spite of having an FPT algorithm for recognition of a fixed k.

9.5.1 Exercises

Exercise 9.5.5 Prove Theorem 9.5.2.

9.6 Another Basic Hardness Class and XP Optimality*

Space considerations preclude us from discussing one important programme in parameterized complexity, called XP-*optimality*. I will give a very brief description. This programme regards the classes like W[1] as artifacts of the basic problem of proving hardness under reasonable assumptions, and strikes at membership in XP.

The engine is a new class(!) of parameterized problems. We define the following.

$k \log n$ MINI-CIRCSAT
Input : A positive integers k and n in unary, and a Boolean circuit C of total description size n.
Parameter : A positive integer k.
Question : Is there any input vector x of weight $\leqslant k \log n$ with $C(x) = 1$?

We are lead to the following (which has higher level extensions to $M[t]$ for $t > 1$), by confining ourselves to boolean circuits of weft t.

---
[10] Now a professor at the University of Chicago.

Definition 9.6.1 (Chen, Flum and Grohe [CFG04]). The class $M[t] =_{\text{def}} k \log n$ Mini-$W[t]$ is the class of problem fpt-reducible to the following core problems.

$k \log n$ MINI-$H[t]$
Input : Positive integers k and n and m, and a weft t Boolean circuit C with $k \log n$ variables and of total description size m.
Parameter : A positive integer k.
Question : Is there any input vector x such that $C(x) = 1$?

It is not hard to see that $M[1] \subseteq W[1]$. Next we will need a very deep result from computational complexity theory:

Theorem 9.6.1 (Sparsification Lemma-Impagliazzo, Paturi and Zane [IPZ01]). *Let $d \geqslant 2$. Then there is a computable function $f : \mathbb{N} \to \mathbb{N}$ such that for each k and every instance φ of k-SAT, with n variables, there is a collection $\psi_1, \ldots, \psi_p$ of 2-CNF formulae such that*

1. $\beta = \vee_{i=1}^{p} \psi_i \equiv \varphi$.
2. $p \leqslant 2^{\frac{n}{k}}$.
3. $|\psi_i| \leqslant f(k) \cdot n$ for all i.

Moreover, the algorithm computing the ψ_i runs in time $2^{\frac{n}{k}} |\varphi|^{O(1)}$.

We can use this lemma to establish lower bounds assuming something called the *exponential time hypothesis*. We need a completeness result:

Theorem 9.6.2 (Cai and Juedes [CJ03], Downey et. al. [DECF+03]). MINI-3CNF SAT *is $M[1]$-complete under parameterized Turing reductions.*

Proof. It is quite straightforward to prove that MINI-3CNF SAT $\in M[1]$. (Exercise 9.6.6)

For the hard direction, it suffices to show that $k \log n$ MINI d-CNF SAT is reducible to MINI-3CNF SAT. Let φ be an instance of $k \log n$ MINI d-CNF SAT with size m, $n' = k \log n$ variables so that $k = \lceil \frac{n'}{\log m} \rceil$. We have f depending on d and can construct $\beta = \vee_{i=1}^{p} \psi_i$ depending on these parameters given by the Sparsification Lemma. We note

$$2^{\frac{k \log n}{k}} = \frac{k \log n}{2^{\lceil \frac{k \log n}{\log m} \rceil}} \leqslant m.$$

We have $p \leqslant m$, and β computed in time $m^{O}(1)$. Note that each ψ_i has at most $f(k) \cdot k \log n$ many clauses. Consequently there is a 3-CNF formula ψ_i' with at most $f(k) \cdot d \cdot k \log n$ many variables of length $O(f(k) \cdot k \log n \cdot m)$ such that ψ is satisfiable iff ψ' is satisfiable.

Therefore φ is satisfiable iff there is an i such that ψ_i' is satisfiable with $1 \leqslant i \leqslant p$. Moreover, $k_i' = \lceil \frac{|\mathrm{var}(\psi_i')|}{\log m} \rceil$ where $\mathrm{var}(\psi_i')$ denoted the variables in ψ_i', and hence $k_i' \leqslant O(\frac{f(k) \cdot d \cdot k \log n}{\log m}) \leqslant O(f(k) \cdot d \cdot k)$. Therefore we can decide if φ is satisfiable by querying the instances of ψ_i', m for $1 \leqslant i \leqslant p$ in MINI 3CNF SAT. $\square$

Lower bounds rely on the following refinement of P$\neq$NP. It reflects the belief that not only does P$\neq$NP, but it requires complete search infinitely often.

Exponential Time Hypothesis(Impagliazzo, Paturi and Zane [IPZ01])

There is no algorithm with running time $2^{o(n)}$ that determines, for a weft 1 boolean circuit C of total description size n, whether there is an input vector x such that $C(x) = 1$.

Equivalently, n-variable 3-SAT $\notin$ DTIME$(2^{o(n)})$.

The key result is the following:

Theorem 9.6.3 (Cai and Juedes [CJ01, CJ03]). MINI-CIRCSAT *is fixed-parameter tractable if and only if the ETH fails. That is, $M[1] =FPT$ iff ETH fails.*

Proof. One direction follows from the fact that a language decidable in time $2^{o(k \log n)}$ is FPT.

For the other direction, suppose we are given a weft 1 Boolean circuit C of size N, and suppose that MINI-CIRCSAT is solvable in FPT time $f(k)n^c$. Take $k = f^{-1}(N)$ and $n = 2^{(N/k)}$. Then, of course, $N = k \log n$. For example, if $f(k) = 2^{2^k}$ then $f^{-1}(N) = \log \log N$. In general, $k = f^{-1}(N)$ will be some slowly growing function of N, and therefore $N/k = o(N)$, and also $cN/k = o(N)$ since c is a constant, and furthermore by trivial algebra $cN/k + \log N = o(N)$. Using the FPT algorithm, we thus have a running time of

$$f(f^{-1}(N))(2^{N/k})^c = N2^{cN/k} = 2^{cN/k+\log N} = 2^{o(N)}$$

to analyze the circuit C. $\square$

It is possible to show that many of the classical reductions work for the miniaturized problems, which miniaturize the *size* of the input *and not some part of the input*; meaning we can often re-cycle classical proofs. Such methods allow us to show the following.

Corollary 9.6.1.

1. *The following are all $M[1]$-complete under fpt Turing reductions:* MINI-SAT, MINI-3SAT, MIN-d-COLOURABILITY, *and* MINI-INDEPENDENT SET.
2. *Hence neither* MINI-VERTEX COVER *nor any of these can have a subexponential time algorithm unless the ETH fails.*

Theorem 9.6.4 (Cai and Juedes [CJ01, CJ03]). *Assuming ETH (equivalently $M[1] \neq FPT$) there is no $O^*(2^{o(k)})$ algorithm for k-PATH, FEEDBACK VERTEX SET, and no $O^*(2^{\sqrt{k}})$ algorithm for PLANAR VERTEX COVER. PLANAR INDEPENDENT SET, PLANAR DOMINATING SET, and PLANAR RED/BLUE DOMINATING SET*

We can even refine this further to establish exactly which level of the M-hierarchy classifies the subexponential hardness of a given problem. For example, INDEPENDENT SET and DOMINATING SET which certainly are in XP. But what's the best exponent we can hope for for slice k?

Theorem 9.6.5 (Chen et. al [CCF$^+$05]). *The following hold:*

(i) INDEPENDENT SET *cannot be solved in time $n^{o(k)}$ unless* FPT$=$M[1].
(ii) DOMINATING SET *cannot be solved in time $n^{o(k)}$ unless* FPT$=$M[2].

In some sense we have returned to the dream we started with for computational complexity in Chapter 6; we can prove algorithms more or less optimal subject to complexity assumptions. We refer the reader to Downey and Fellows [DF13], Chapter 29, for more on this fascinating story.

9.6.1 Exercises

Exercise 9.6.6 Prove that MINI-3CNF SAT $\in M[1]$.

Chapter 10
Other Approaches to Coping with Hardness*

Abstract We look at several other approaches to coping with intractability. They include approximation algorithms, PTAS's, average case complexity, smoothed analysis, and generic case complexity. We look at both the positive techniques and the hardness theories.

Key words: Approximation algorithms, PTAS, DistNP, average case complexity, parameterized approximation, generic case complexity, smoothed analysis, APX completeness

10.1 Introduction

The apparent ubiquity of intractability led us to the theory of parameterized complexity in the last chapter. There are, of course, many other approaches to dealing with the problem of intractability. In this chapter we will give a brief survey of some of these methods. Each of them would deserve a book to themselves. Indeed, we will give references to such books. But we plan to give the basic ideas.

10.2 Approximation Algorithms

We have already met approximation schemes in Chapter 9, specifically §9.3.4, where we use parameterized complexity to analyse the existence of fully polynomial time approximation schemes. As we mentioned there, polynomial time approximation schemes (PTAS's) are one of the main traditional methods of coping with intractability. If you cannot solve the problem exactly, try to get a solution to a minimization or maximization optimization problem whose solution is *close*. As we did in Chapter 9, we will mainly look at minimization problems for concreteness. *Maximization* is dual. Thus, our goal is to minimize some relation R on the instance. For example, MINIMAL VERTEX COVER is the problem of finding the smallest vertex cover in a graph G.

Formally, a solution to a minimization problem is specified by a set $\mathcal{I}$ of instances I, a function $f : \mathcal{I} \to \mathcal{S}$ where $\mathcal{S}$ is a set of *feasible solutions* S, and a value function $c(I, S)$ saying what the value of the solution S is for instance I. For example, for MINIMAL VERTEX COVER I would be a graph, $f(I)$ a vertex cover of I, and $c(I, S)$ the number of vertices of the cover.

In the case of VERTEX COVER, a silly solution is to list all the vertices of G. Evidently, for this to be a useful concept, it would be reasonable to ask that the solution we find is "close" to the optimal one, OPT(I). Thus, for a good solution we want a "good" f picking a solution for each instance $I \in \mathcal{I}$.' If the problem is in P then we can compute the exact solution, but if the problem is NP-hard, then what do we mean by close?

10.2.1 Constant Distance Approximation

The strongest requirement we might ask is that we can compute a solution in polynomial time which is within a fixed constant distance from OPT(I). The reader might imagine that there are very few such examples, and that they are all trivial. However there are some notable examples.

PLANAR GRAPH k-COLOURING is one such example. For $k = 3$ we know the problem is NP-complete (Theorem 7.2.16). But famously, for $k = 4$ we simply say *yes* to the decision problem, and it has been shown that there is a polynomial time algorithm for computing a 4-colouring (Haken and Appel [HA77], and Haken, Appel and Koch [AHK77].) Using variations of the complex Haken and Appel methods, it has also been shown that planar graphs can be 4 coloured in quadratic time (i.e. on a register machine-see Thomas [Tho98]). However, if we are only interested in a constant approximation algorithm then using the 5-colour theorem is enough. By some basic graph theory due to Euler, such a planar G must have a degree 5 vertex and this allows an easy 5 colouring using the "Kempp Chain" Technique. The Kempp chain technique is concerned with connectivity and hence can be implemented in polynomial time. That is, use a recursive algorithm to find a vertex of degree 5, then 5-colour the remainder recursively, add the degree 5 vertex back and show either the neighbours only used 4 colours or one can be recoloured so that there are only 4. This is an approximation algorithm within a constant 2.

Another example is EDGE COLOURING of a graph G. If Δ is the largest degree, Vizing's Theorem [Viz64] says says that the number of colours needed to edge colour G is either Δ or $\Delta + 1$. Misra and Gries [MG92] showed that there is a polynomial time algorithm which $\Delta + 1$ colours any graph.

It is relatively easy to see that most problems which are NP-complete cannot have approximation algorithms to within a constant unless P=NP. Some such problems can be found in the exercises below.

10.2.2 Exercises

Exercise 10.2.1 Show that MAXIMAL INDEPENDENT SET cannot have an approximation algorithm to within a constant unless P=NP.

Exercise 10.2.2 Show that MINIMAL VERTEX COVER cannot have an approximation algorithm to within a constant unless P=NP.

Exercise 10.2.3 (Moret [Mor98]) Consider the following problem:

SAFE DEPOSIT BOXES
Instance : A collection of boxes $\{B_1, \ldots, B_n\}$ such that for each $j \in \{1, \ldots, n\}$, B_j contains $s_{i,j}$ of currency i for $i \in \{1, \ldots, m\}$, together with target amounts b_i for $1 \leqslant i \leqslant m$ and a target bound $k > 0$.
Question : Does there exists a collection of k or fewer boxes that contain sufficient currency to meet the targets b_i for all i?

Prove:
(i) For $m \geqslant 2$, the problem is NP-complete.
(ii) The problem has an approximation algorithm to within a constant.

10.2.3 Constant Approximation Ratios

Since problems seem to be unlikely to have approximation algorithms to within an additive constant, we need to look for weaker approximation guarantees. One method is to look at approximation to within a *multiplicative* constant, reflected in the definition of approximation ratio below.

Definition 10.2.1. If f is an algorithm solving a minimization problem, then for each instance I, the *approximation ratio* is the quantity

$$\frac{c(I, f(I))}{c(I, OPT(I))}.$$

If this is bounded by a constant for all instances $I \in \mathcal{I}$, then we say that f is a *constant approximation ratio* algorithm for the optimization problem.

The idea is that even if we cannot solve the problem exactly, maybe we can do no worse than (e.g.) double the true solution. One of the classic constant ratio approximation algorithms is for the problem BIN PACKING which we proved NP-complete in Theorem 7.2.25. We recall that it is the following problem:

BIN PACKING
Input : A finite set S, volumes $w : S \to \mathbb{N}$, bin size $B \in \mathbb{N}$, $k \in \mathbb{N}$.
Question : Is there a way to fit all the elements of S into k or fewer bins?

The optimization problem asks us to minimize the number of bins, the value being the number of bins needed. We will normalize and regard all the bins of size 1. Thus the objects u will have sizes $w(u) \leqslant 1$. A classic algorithm for this problem is *first fit*. This is the simplest algorithm imaginable. We start with a potentially infinite set of bins $\{B_i : i \in \mathbb{N}^+\}$. That is, we start with one bin and a bin making machine. The items are considered in order $u_1, u_2, \ldots$ and when they arrive, we simply put them into the first open bin that they fit into. If there is no open bin it will fit into, we manufacture another bin and place the object into that bin. We denote this as FF(I) for an instance I. Notice that with this protocol, there can be at most one open bin with contents of total size $\leqslant \frac{1}{2}$ at any stage of the algorithm. It follows that we can do at worst a constant approximation ration of 2. Actually with cleverer algorithms we can improve the ratio of 2 to $\frac{11}{9}$ (Johnson [Joh73]) and this has been an area of much research.

The reader should note also that the algorithm given is an *online algorithm*. For online algorithms we imagine the input as a stream of instances and at each stage, without any knowledge of the future, we have to construct a feasible solution. As data becomes ever larger, and situation more dynamic, online algorithms are becoming ever more important. We don't have space in this book to consider them further.

We will look at another approximation algorithm which does not use an online algorithm. It is for MINIMUM VERTEX COVER, and uses (at least in this proof) linear programming. Recall that a *kernel* for a parameterized (graph) problem takes an instance (G, k) and in polynomial time produces an instance (G', k') where $k' \leqslant k$ and $|G'| \leqslant f(k)$ such that (G, k) is a yes instance iff (G', k') is a yes instance. $f(k)$ is the *size* of the kernel.

Theorem 10.2.4 (Nemhauser and Trotter [NT75]). *Let* $G = (V, E)$ *have n vertices and m edges, with parameter k. In polynomial time[1] we can compute a graph (G', k) such that G has a vertex cover of size $\leqslant k$ iff G' does, and G' is a size $2k$ kernel. Thus* MINIMUM VERTEX COVER *has an approximation algorithm with approximation ratio 2.*

---

[1] The actual running time of the best algorithm is $O(\sqrt{n}m)$.

Proof (Bar-Yehuda and Even [BYE85]). We can formulate VERTEX COVER as an integer programming problem. Let $G = (V, E)$ be a graph. The problem is then formulated as follows. For each $v \in V$ we have a variable x_v with values $\{0, 1\}$. VERTEX COVER is then :

- Minimize $\sum_{v \in V} x_v$ subject to
- $x_u + v_v \geqslant 1$ for each $uv \in E$.

The classical relaxation of this is to make it a linear programming problem by having x_v rational valued with $x_v \in [0, 1]$.

We can define the following sets, by asking that the values be in the set $\{0, \frac{1}{2}, 1\}$.

$$S_1 = \{v \in V \mid x_v > \frac{1}{2}\}$$

$$S_{\frac{1}{2}} = \{v \in V \mid x_v = \frac{1}{2}\}$$

$$S_0 = \{v \in V \mid x_v < \frac{1}{2}\}.$$

We claim that *there is a minimal sized (integral)* VERTEX COVER S *with* $S_1 \subseteq S \subseteq S_1 \sqcup S_{\frac{1}{2}}$, $S \cap S_0 = \emptyset$ *and that* $(G[S_{\frac{1}{2}}], k - |S_1|)$, *the subgraph induced by* $S_{\frac{1}{2}}$ *with parameter* $k - |S_1|$, *is a kernel of size at most* $2k$. To see this, let S be a minimal vertex cover. Now, we consider $(S_1 - S) \cup (S - S_0)$. Now since we need to make $x_u + x_v \geqslant 1$ for all $uv \in E$, we conclude that $N(S_0) \subseteq S_1$, and hence $|S_1 - S| \geqslant |S \cap S_0|$, since S is minimal. We claim that $|S_1 - S| \leqslant |S \cap S_0|$. Suppose otherwise. Then if we define $\delta = \min\{x_v - \frac{1}{2} \mid v \in S_1\}$, and for all $u \in S \cap S_0$ set $x_u := x_u + \delta$ and for $v \in S_1 - S$, set $x_v := x_v - \delta$, then $\sum_{v \in V(G)} x_v$ would then be smaller if $|S_1 - S| > |S \cap S_0|$, which is impossible. Therefore $|S_1 - S| = |S \cap S_0|$ and we can substitute $S_1 - S$ for $S \cap S_0$ which is allowable since $N(S_0) \subseteq S_1$. Thus we obtain a vertex cover of the same size.

To see that $|S_{\frac{1}{2}}| \leqslant 2k$, as with many relaxations we note that the solution to the (objective function for the) linear programming problem is a lower bound for corresponding integer programming problem. Thus applying this to S_1, we see that any solution is bounded below by $\sum_{v \in S_{\frac{1}{2}}} x_v = \frac{|S_{\frac{1}{2}}|}{2}$. Thus if $|S_1| > 2(k - |S_{\frac{1}{2}}|)$ then this is a no instance. Thus we can take $G[S_{\frac{1}{2}}]$ as the kernel. $\square$

A more precise statement of the Nemhauser-Trotter Theorem can be extracted from the proof above as we invite the reader to prove in Exercise 10.2.7.

As we stated, a consequence of the Nemhauser-Trotter Theorem is that if we have a minimal vertex cover of size k then the proof above gives an approximate solution of size $2k$; that is the approximation ratio is 2. Can we do better? Clearly if P equals NP then the answer is yes. Assuming P$\neq$NP the answer will depend on how strong a complexity hypothesis we might consider.

There is a classical hypothesis called "The Unique Games Conjecture" (Khot [Kho02]) which implies that you cannot do better than 2. Anyway, whether you can do better than 2 is a very longstanding open question. Hence, showing that the answer is yes (or no!) will resolve an important conjecture in computational complexity theory.

10.2.4 APX Completeness

As with most (?all?) concepts in computational complexity which seek to understand problems whose decision versions are in NP, there are *hardness* classes. We have seen this in parameterized complexity with the W-hierarchy and the M-hierarchy. Later we will see it when we look at average case complexity in §10.3. In this section we look at hardness for approximation. Because the material is rather technical, we confine ourselves to briefly sketching the main ideas, and refer the reader to the books Vazirani [Vaz01] or Ausiello et. al. [ACG⁺99] for a detailed exposition.

As usual, the key ingredient is a notion of *reduction*. For approximation theory, we need a reduction that preserves approximations. It turns out that there are several notions of reduction: AP-, APX- and PTAS- reductions. One such reducibility is defined in Exercise 10.2.9.

Then for a notion of reduction R, what is needed is that the following lemma holds:

Lemma 10.2.1. *If $(\mathcal{I}_1, \mathcal{S}_1) \leqslant_R (\mathcal{I}_2, \mathcal{S}_2)$ and $(\mathcal{I}_2, \mathcal{S}_2)$ has a constant ratio approximation algorithm, so does $(\mathcal{I}_1, \mathcal{S}_1)$. The same holds for PTAS's for the next section.*

In Exercise 10.2.9, we invite the reader to verify that this lemma holds for one of the reducibilities $\leqslant_{AP}$.

Using these notions of reduction, we can now define APX-R-complete optimization problems as those whose decision problem is NP-complete, have constant approximation algorithms, and every other such language R-reduces to them.

The following problem is APX-complete for appropriate reductions R. Detailed proofs appear in for example, Vazirani [Vaz01] or Ausiello et. al. [ACG⁺99].

MAXIMUM WEIGHTED SATISFIABILITY
Input : A boolean formula φ
Question : What is the solution which maximizes the Hamming weight of the satisfied clauses?

The 3-SAT version has φ of maximum clause size 3.

Theorem 10.2.5. *For appropriate reducibilities R-,* MAXIMUM WEIGHTED SATISFIABILITY *and* MAXIMUM WEIGHTED 3-SATISFIABILITY *are both APX-R-complete.*

The proofs are quite technical. One major consequence is the use of APX-completeness when combined with what is called the PCP Theorem. The PCP theorem gives a complicated reduction which achieves many non-approximability results using P$\neq$ NP. For example, using it we can show that if P$\neq$ NP then there is no $\frac{1}{2}$-ratio polynomial time approximation algorithm for MAXIMUM WEIGHTED 3-SATISFIABILITY. This uses a kind of "gap creation" argument much more complicated than the simpler one we give below for TSP. Using this, if we R-reduce MAXIMUM WEIGHTED 3-SATISFIABILITY to some other problem we can deduce lower bounds on the "best" approximation ratio achievable assuming P$\neq$ NP. In §10.2.8, we will also give another method of proving non-approximability using the stronger hypothesis that $W[1] \neq$ FPT. We refer the reader to Vazirani [Vaz01] for a detailed use of this technique, and for many of the beautiful techniques which have been used in approximation. We also refer the reader to Ausiello et. al. [ACG$^+$99] particularly for the use of randomization in the design of approximation algorithms.

We remark that some early classical results allowed certain problems to be shown not to have constant ratio approximation schemes. We finish this section with one easy example (as usual) proven by "gap creation", which gives some intuition behind the use of the PCP Theorem.

Theorem 10.2.6 (Sahni and Gonzalez [SG76]). *If P $\neq$ NP then TSP cannot have a constant ratio approximation scheme.*

Proof. Suppose not and assume that TSP has a constant ratio approximation scheme with ratio c. Take a standard instance of HAMILTON CYCLE (which is NP-complete by Theorem 7.2.27). Now construct a new graph H with $V(G) = V(H)$, the edges of H forming a clique, but defining the weights of the edges of H via

$$w(x,y) = \begin{cases} 1 \text{ if } xy \in E(G) \\ c \cdot |V(G)| \text{ otherwise} \end{cases}$$

If G has a Hamilton cycle then the optimal solution is $|V(G)|$, and if not then the optimal solution must have weight $\geqslant c \cdot |V|$. The competitive ratio allows us to use this show HAMILTON CYCLE is in P. $\square$

10.2.5 Exercises

Exercise 10.2.7 The original statement of the Nemhauser-Trotter Theorem is the following.

Theorem 10.2.8 (Nemhauser and Trotter [NT75]). *Let* $G = (V, E)$ *have* n *vertices and* m *edges. In time* $O(\sqrt{n}m)$ *we can compute two disjoint sets of vertices,* C *and* H *such that*

1. *If* $D \subseteq H$ *is a vertex cover of* $G[H]$*, then* $X = D \sqcup C$ *is a vertex cover of* G.
2. *There is a minimum sized vertex cover of* G *containing* C.
3. *There is a minimum vertex cover for the induced subgraph* $G[H]$ *has size* $\geqslant \frac{|H|}{2}$.

Show that this can be obtained from the given proof.

Exercise 10.2.9 We say that $(\mathcal{I}_1, \mathcal{S}_1) \leqslant_{AP} (\mathcal{I}_2, \mathcal{S}_2)$ (*AP-reducible*) if there are functions f and g, and a constant d such that

(i) For each instance I of $\mathcal{I}$ and rational $r > 1$, $f(I, r) \in \mathcal{I}_2$.
(ii) For each $I \in \mathcal{I}$ and $r > 1$, if I has a feasible solution then $f(I, r)$ also has a feasible solution.
(iii) For each $I \in \mathcal{I}$ and $r > 1$, and any solution S_2 of $f(I, r)$, $g(I, S_2, r)$ is a solution to I in $\mathcal{S}_1$.
(iv) For each r, f and g are polynomial time computable by algorithms A_g and A_f.
(v) For any $I \in \mathcal{I}_1$, and any $r > 1$, and any solution S_2 of $f(I, r)$, the performance ration of $(f(I, r), S_2) \leqslant r$ implies the performance ratio of $(I, g(I, S_2, r)) \leqslant 1 + d(r - 1)$.

Prove that Lemma 10.2.1 holds for $\leqslant_{AP}$.

10.2.6 PTAS's

If we can have an approximation scheme with a constant approximation ratio, perhaps we can do even better and have ones with *arbitrarily good* approximation ratios.

We recall the following definition.

Definition 10.2.2. For a minimization problem of minimizing R on a class of instances $\mathcal{C}$, a *polynomial time approximation scheme* (PTAS) is an algorithm $A(\cdot, \cdot)$ which, fixing $\varepsilon > 0$, runs in polynomial time; and given and $G \in \mathcal{C}$, will compute a solution S, with

$$(1 + \varepsilon)|Q| \leqslant |S|.$$

Here Q is a minimal sized solution for the instance G.

In §9.3.4, we met these are showed that some problems have PTAS's which were quite infeasible assuming $W[1] \neq$ FPT. As we mentioned there some problems do have quite feasible PTAS's, and there is quite a rich literature around this area. We will limit ourselves to an analysis of the following which problem. It is a variation of PARTITION we met in Chapter 7.

MINIMUM PARTITION
instance : A finite set X of items and for each $x_i \in X$ a *weight* a_i.
Solution : A partition $Y_1 \sqcup Y_2 = X$.
Cost : $\max\{\sum_{x_i \in Y_1} a_i, \sum_{x_i \in Y_2} a_i\}$.

The following result is folklore, or at least very hard to track who proved it first.

Theorem 10.2.10. MINIMUM PARTITION *has a* PTAS.

Proof. We begin with an instance of MINIMUM PARTITION and a rational number $r > 1$. If $r \geqslant 2$ simply give $X = Y_1$ and $Y_2 = \emptyset$. If $r < 2$, then first sort the items as $\{x_1, \ldots, x_n\} = X$ in non-increasing order of weight. Define $k(r) = \lceil \frac{2-r}{r-1} \rceil$. Now find an optimal partition $Y_1 \sqcup Y_2 = \{x_1, \ldots, x_{k(r)}\}$. Thereafter, for $j = k(r)+1, \ldots, n$, in order of j if $\sum_{x_i \in Y_1} a_i \leqslant \sum_{x_i \in Y_2} a_i$, add x_j to Y_1, and otherwise add it to Y_2. This rule says add x_j to the part of the partition of least overall weight.

The verification that this process works is as follows. If $r \geqslant 2$, then the solution is valid since any feasible solution has measure at least half the total weight. For $r < 2$, let $a(Y_k) = \sum_{x_i \in Y_k} a_i$, and let $d = \frac{a(X)}{2}$. We can assume that $a(Y_1) \geqslant a(Y_2)$ without loss of generality, and we can suppose that the last action is of x_q added to Y_1. Then $a(Y_1) - a_q \leqslant a(Y_2)$. Therefore $a(Y_1) - d \leqslant \frac{a_q}{2}$, by adding $a(Y_1)$ to both sides and dividing by 2 If a_q has been inserted into Y_1 during the algorithm where we find an optimal partition $Y_1 \sqcup Y_2 = \{x_1, \ldots, x_{k(r)}\}$, then the solution is, in fact, optimal. If this is not the case, then $a_q \leqslant a_j$ for all j with $1 \leqslant j \leqslant k(r)$, and $2d \geqslant a_q \cdot (k(r)+1)$. As $a(Y_1) \geqslant d \geqslant a(Y_2)$, and the optimal solution has cost $c \geqslant d$, the performance ratio is

$$\frac{a(Y_1)}{c} \leqslant \frac{a(Y_1)}{d} \leqslant 1 + \frac{a_q}{2d} \leqslant 1 + \frac{1}{k(r)+1} \leqslant \frac{1}{\frac{2-r}{r-1}+1} = r.$$

The running time of the algorithm is polynomial since, first we nee to sort the items, taking $O(n \log n)$ (as is well-known). The preliminary optimal solution takes $n^{k(r)}$ since $k(r)$ is $O(\frac{1}{r-1})$, we get that the running time is $O(n \log n + n^{k(r)})$. $\square$

I believe it is open whether MINIMUM PARTITION can have an EPTAS or whether, for instance, parameterizing with $k = \frac{1}{r}$ will allow us to show that under the assumption $W[1] \neq$ FPT, there is no such EPTAS.

10.2.7 Parameterized Approximation

We have already seen the interactions of approximation and parameterized complexity in §9.3.4. Parameterized complexity also has another version of having an approximate solution. The idea is that for a given parameter, we have a method that either gives a "no" guarantee or an approximate solution. That is, we either see a certificate that there is no solution of size k or we find an approximate solution of size $g(k)$.

Definition 10.2.3. A parameterized approximation (minimization for definiteness) algorithm (FPT-approximation algorithm) for a problem $(\mathcal{I}, \mathcal{S})$ is an FPT algorithm A such that for each instance (I, k)

$$A(I, k) = \begin{cases} \text{a solution } S \in \mathcal{S} \text{ with } c(I, S) \leqslant g(k) \\ \text{"no", if } (I, S) \text{ has no solution of size } \leqslant k \end{cases}$$

The online algorithm FF for BIN PACKING is a good example of an FPT-approximation algorithm in that for any fixed k, would say you cannot pack into k bins or gives a packing into at most $g(k) = 2k$ bins. There are a number of examples from the theory of "graph minors" which give FPT-approximation algorithms, and one nice example is the following. A graph G is said to have an *interval representation* by specifying a function $f : V(G) \to P([0,1])$ where each $x \mapsto I_{f(x)}$ where I_x is an subinterval of $[0,1]$, such that for all x, y if $xy \in E(G)$ then $I_{f(x)} \cap I_{f(y)} \neq \emptyset$. Clearly this can be achieved by mapping $x \mapsto [0,1]$ for all x, but the definition gains traction by looking at the "maximal intersections" and bounding those.

Definition 10.2.4.

(i) A graph G is called an *interval graph* of width $\leqslant k$ if it has an interval representation such that at most k intervals have common intersection.
(ii) The minimum width of any interval representation of an interval graph is called the *pathwidth* of G.
(iii) G is called a graph of *pathwidth* k (or a partial interval graph) if G is a subgraph of an interval graph of width k.

Graphs of bounded pathwidth have many nice algorithmic properties, roughly speaking because you can work dynamic programming on the in-

terval representation. It would take us too far afield for look into properties
of interval graphs and bounded pathwidth graphs, so we refer the reader to either Diestel's [Die05] excellent book for a good treatment or perhaps Downey
and Fellows [DF13] which gives a reasonable source for the main ideas. It is
known that determining the pathwidth of a graph is NP-complete, but also
FPT for a fixed k. (Bodlaender [Bod96]). However, the algorithm is horrible,
and for a fixed k takes $O(2^{35k^3}|G|)$. Cattel, Dinneen, and Fellows [CDF96]
produce a very simple algorithm based on "pebbling" the graph using a pool
of $O(2^k)$ pebbles, that in linear time (for fixed k), either determines that the
pathwidth of a graph is more than k or finds a path decomposition of width
at most the number of pebbles actually used. The main advantages of this
algorithm over previous results are (1) the simplicity of the algorithm and (2)
the improvement of the hidden constant for a determination that the pathwidth is greater than k. The algorithm works by trying to build a binary tree
T as a topological minor (cf. Exercise 7.2.32) of G (represented by pebbles
which are slid along the edges) or builds a path decomposition of width 2^k.
It is known that G has pathwidth k iff it does not have the complete binary
tree as a topological minor. Again it would take us a bit too far from our
main goals to look at the detailed proof of this result. Hence we only state it
below:

Theorem 10.2.11 (Cattel, Dinneen, and Fellows [CDF96]). *Given a
graph H of order n and an integer k, there exists an $O(n)$ time algorithm
that gives a certificate which demonstrates that the pathwidth of H is greater
than k or finds a path decomposition of width at most $O(2^k)$.*

10.2.8 Parameterized Inapproximability

It is often possible to show that, for a given function f, a parameterized problem has no FPT approximation algorithm using f, at least assuming that
$W[1] \neq$ FPT or something of that ilk. A beautiful example of this was given
by Chen and Lin [CL19] who showed that, unless $W[1] =$FPT, DOMINATING
SET has no FPT-approximation algorithm with a constant multiplicative ratio; that is, no algorithm which gives a dominating set of size $p \cdot k$ or says
there is no size k dominating set. The authors remark that this result also
shows that the unparameterized DOMINATING SET has no classical constant
ratio approximation algorithm using elementary methods which are far simpler than the PCP machinery. The proof of this result is beyond the scope of
this text.

Sometimes it is possible to show that problems can be *very bad indeed* in
terms of parameterized approximation. In this section, we will look at problems that have the property that *any* parameterized approximation algorithm

would entail collapse of the W-hierarchy. We consider the following problem. Let g be a computable (increasing) function.

$g(k)$-APPROXIMATE INDEPENDENT DOMINATING SET
Input : A graph $G = (V, E)$
Parameter : An integer k.
Output : 'NO', asserting that no independent dominating set $V' \subseteq V$ of size $\geqslant k$ for G exists, or an independent dominating set $V' \subseteq V$ for G of size at most $g(k)$.

Theorem 10.2.12 (Downey, Fellows and McCartin [DFM06]). *There is no* FPT *algorithm for k-*APPROXIMATE INDEPENDENT DOMINATING SET* *unless* $W[2]$ =FPT. *That is, unless* DOMINATING SET *is FPT.*

Proof. We reduce from DOMINATING SET. Let (G, k) be an instance of this. We construct $G' = (V', E')$ as follows. V' is $S \cup C \cup T$:

1. $S = \{s[r, i] \mid r \in [k] \wedge i \in [g(k) + 1]\}$ (the *sentinel* vertices).
2. $C = \{c[r, u] \mid r \in [k] \wedge u \in V\}$ (the *choice* vertices).
3. $T = \{t[u, i] \mid u \in V \wedge i \in [g(k) + 1]\}$ (the *test* vertices).

E' is the set $E'(1) \cup E'(2) \cup E'(3)$:

1. $E'(1) = \{s[r, i]c[r, u] \mid r \in [k] \wedge i \in [g(k) + 1] \wedge u \in V\}$.
2. $E'(2) = \{c[r, u]c[r, u'] \mid r \in [k] \wedge u, u' \in V\}$.
3. $E'(3) = \{c[r, u]t[v, i] \mid r \in [k], u \in V, v \in N_G[u], i \in [g(k) + 1]\}$.

The idea here is that the $E'(2)$ are k groups of *choice* vertices. They form a clique. To each of these k cliques are *sentinel* vertices S, and the edges $E'(1)$ connect each sentinel vertex to the corresponding choice clique. Note that the sentinel vertices are an independent set. The other independent set is the *test* vertices T, and the edges $E'(3)$ connect the choice vertices to the test vertices according to the structure of G.

First we note that if G has a k-dominating set, $v_1, \ldots, v_k$, then the set $\{c[i, v_i] \mid i \in [k]\}$ is an independent dominating set for G'.

Second, suppose that g' has an independent dominating set D' of size $\leqslant g(k)$. Then, because of the enforcement by the sentinels, there must be at least one $c[i, u_i]$ per C, and hence exactly one as D' is independent. Let $D = \{u_i \mid c[i, u_i] \in D'\}$. We need to argue that D is a dominating set in G. Note that for all $v \in V$, there is a $t[v, i] \in T$, and by cardinality constraints, at least one is not in D'. This must correspond to a domination in G. $\square$

Many non-approximability results were given by Marx who proved the following. For simplicity, we will say that a minimization problem is FPT *cost inapproximable* there is no computable function g and FPT algorithm giving a no for one of size k or a solution of size $\leqslant g(k)$. Marx [Mar10] established that many problems complete for the parameterized complexity class $W[P]$ are FPT cost inapproximable. The method is to use certain gap inducing reductions. We refer the reader to [DF13].

10.2.9 Exercises

Exercise 10.2.13   (i) Show that G is an interval graph of width k iff its maximal cliques have size $\leqslant k + 1$ and can be ordered $M_1, \ldots, M_p$ so that every vertex belonging to two of these cliques also belongs to all cliques between them.

(ii) Hence, or otherwise, show that being an interval graph of width $\leqslant k$ can be decided in polynomial time.

Exercise 10.2.14 (Kierstead and Qin [KQ95]) Show that if G is an interval graph of width k, then First Fit colours G with at most $40k$ many colours[2] This is not easy. The method is to take an interval decomposition of the graph and imagine first fit interacting with this.

Exercise 10.2.15 (Askes and Downey [AD22]) * Show that there is an online approximation algorithm which colours a graph of pathwidth k with at most $3k + 1$ many colours[3] (This is a difficult research level result. There is a solution in the appendix.)

10.3 Average Case Complexity

The idea behind this approach, is to figure out which problems have algorithms which are hard/easy for "typical" instances. Most problems seem, in practice, to have lots of easy instances. How to capture this idea? Problems which are not in P will have an infinite complexity core of hard instances (Exercise 10.3.3), but they can be distributed in many different ways. We have already seen this in the the Buhrman and Hitchcock Theorem which showed that under the assumption that co-NP $\not\subseteq$ NP/Poly, then NP m-complete problems have exponentially dense hard instances.

Pioneered by Leonid Levin [Lev86], authors looked at what it meant for a problem to be easy *on average*. By way of motivation, consider the problem of graph colouring in k colours. Suppose that we posit that the graphs with n vertices are distributed uniformly at random. (That is, if G is one of the $2^{\binom{n}{2}}$ graphs with n vertices the probability of G be randomly chosen is $2^{-\binom{n}{2}}$. This is the standard "coin flip" probability.) Then, if we simply use backtracking,

---

[2] Naranaswamy and Subhash Babu [NSB08] gave an approximation ratio of 8. We remark that in 1988, Chrobak and Ślusarak [CS88] gave a lower bound of 4.4. Finally, in 2016 Kierstead, Smith, and Trotter [KST16] showed that the performance ratio of first fit on interval graphs is at least 5. The most recent word is by Dujmovic, Joret and Wood [DJW12], who have shown that first-fit will colour graphs of pathwidth k (i.e. not just interval graphs) with at most $8(k + 1)$ many colours. The precise best bound is not yet known.

[3] After Kierstead and Trotter [KT99] who proved this for interval graphs.

with overwhelming probability we will discover a k-clique for large values of n, since most graphs have *many* edges compared to the number of vertices. Thus there is a *constant* time algorithm for this problem. The precise analysis of this is quite hard. For $k = 3$ Wilf [Wil85] showed that, on average, the size of a backtracking (for any n) tree is 197. (Technically, the distribution is biased towards the negative solution.)

10.3.1 Polynomial Time on Average

We first need to define what we even mean by "polynomial time on average". We will assume that we have some probability distribution $\mu = \mu_n$ over instances of size n. That is, if we have $\text{Prob}(y)$ being the probability of y occurring, then the distribution is

$$\mu(x) = \sum_{y \leqslant_L x} \text{Prob}(y),$$

where $\leqslant_L$ is lexicographic ordering. As we discus below, this will be *computable in polynomial time* in a sense we define in Definition 10.3.2. the present we will treat this as intuitive. Our first guess for "polynomial time on average" might be the following. If $t(x)$ is a running time for an algorithm, the first guess would be to to use the sum

$$\sum_{|x|=n} t(x)\mu(x).$$

Unfortunately, this fails to be machine independent. For example, we might have an algorithm running in time in linear time on all but $2^{-0.1n}$ of instances of length n and runs in time $2^{0.9n}$ on the remainder. Then the sum above is polynomial, so we seem to run in polynomial time, on average. If we translated, for example, from a register machine to a Turing machine with quadratic overhead, then on the exceptional instances, now we run in time $2^{0.18n}$ time, and now the sum evaluates to be exponential. To overcome this problem, we declare that the longer an instance takes to compute, the rarer it should be. There should be a polynomial time trade-off between the running times $t(x)$ and fractions of hard instances.

Definition 10.3.1. We say that an algorithm is *polynomial time on average* (relative to a distribution μ) if there is an algorithm A running in time $t_A = t$, a constant c, and a constant $\varepsilon > 0$ such that for all d,

$$\mu(t(x) > d) \leqslant \frac{|x|^c}{d^\varepsilon}.$$

That is, after running for $t(x)$ many steps, the algorithm solves all but a $\frac{|x|^c}{t(x)^\varepsilon}$ fraction of the inputs of length $|x|$. The reader should note that the use of ε means that we make the definition machine independent.

To complete our notion of average case, we need to say what typical means for inputs. Thus we will need specify a pair (L, μ) where μ is a probability distribution with associated probability $\text{Prob}_\mu = \text{Prob}$. Since it seems foolish to define this for distributions which take a long time to compute, we use distributions that we can compute, and compute quickly on average. The reader should note that most distributions are, in fact, real-valued, but we can ask that they be approximable in polynomial time.

Definition 10.3.2. $f : \Sigma^* \to [0,1]$ is polynomial time computable iff there is an algorithm A and a constant c such that, in time $O(\langle |x|^c, d \rangle)$, the algorithm computes $q \in \mathbb{Q}$ (i.e. as $\frac{p}{r}$) with $|f(x) - q| < 2^{-d}$.

Thus we will be asking that μ is polynomial time in the sense of Definition 10.3.2. Note that asking that $\mu(x)$ is polynomial time is at least as strong as asking that $\text{Prob}(x)$ is computable in polynomial time since

$$\text{Prob}(x) = \mu(x) - \mu(\widehat{x})$$

where $\widehat{x}$ is the lexicographic predecessor of x. Sometimes people have used what are called *P-samplable* distributions where instead we only ask that we can draw random samples in polynomial time, that is, $\text{Prob}(x)$ is in polynomial time. (These two notions may not be the same, but can coincide under certain complexity assumptions and can differ under others.)

One of the principal troubles with this theory is that it is quite hard to figure out the "correct" distribution for natural problems. Generally, we seem to work with the uniform distribution which assumes that all instances of size n are equally likely. Since we will need to sample from an infinite set, we will need a way of sampling from Σ^*. To do this we say a polynomial time $\mu(x)$ computable distribution is called *uniform* if there exists a c and a distribution ν on $\mathbb{N}$ such that we can write $\mu(x) = \nu(|x|)2^{-|x|}$ and $\nu(n) \geqslant \frac{1}{n^c}$ almost always. The canonical distribution uses $\nu(|x|) = \frac{6}{\pi^2}|x|^{-2}$. This is certainly uniform, in this sense. (Exercise 10.3.4.)

To say why this is reasonable, we need a way of comparing distributions in polynomial time.

Definition 10.3.3.

(i) If μ_1 and μ_2 are two distributions. Then we say that μ_2 (polynomially) *dominates* μ_1, if there is a d such that $\mu_1(x) \leqslant |x|^d \mu_2(x)$.

(ii) If (L_1, μ_1) and (L_2, μ_2) are two distributional problems, and f is a polynomial time m-reduction from L_1 to L_2, then we say that μ_1 is dominated by μ_2 via f, if there exists a distribution ρ_1 dominating μ_1 with $\mu_2(y) = \sum_{f(x)=y} \rho_1(x)$.

The idea of (ii) is the following. If we consider the set of all x mapped to the same y under f, this will account for $\mu_2(y) = \sum_{f(x)=y} \rho_1(x)$. But since μ_1 is dominated by ρ_1, we see $\mu_2(y) \geqslant \frac{\sum_{f(x)=y} \rho_1(x)}{|x|^d}$. That is, the probability of y cannot be more than a polynomial factor smaller than the set of instances mapping to it.

The following lemma says that any polynomial time distribution is dominated by a uniform one. It also says that if the distribution has reasonable probabilities for all $x \in \Sigma^*$, then the distribution is closely approximated by a uniform one.

Lemma 10.3.1. *Suppose that μ is a polynomial time computable distribution. Then there exists a constant c and a 1-1, invertible polynomial time computable $g : \Sigma^* \to \Sigma^*$ such that, for all x, $\mu(x) \leqslant c \cdot 2^{-|g(x)|}$. Moreover, if there is a constant d, such that for all x, if $\mu(x) \geqslant 2^{-|x|^d}$, then there is some $c' \in \mathbb{N}$ such that for all x,*

$$c' \cdot 2^{-|g(x)|} \leqslant \mu(x) \leqslant c \cdot 2^{-|g(x)|}.$$

We will omit the somewhat technical proof, although we invite the reader to prove Lemma 10.3.1 in Exercise 10.3.5.

Definition 10.3.4. If (L, ρ) is a distribution problem with a polynomial distribution, we say that it is solvable in AvgP if there is an algorithm running in polynomial time on ρ-average accepting L.

The art here is finding the relevant distributions where a problem becomes tractable. Alternatively, we can formulate the problem so that it is only concerned with instances where the problem becomes average case tractable. An illustration of this is the paper [DTH86] by Dieu, Thanh and Hoa. They consider the problem below.

$\text{CLIQUE}(n^e)$

Input : A Graph G and a integers n, k.

Question : Does G have n vertices and $\leq n^e$ edges and a clique of size k?

They prove the following.

Theorem 10.3.1 (Dieu, Thanh and Hoa [DTH86]). *For* $0 < e < 2$, $\text{CLIQUE}(n^e) \in AvgP$.

Dieu, Thanh and Hoa also show that the problem is NP-complete. Their proof actually shows that the the algorithm of Tsukiyama, Ide, Ariyoshi and Shirakawa [TMAS77] runs in polynomial time on average. Unsurprisingly, the methods are taken from probability theory and are sufficiently complex that we will omit them and refer the reader to that paper.

Actually, this example shows one of the main drawbacks of the theory. It seems that applying the theory involves some complicated mathematics, and there are not too many genuine applications in the literature. Although average case complexity seems a fine idea in theory, as a practical coping strategy, it has proven quite unwieldy.

Average case complexity also has a hardness theory. The next subsection looks at this theory: when things seem hard on average.

10.3.2 DISTNP

We have introduced AvGP which is a average case version of P. There is an analog of NP; a hardness class. To define it we will need a notion of reduction.

Definition 10.3.5. We say that a distributional problem (L_1, μ_1) m-reduces to a distributional problem (L_2, μ_2), $(L_1, \mu_1) \leq_m (L_2, \mu_2)$, iff there is a polynomial time computable f and a constant d, such that

(i) $x \in L_1$ iff $f(x) \in L_2$.

(ii) There is an $\varepsilon > 0$ such that for all x, $|f(x)| \geq |x|^\varepsilon$. (This says that f is "honest".)

(iii) μ_2 dominates μ_1 via f.

The principal new ingredient is (iii) in this definition. The rationale is the following. Suppose we had an algorithm A_2 for (L_2, μ_2) and choose x randomly according to μ_1. Now we compute $f(x)$, and compute A_2 on $f(x)$. The idea is that we want that A_2 to run quickly on $f(x)$, and the condition (iii) says that slow running inputs are unlikely when we sample x and compute $f(x)$ because of the domination property discussed above.

Definition 10.3.6 (Levin [Lev86]). The class DISTNP is the class of distributional NP languages (L, μ) where μ is dominated by a polynomial time computable distribution.

Levin proved that there are problems complete under this notion of membership and reduction. It is an analog of our old friend we found in Theorem 7.2.1, which had echoes for other complexity classes: $L = \{\langle e, x, 1^n \rangle :$ some computation of N_e accepts x in $\leqslant n$ steps$\}$. Now we need to get the distribution involved, and this seems somewhat artificial.

Theorem 10.3.2 (Levin [Lev86]). *The following distributional problem is* DISTNP *complete:* (L, μ) *where L is as above, and*

$$\mu(\langle e, x, 1^n \rangle) = c \cdot n^{-2} |x|^{-2} |M_e|^{-2} 2^{-|M_e| - |x|}.$$

Proof. If $(L, \mu) \in$ DISTNP, then let M_e be a nondeterministic polynomial time machine accepting L. Now apply Lemma 10.3.1 and choose g with $\mu(x) \leqslant 2^{-|g(x)|}$. Define a new machine $N = M_{e'}$ as follows. On input σ, if $g^{-1}(\sigma)$ is defined, N simulates $M_e(g^{-1}(\sigma))$, and otherwise N rejects σ. Clearly M_e accepts x iff $M_{e'} = N$ accepts $g(x)$. Also there is some d such that N run on $g(x)$ completes in time $|x|^d$ for all x. Then we will map $\langle e, x, \mu \rangle$ to $\langle e', g(x), 1^{|x|^d} \rangle$. The mapping is injective, polynomial time computable, and g gives the domination property. $\square$

10.3.3 Livne's "Natural" Average Case Complete Problems

The construction in Theorem 10.3.2 above seems rather contrived. However, it seems that all "natural" NP-complete problems have polynomial time computable distributions which are DistNP-complete. This was more or less proved by Livne [Liv10] where he proved this assertion for SAT, CLIQUE, and HAMILTON CYCLE, and observed that his methods were sufficiently generic to work for any reasonable NP-complete problem. In this subsection, we will give a brief treatment of Livne's methods.

Livne gave a simple sufficient condition to the *decision* version of a language that it would have a *distributional* version which was DISTNP complete. Suppose that we wanted to do this for SAT and let (L, μ) be a DISTNP-complete problem such as that given via Theorem 10.3.2. Choose a polynomial time m-reduction h from L to SAT which is length preserving:

$$|y| \leqslant |x| \text{ iff } |h(y)| \leqslant |h(x)|.$$

We invite the reader to prove that such an h exists in Exercise 10.3.7. Livne then defines a new polynomial time m-reduction f, in place of h which explicitly codes z into $f(z)$. For example, if we use an underlying lexicographic ordering on formulae, and consider the formulae $e_0 = (x_0 \vee \neg x_0), e_1 = (x_1 \vee \neg x_1)$, so that (the coding of) $e_0 \leqslant_{lex} e_1$, with $|e_0| = |e_1|$, for $w \in \{0,1\}^*$, Livne defines

$$f(w) = e_{w_1} \wedge e_{w_2} \wedge e_{w_{|w|}} \wedge h(w),$$

where w_i is the i-th bit of w. Then

- Given $f(w)$ we can compute w.
- $w \leqslant_{lex} \widehat{w}$ implies $f(w) \leqslant_{lex} f(\widehat{w})$.
- Since e_0 and e_1 are tautologies, $f(w)$ preserves the truth values of $h(w)$.

We call f an *order preserving and polynomial invertible* reduction from L to SAT. Now we want to make a distributional version of SAT. We give it the distribution ν via

$$\nu(x) = \begin{cases} \mu(f^{-1}(x)) \text{ if } x \in \text{ra}(f). \\ 0 \text{ otherwise.} \end{cases}$$

Then $(L, \mu) \leqslant_m (\text{SAT}, \nu)$, because it maps each instance of L to one of SAT, and the probabilities are preserved[4]. It is also clear that ν is P-computable as μ is.

As Livne observes, the proof above utilizes padding, and lexicographic order. So all that is needed are similar methods for other problems to define length preserving invertible reductions. He provides details for CLIQUE and HAMILTON CYCLE. It is relatively clear that we could also do this for all of, for example, Karp's original 21 problems.

10.3.4 Exercises

Exercise 10.3.3 (Folklore) Show that if a computable language $L \subset \Sigma^*$ is not in P, then there is a computable infinite subset $X \subset L$, such that any decision algorithm for X must take more than a polynomial number of steps for almost all instances of $x \in X$?. (Hint: This is a straightforward diagonalization argument. Meet requirements R_e diagonalizing against φ_e accepting X in time $|x|^e$. Note that simulating L long enough will guarantee finding some witness x.)

Exercise 10.3.4 Prove that the distribution $\mu(x) = \frac{6}{\pi^2}|x|^{-2}2^{-|x|}$ on $\{0,1\}^*$ is computable and uniform.

---

[4] The reader might be somewhat dubious about whether this says a lot about "normal" distributional versions of the combinatorial problems; but the point is that the methods show that there is some *some* distributional version.

Exercise 10.3.5 Prove Lemma 10.3.1.

Exercise 10.3.6 Show that if $(L, \mu) \leqslant (L', \mu')$ and $(L', \mu') \in \mathrm{AvgP}$, then $(L, \mu) \in \mathrm{AvgP}$.

Exercise 10.3.7 Prove that if $L \in \mathrm{NP}$ then there is a reduction f from L to SAT such that $|y| \leqslant |x|$ iff $|h(y)| \leqslant |h(x)|$.

10.4 Generic Case Complexity

One relatively recent initiative is an area called generic case complexity which resembles average case complexity, but has a somewhat different flavour. Generic case complexity is still largely undeveloped. Its motivation comes from the idea of an algorithm being essentially good provided that it works well *given that* that we *completely ignore* a small collection of inputs[5]. Its origin was combinatorial group theory. We have seen in §4.3.5 that there are finitely presented groups with unsolvable word problems. However, in practice, it seems that you can solve the word problem in a typical finitely presented group on a laptop. In particular, the reader might imagine that we might have a partial algorithm for a problem which, for instance, *might not halt* on a very small set of inputs, but might run very quickly on the remainder. This partial algorithm would have *no* "average" running time as the run time would not even be defined for such a partial algorithm. This is the idea for generic case complexity, defined formally below. So far, it has proven that generic case complexity is rather simpler to apply than average case complexity. This is because generic case complexity feasibility proofs rely on less heavy duty probability theory, at least to this author's knowledge.

10.4.1 Basic Definitions

Let $\mathcal{I}$ be a set of inputs for some computational problem.

Definition 10.4.1. A *size* function is a function $s : \mathcal{I} \to \mathbb{N}^+$ with infinite range. For each $n \in \mathbb{N}^+$, $\mathcal{I}_n = s^{-1}(n)$ and $B_n = \cup_{j \leqslant n} \mathcal{I}_n$, is called the *ball of radius n*.

---

[5] There have been other examples of this in numerical analysis, such as Amelunxen and Lotz [AL17], which is entitled "Complexity theory without the black swans" which comes across as amusing to me, as I originate from Australia, where for the most part swans are black.

For each n we can think of B_n as having a ("atomic") probability distribution, μ_n generating what is called an *ensemble of distributions* $\{\mu_n\}$. The precise μ_n used varies with the application, but as a natural illustration, think of B_n as the ball of binary strings of length $\leqslant n$, and the distribution being the usual counting one. Another example would be that $\mathcal{I} = \mathbb{N}$ and $B_n = \{0, \ldots, n\}$. For this section, we will assume that the B_n are all finite and there is such and underlying μ_n.

Definition 10.4.2. For $X \subset \mathcal{I}$, $\rho(X) = \lim_n \mu_n(X \cap B_n)$ is called the *volume density of X* or *asymptotic density of X* if the limit exists.

Definition 10.4.3. We will say that $X \subset \mathcal{I}$ is *generic* if $\rho(X) = 1$, and *negligible* if $\overline{X}$ is generic.

The standard example would be where $\mathcal{I} = \mathbb{N}$ with counting density. Then X would have density 1 iff

$$\lim_n \frac{|\{X \cap \{0, \ldots, n\}|}{n} = 1.$$

For $\mathcal{I} = \{0,1\}^*$ with B_n the strings of length $\leqslant n$, then X has density 1 iff

$$\lim_n \frac{|\{\sigma \in X : |\sigma| \leqslant n\}|}{2^n} = 1.$$

Lemma 10.4.1. *Suppose that ρ is an asymptotic density defined on $\mathcal{I}$.*

(i) $\rho(\mathcal{I}) = 1$.
(ii) If X and Y are disjoint and both $\rho(X)$ and $\rho(Y)$ are each defined, then $\rho(X \sqcup Y) = \rho(X) + \rho(Y)$, and hence $\rho(\overline{X}) = 1 - \rho(X)$.
(iii) If X is generic and $X \subseteq Y$ then Y is generic.

We leave the proof to the reader (Exercise 10.4.3). It is important to understand how fast $\rho(X) \to 1$. The point is that for generic complexity to be most useful is that halting instances should happen most of the time for reasonably small problem instances.

Definition 10.4.4. If $\rho(X)$ exists for $X \subset \mathcal{I}$, let $\delta(n) = 1 - \mu_n(X \cap B_n)$.

(i) We say that X has *super-polynomial convergence* if $\delta(n) = o(n^{-k})$ for all $k \in \mathbb{N}$.

(ii) We say that X is *strongly generic* if it exhibits super-polynomial convergence.

(iii) We say that it has *exponential convergence* if $\delta(n) = 2^{-c \cdot n}$ for some $c \in \mathbb{N}^+$.

(iv) We say that X is *exponentially strongly generic* if it has exponential convergence.

We can finally give the main definitions.

Definition 10.4.5. Let $\mathcal{I}$ be a yes/no problem with size function s and let the language L represent the "yes" instances of $\mathcal{I}$. Then a partial computable function A *solves $\mathcal{I}$ generically* if

(i) $X(A) = \{x : A(x) \downarrow\}$ is a generic subset of $\mathcal{I}$, and

(ii) $x \in X(A)$ implies $A(x) = L(x)$.

That is, A decides $x \in L$ on a density 1 set of inputs, and whenever $A(x)$ halts, it is correct.

Definition 10.4.6 (Kapovich, Myasnikov, Schupp and Shpilrain [KMSS03]).

(i) We say that a language L is in GenP, if there exists and algorithm A is generically computing L in the sense of Definition 10.4.5, and there is a constant c such that, for all x,

$$\{x : A(x) \text{ halts in } \leqslant |x|^c \text{ many steps}\},$$

is generic in $\mathcal{I}$.

(ii) L is said to be in *SGP*, STRONGLY GenP, if

$$\{x : A(x) \text{ halts in } \leqslant |x|^c \text{ many steps}\},$$

is strongly generic in $\mathcal{I}$.

Here is one example drawn from the unpublished note [GMMU07]. Recall that SUBSET SUM is NP-complete by Theorem 7.2.24. SUBSET SUM is theoretically difficult as it is NP-complete, but easily solved in practice. We will use a variation of the definition from Chapter 7, which is easily seen to be equivalent. (Exercise 10.4.4.)

SUBSET SUM 2
Input : A finite set $S = \{w_1, \ldots, w_n\} \subset \mathbb{N}$, and a target integer c
Question : Is there a subset $S' \subseteq S$ such that $\sum_{w \in S'} w = c$?

We need to make this a distributional problem. If we fix a parameter b, and choose weights for S uniformly at random from $\{1, \ldots, b\}$, and c uniformly at random from $\{1, \ldots, c \cdot b\}$, then

$$\mathcal{I}_n = [1, nb] \times [1, b]^n.$$

Theorem 10.4.1 (Gilman, Miasnikov, Myasnikov, Ushakov [GMMU07]).
With parameter b, SUBSET SUM 2 is in GENP.

Proof. The polynomial time algorithm computes $w_1 + w_2 + w_3 + \ldots$ until one of the following occurs

1. $\sum_{i=1}^{j} w_i = c$. Accept.
2. $\sum_{i=1}^{n} w_i < c$. Reject.
3. $\sum_{i=1}^{j-1} w_i < c < \sum_{i=1}^{j} w_i$, then if there is some $k \in (j, n]$ with

$$\sum_{i=1}^{j-1} w_k = c, \text{ Accept.}$$

Otherwise, say *Don't Know*.

Clearly the algorithm is correct except in the case that it returns "Don't Know." Since c is sampled uniformly from $\{1, \ldots, nb\}$

$$\text{Prob}(\sum_{i=1}^{j-1} w_i < c < \sum_{i=1}^{j} w_i) = \frac{w_j - 1}{nb} \leqslant \frac{1}{n}.$$

The probability that there is no good w_k is $(\frac{b-1}{b})^{(n-j)}$. Therefore the probability we return "Don't Know" is

$$\leqslant \sum_{j=1}^{n} (\frac{1}{n})(\frac{b-1}{b})^{(n-j)} \leqslant \frac{b}{n} \to 0,$$

as $n \to \infty$. $\square$

With a wee bit more work, it is possible to eliminate the use of parameters to see that SUBSET SUM 2 is in GENP. This is an interesting example, in that it is observed in [GMMU07] that the measure of negligible strings only goes to 0 rather slowly.

The study of generic case complexity began in group theory; specifically finitely presented groups. On these there is a natural definition of size. Specifically if $\langle x_1, \ldots, x_n | R_1, \ldots, R_k \rangle$ is a finite presentation of a group G, then words in G are naturally presented as concatenations of the generators $\Sigma = \{x_1, \ldots, x_n\}$, that is members of Σ^*. Modulo some length considerations, the density of $X \subseteq \Sigma^*$ would now be

$$\rho(X) = \frac{|\{w \in X : |w| \leqslant n\}|}{|B_n|}.$$

Here B_n is the collection of words of length $\leqslant n$. This works, in fact, for any finitely generated group.

Here is an example of a strongly generic algorithm in this area. To make sense of this example, the reader will need to be familiar with some combinatorial group theory for the correctness proof. The reader unfamiliar with the concepts below should simply skip the details. Let $G = \langle a, b | R \rangle$ be any 2-generator group. It is a fact from combinatorial group theory (See [LS01]) that any countable group is embeddable in a 2-generator group. Thus, there are uncountably many such G! Let $F = \langle x, y \mid \ \rangle$ be the free group of rank 2. Define $H = G * \langle x, y \rangle := \langle a, b, x, y; R \rangle$, that is, the free product of G and F.

Theorem 10.4.2 (Kapovich, Myasnikov, Schupp, Shpilrain [KMSS03]). *The word problem for H is strongly generically solvable in linear time.*

Proof. (Sketch) Take a long word w on the alphabet $\{a, b, x, y\}^{\pm 1}$, e.g. $abx^{-1}bxyaxbby$.

Erase the a, b symbols, freely reduce the remaining word on $\{x, y\}^{\pm 1}$, and if any letters remain, *Reject*.

This partial algorithm gives no incorrect answers because if the image of w under the projection homomorphism to the free group F is not 1, then $w \neq 1$ in H.

$$abx^{-1}bxyaxbby \to x^{-1}xyxy \to yxy \neq 1$$

The successive letters on $\{x, y\}^{\pm 1}$ in a long random word $w \in H$ is a long random word in F which is not equal to the identity. So the algorithm rejects on a strongly generic set and gives no answer if the image in F is equal to the identity. □

The above is called the *quotient method* and can be used to show that G has a generic case solvable word problem for any $G = \langle X, R \rangle$, G subgroup of K of finite index for which there is an epimorphism $K \to H$, such that G is hyperbolic and not virtually cyclic. (Don't worry about these terms if they are not familiar. Enough to know that these are important classes of geometric groups.)

The methods of the paper [KMSS03] and follow up papers allowed us to show that the word problems are generically linear time decidable for 1-relator groups with $\geqslant 3$ generators. 1-relator finitely presented groups famously have solvable word problems by Magnus (see [LS01]). However, all known algorithms use double exponential time. The paper [KMSS03], also gives generic case algorithms for isomorphism problems and algorithms for braid groups. (Again, don't worry if you are not familiar with these concepts from group theory as they only serve as illustrations of important problems which have good generic case decidable solutions.) Certain finitely presented groups with undecidable work problems have generic case decidable word problems. For example, the Boone group, we constructed to give one with an unsolvable word problem, has a strongly generic case decidable word problem ([KMSS03]).

We remark that the algorithms above (in some sense) show the limitations of the concept in that *most* of the generic case algorithms are *trivial*. Also generic case complexity seems hard to apply to settings outside of group theory. For example, we can define what it means for a first order theory under a reasonable coding to be generically decidable with the obvious meaning. One way to do this would be to count the number of possible well-formed sentences of a given length (determined by counting the symbols in the expressions), and asking for an algorithm which generically decides if a sentence is valid, according to this notion of density. However, many workers have independently observed that a generic case decidable theory with this notion of density is decidable. (This is reasonably clear. If we have a natural size measure for formulae, then to decide if a sentence X is true, we can keep padding the sentence with $X \land (Y \lor (a \lor \neg a))$ for all longer sentences Y. This will have positive density and hence at some stage the generic case algorithm run upon these padded instances will need to answer yes or no, and it must be correct. All we need to do is wait.)

Generic case complexity also has a hardness theory. Not surprisingly, a "dense" variation of the halting problem has no algorithm which can generically decide it; and there is a GENP version of this.

We also remark that there are other variations on the theme of ignoring inputs. For example, X is called *coarsely* computable if there is an algorithm A halting on *all inputs* and *is correct* on a density 1 set, and combining the two, there is a notion of *coarsely generic*. All finitely presented groups have coarsely decidable word problems, exponentially fast (in terms of the density upon which they are correct) and in linear time. ([KMSS03])

It is not yet clear how important this area is, as it remains in a state of formation.

10.4.2 Exercises

Exercise 10.4.3 Prove Lemma 10.4.1.

Exercise 10.4.4 Prove that SUBSET SUM from Chapter 7 is polynomially equivalent to SUBSET SUM 2 above.

10.5 Smoothed Analysis

In this last short section, we will give a brief account of one recent area of computational complexity, called smoothed analysis, which has delivered techniques which (sometimes) seem to explain why *algorithms* seem to work better in practice than we would expect. We have seen some other explanations in the two preceding sections on average case and generic case complexity.

It is fair to say that the motivating example for smoothed analysis was an analysis of the SIMPLEX ALGORITHM for LINEAR PROGRAMMING. The SIMPLEX ALGORITHM was created in 1947 by George Dantzig for solving linear programming questions. These questions seek to optimize a set of solutions to multivariable linear inequalities over $\mathbb{R}$. The algorithm works by creating a simplex of feasible solutions and then systematically searches for a a solution by traversing the simplicial edges[6]. Because of this it is known to take exponential time in worst case. But, *in practice*, it seems to almost always run quickly, and for many years researchers were puzzled by that fact.

There were several attempts to explain the behaviour by average case analysis, such as Smale [Sma83]. In 2001, Spielman and Teng [ST01] and [ST09] suggested an explanation they called *smooothed analysis*. This suggestion has attracted considerable interest in the last two decades. Spielman and Tang's idea is that not only is the problem average case feasible, but also the distribution of hard instances has a certain very nice feature: *A hard instance is "very close" to an easy one.* That is, if we randomly perturb the problem then almost surely we will hit an easy instance.

The relevant definition if the following.

Definition 10.5.1 (Spielman and Teng [ST01]). Given an algorithm A for a problem $\mathcal{I}$, with running time (cost) $t(A, I)$ for an instance I. Then the *smoothed complexity* of $(A, \mathcal{I})$ is

$$Sm(A, \mathcal{I}, \varepsilon, n) = \sup_{|I|=n} \mathbb{E}_{y \in U_\varepsilon(I)}[\{t(A, y)\}],$$

where $U_\varepsilon(I)$ denotes an ε neighbourhood of I as determined by some distribution, and $\mathbb{E}$ is the expectation.

Notice that this is a mixture of worst case analysis since we look at all inputs I of length n, and average case by looking the expected behaviour within neighbourhoods of each I.

---

[6] It is not important for this motivating example to know the precise details, only to know that this is one of the most important and used algorithms for optimization in existence.

Definition 10.5.2. If the function $Sm(A, \mathcal{I}, n)$ is polynomial in n, we say that $(A, \mathcal{I})$ is in *smoothed polynomial time* with factor ε.

In Spielman and Teng's paper on the simplex method, U_ε is $\{I + r(\sigma)\}$ where $I \in [-1, 1]^n$ and $r(\sigma)$ ranges over random permutations given by the Gaussian distribution with standard deviation σ. The case of the SIMPLEX METHOD is reasonably complex, so we will try to illustrate the smoothed analysis method with a somewhat simpler example.

This concerns our old friend TSP, now in the unit square $[0, 1] \times [0, 1]$. There is a very successful heuristic called *local search* which looks at local perturbations to get an nearly optimal solution, and here we get what is called 2-OPT heuristic.

We are given n points $\{x_1, \ldots, x_n\}$ in $[0, 1] \times [0, 1]$, considered now as "real numbers" (obviously in practice they would be rational numbers). We want to find a tour of shortest length and are using the "taxi-cab" norm $\|x\| = |x^1| + |x^2|$, the sum of the length of the components of x. So we will need to find a permutation j so that $\{x_1 = x_{j(1)}, x_{j(2)}, \ldots, x_{j(n)}, x_1\}$ to we minimize

$$\sum_{i=1}^{n} \|x_{j(i+1)} - x_{j(i)}\|.$$

Here we regard the members of the cycle as connected by edges.

The 2-OPT Heuristic works as follows. Given a cycle, we will remove two edges which leaves two connected components. We can then reconnect the components to form a different cycle (this is unique). If this drops the value of $\sum_{i=1}^{n} \|x_{j(i+1)} - x_{j(i)}\|$, then the swap improves the solution. This method will require only a polynomial number of swaps as there are only n^2 possible swaps for each cycle. It is known that this strategy will need an exponential number of iterations to find an optimal solution in the worst case (Englert, Röglin, and Vöcking [ERV14]).

To understand this heuristic from a smoothed point of view, as with all distributional problems, we need to describe what we mean by a random perturbation. For simplicity we will assume that we will be using the uniform distribution over a subset of the unit square with area d, but as pointed out in [ERV14], the analysis below is robust for any reasonable distribution so long as measure is not concentrated.

Theorem 10.5.1 (Englert, Röglin, and Vöcking [ERV14]). *The smoothed complexity of the 2-OPT heuristic above is polynomial*[7]

Proof. The protocol begins with a cycle and removes edges uv, xy and replaces them with xu, vy. The decrease in the TSP value is

---

[7] The expected number of iterations is $O(d^{-1} n^6 \log n)$.

$$||u - v|| + ||x - y|| - ||u - x|| - ||v - y||.$$

Coordinatizing, this yields

$$|u^1-v^1|+|u^2-v^2|+|x^1-y^1|+|x^2-y^2|-|u^1-y^1|-|u^2-y^2|-|v^1-x^1|-|v^2-x^2|.$$

Assuming this is chosen by the algorithm, it will be a decrease, one that is independent of the rest of the cycle. We want this decrease to be significant, so we will call the decrease ε-*small* if it is between 0 and δ.

Claim. For a perturbed instance and $\varepsilon > 0$, the probability that there is an ε-small swap is $O(\varepsilon \cdot \frac{n^4}{d})$.

We first prove the claim. Considering all improving swaps, there are only $O(n^4)$ distinct swaps. We first establish a bound of $O(\frac{\varepsilon}{d})$ on the probability that a *fixed swap* is ε-small. Consider the swap xy, uv as above. Then $|u^1 - v^1|+|u^2-v^2|+|x^1-y^1|+|x^2-y^2|-|u^1-y^1|-|u^2-y^2|-|v^1-x^1|-|v^2-x^x|$, is linear in 8 variables with each variable have a coefficient from $\{-2,0,2\}$. There are $576 = (4!)^2$ linear combinations (i.e. all pairs in differing orders). Thus, the expression is ε-small iff one of these combinations lies in $(0, \varepsilon)$. For each fixed linear combination, we claim that the probability that the linear combination lies in $(0,\varepsilon)$ is $\frac{\varepsilon}{2d}$. Since all the coefficients cannot be 0 to give a ε-small improvement, as can assume that at least one is ± 2. The values of u^1, plus value of the linear combination of the other 7 variables together with $2u^1$ or $-2u^1$ to lie in $(0, \varepsilon)$ is at most $\frac{\varepsilon}{2d}$ since we are using the uniform distribution. Finally, note that the $O(n^4)$ swaps are all independent events, and hence we get a bound of $O(n^4 \frac{\varepsilon}{2d}) = O(d^{-1}n^4 \cdot \varepsilon)$, as required to establish the claim.

Now we can finish the proof of the Theorem. We first note that in $[0,1] \times [0,1]$, the maximal distance between any two points using the norm $||$ is 2, and hence any cycle has length at most $2n$. If there are no ε-small swaps, then the local search will halt within $\frac{2n}{\varepsilon}$ many iterations. The worst case number of iterations is bounded above by $n!$ as this is the number of different cycles. We estimate how many iterations N are needed. Let $\varepsilon = \frac{2n}{N}$. The expected number of iterations is $\sum_{N=1}^{n!}$ Prob[The number of iterations $\geq N$]. This sum is bounded above by $\sum_{N=1}^{n!}$ Prob[There is a $\frac{2n}{N}$-small swap]. By Claim 10.5, this is

$$\leq \sum_{N=1}^{n!} O(\frac{2n}{N} \cdot \frac{n^4}{d}) = \sum_{N=1}^{n!} O(\frac{n^5}{N \cdot d}).$$

Now,

$$\sum_{N=1}^{n!} O(\frac{n^5}{N \cdot d}) = O(d^{-1} \cdot n^5(1 + \frac{1}{2} + \frac{1}{3} + \cdots + \frac{1}{n!})).$$

The growth rate of the series $1 + \frac{1}{2} + \frac{1}{3} + \cdots + \frac{1}{n!}$ is known from some basic calculus, and is $O(n \log n)$. That is, the expected number of iterations is bounded by $O(d^{-1} n^6 \log n)$. $\quad \square$

With improved analysis, it is shown in [ERV14], that the bound n^6 can be improved to n^4.

Smoothed analysis has been applied in a number of domains. Spielman has a "Smoothed Analysis Homepage" with links to a number of such papers. Some I would suggest are Beier and Vöcking's [BV04] smoothed analysis of KNAPSACK, the smoothed analysis of three combinatorial problems by Banderier, Beier, and Mehlhorn [BBM03], (although there is some question about the distribution chosen for QUICKSORT) and the smoothed analysis of Gaussian Elimination by Sankari, Spielman and Teng [SST06]. Gaussian Elimination and Condition Numbers are also looked at for another kind of smoothed analysis in Amelunxen and Lotz [AL17].

Whilst she was an undergraduate, Abigail See gave a nice, if perhaps dated, summary in her essay on smoothed analysis and its applications to machine learning [See14].

> "Smoothed analysis appears to be very difficult to carry out; most papers proving the smoothed complexity of an algorithm are quite long and intricate. It would be easier if smoothed complexity was more integrated into existing complexity theory. For example, is there a notion of completeness for smoothed complexity? Can we relate smoothed complexity to hardness of approximation?"

10.6 Summary

In this chapter we gave brief and often sketchy snapshots of several methods of coping with intractability. These were not meant to be complete treatments and we hope that the reader will be inspired to seek out fuller details for themselves. One slightly dated but very interesting article is Impagliazzo [Imp95], where the Impagliazzo discusses the implications of the validity of various complexity hypotheses about average case complexity, and proposes "Five Worlds" and as he says:

> "In each such "world", we will look at the outcomes of these questions on algorithm design for such areas as articicial intelligence and VLSI design, and for cryptography and computer security".

I believe that we are far from understanding the algorithmic behaviour of natural problems, and why things work so well in practice. We are far from understanding, for example, the unreasonable efficiency of SAT SOLVERS (see Ganesh and Vardi [GV21]), or the amazing efficacy of modern machine learning tools like transformers (something beyond the scope of this book). The reader should feel inspired to solve these vexing questions. Good luck!

Part IV
Selected Solutions to Exercises

Chapter 11
Solutions

11.1 Chapter 1

Exercise 1.2.2.

1. Use, for example, Gödel numbers. If $\Sigma = \{a_1, \ldots, a_n\}$ then let a_i be represented by i, i.e. $\#(a_i) = i$, and then a string $x_1 \ldots x_n$ could be represented by $2^{\#(x_1)} 3^{\#(x_2)} \ldots (p_n)^{\#(x_n)}$, where p_n denotes the n-th prime.
2. The same technique works for $\Sigma = \{a_1, a_2, \ldots\}$.
3. To specify the periodic sequence, you only need to specify the finite subsequence which generates it.
4. For eventually periodic, you'd need to specify the finite initial segment, and the the finite periodic part.

Exercise 1.2.3 and 1.2.4 both can be done with Gödel numbers, the first one speficying $\overline{A}$ for cofinite A, and the second the coefficients of the polynomial.

Exercise 1.2.5. This is purely algebraic. Is f onto? If $x = 0$, $y = 0$, then

$$f(x,y) = \frac{1}{2}(0^2 + 2.0.0 + 0^2 + 3.0 + 0) = 0$$

So if $S \subseteq \mathbb{N}$, and $S = \{m \in \mathbb{N} : \exists x', y' \text{in} \mathbb{N}, F(x', y') = m\}, 0 \in \mathbb{N}$.

Suppose that $k \in S$, so $\exists x', y' \in \mathbb{N} : f(x', y') = k$. Then what values for x and y do we need so that $f(x,y) = k+1$?

Either $y' = 0$, or it isn't. If it is, then $2k = (x')^2 + 3x'$. What values for x and y are needed so that

$$2(k+1) = x^2 + 2xy + y^2 + 3x + y?$$

We can choose $x = 0$ and $y = x' + 1$. Then

$$
\begin{aligned}
(x+y)^2 + 3x + y &= (0 + x' + 1)^2 + 3.0 + x' + 1 \\
&= (x')^2 + 2x' + 1 + x' + 1 \\
&= (x')^2 + 3x' + 2 \\
&= 2k + 2 \\
&= 2(k+1)
\end{aligned}
$$

If $y' \neq 0$, then $2k = (x' + y')^2 + 3x' + y'$. So choose $x = x' + 1$, $y = y' - 1$ to get

© The Author(s), under exclusive license to Springer Nature Switzerland AG 2024
R. Downey, *Computability and Complexity*, Undergraduate Topics
in Computer Science, https://doi.org/10.1007/978-3-031-53744-8_11

$$(x+y)^2 + 3x + y = (x' + 1 + y' - 1)^2 + 3(x' + 1) + y' - 1$$
$$= (x' + y')^2 + 3k' + 3 + y' - 1$$
$$= (x' + y')^2 + 3x' + y' + 2$$
$$= 2k + 2$$
$$= 2(k + 1)$$

$k \in S \Rightarrow k + 1 \in S$, and so by the principle of mathematical induction, $S = \mathbb{N}$. Therefore f is onto.

Is f 1-1? Suppose that for some $k \in \mathbb{N}$, the above process gives $a, b \in \mathbb{N}$ such that $f(a, b) = k$, and there exist $c, d \in \mathbb{N}$, where $(a, b) \neq (c, d)$ such that $f(c, d) = k$ also.

Suppose $a + b = c + d$. Then

$$2k = (a + b)^2 + 3a + b = (c + d)^2 + 3c + d$$

$a + b = c + d$ means that $(a + b)^2 = (c + d)^2$, so $3a + b = 3c + d$. Rewriting this, we get that $2a + (a + b) = 2c + (c + d)$, so $a = c$ and therefore $b = d$.

If $a + b \neq c + d$, it must be that either $a + b > c + d$ or $a + b < c + d$. Without loss of generality, suppose that $a + b > c + d$. By our proof that f is onto, we say that f maps ordered pairs (x, y) to $\mathbb{N}$ by first mapping all pairs that add to 0, then all pairs that add to 1, 2, ... $a + b$. $c + d < a + b$ so (c, d) must have occurred earlier in this list than (a, b). Therefore $f(c, d) < f(a, b)$. This gives a contradiction.

It follows that there is no such other pair (c, d), so f is indeed 1-1. And since f is onto and 1-1, it is a bijection.

Exercise 1.3.4 Use diagonalization: Suppose that the collection is countable, and let $b_i = \sum_{n=0}^{\infty} a_{i,n} X^n$ be the i-th binary power series. Now define $c(X) = \sum_{n=0}^{\infty} c_n X^n$ with $c_n = 0$ if $a_{n,n} = 1$ and $c_n = 1$ if $a_{n,n} = 0$. Then $c(X)$ is a binary power series and it does not equal any b_i, since it will differ in position i.

Exercise 1.3.5 Suppose that $S = \{n \mid x_n \notin Y_n\}$ is finite. Let $Q = \{x_n \mid n \in S\}$. Then Q is finite and hence $Q = Y_n$ for some n. This leads to a contradiction.

Exercise 1.3.7 We know that $|\mathbb{R}|$ is the the same as that of the collection C of infinite binary sequences. So it is enough to show that $C \times C$ has the same cardinality as C. For two binary sequences $x = x_0, x_1 \ldots$, and $y = y_0 y_1 \ldots$, map $(x, y) \mapsto x_0 y_0 x_1 y_1 \ldots$. This is injective and shows that $|C \times C| \leqslant |C|$.

11.2 Chapter 2

Exercise 2.2.2 1. $L = L(\alpha)$ with $\alpha = b^* \cup (b^* ab^*) \cup (b^* ab^* ab^*) \cup (b^* ab^* ab^* ab^*)$.
2. $L = L(\beta)$ with $\beta = (a^* ba^* ba^*)^*$.

Exercise 2.2.6 1. $(a^* b^*)^* (b^* a^*)^* = (a \cup b)^*$. Prove this by induction. Clearly λ is in both. if $\sigma \in (a \cup b)^*$, then $\sigma a \in (a^* b^*)^* (b^* a^*)^*$, since the $(b^* a^*)^*$ allows one more a, and similarly b.

Exercise 2.4.5 I strongly suggest you draw a diagram:

$Q_0 = \{q_0\}, (Q_0, a) \vdash \{q_0, q_1\} = Q_1, (Q_0, b) \vdash \{q_3\} = Q_2, (Q_1, a) \vdash \{q_0, q_q, q_2, q_3\} = Q_3, (Q_1, b) \vdash \{q_1, q_3\} = Q_4, (Q_2, a) \vdash \{q_0\} = Q_0, (Q_2, b) \vdash Q_2, (Q_3, a) \vdash Q_3, (Q_3, b) \vdash \{q_3, q_1, q_2\} = Q_5, (Q_4, a) \vdash \{q_0, q_3, q_2\} = Q_6, (Q_4, b) \vdash Q_4, (Q_5, a) \vdash Q_6, (Q_5, b) \vdash Q_5, (Q_6, a) \vdash \{q_0, q_1, q_2\} = Q_7, (Q_6, b) \vdash \{q_3, q_2\} = Q_8, (Q_7, a) \vdash Q_3, (Q_7, b) \vdash Q_5, (Q_8, a) \vdash \{q_0, q_2\} = Q_9, (Q_8, b) \vdash Q_8, (Q_9, a) \vdash Q_7, (Q_9, b) \vdash \{q_3, q_2\} = Q_8$. $K = \{Q_0, \ldots, Q_9\}$ $F = \{Q_0, \ldots, Q_7\}$.

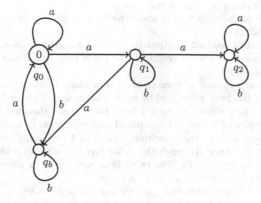

Fig. 11.1: Diagram for Exercise 2.4.5

11.3 Chapter 3

Exercise 3.2.3 $M = \{\langle q_0, B, 0, q_1 \rangle, \langle q_1, 0, L, q_1 \rangle, \langle q_1, B, 0, q_2 \rangle, \rangle q_2, 0, R, q_2 \rangle, \langle q_2, B, 0, q_1 \rangle\}$
Exercise 3.2.5

(i) $f(x) = 2x^2$.

(1) If the input is 0 then stop.
(2) Read the current square. If it is a 1: change the 1 to 0; move to the second blank square to the right and write a 1; move left until finding a 0 then move right one square; return to step (2). If it is a blank square then move left until finding a blank square then move right one square and go to step (3).
(3) Read the current square. If it is a 0: change the 0 to a blank square; move right until finding a 1 then go to step (4). If it is a blank square: go to step (5).
(4) Read the current square. If it is a 1: change the 1 to a 0; move to the second blank to the right and write a 1; move left until finding a 0 then move right one square; return to step (3). If it is a blank: move left one square at a time restoring the string of x consecutive 0's to 1's; move left until reaching the second blank square then move right one square; return to step (3).
(5) Move right one square at a time changing 1's to blank squares until reaching a blank square; move right one square.
(6) Read the current square. If it is a 1: change the 1 to 0; move to the second blank square to the right and write a 1; move left until finding a 0 then move right one square; return to step (6). If it is a blank square then change to a 1; move left one square at a time changing 0's to 1's until finding a blank square then move right one square; change the 1 to a blank square; move right one square and halt.

The steps above compute $2x^2$ as follows:
Step (1) checks if the input is zero and halts if it is. Step (2) copies the input x and writes a string of 1's to the right of the input on the tape. This leaves behind a string of 0's where the input was. Steps (3), (4) and (5) produce a string of x^2 1's on the tape to the right of the copy of x made in step (2). Finally, Step (6) copies the x^2 many 1's and then tidies up the tape to leave a consecutive string of $2x^2$ 1's with the tape head reading the leftmost 1.

(ii) $f(x) = 0$ if $x = 0$ and $f(x) = x^x$ otherwise.

We give a sketch of the algorithm to show the idea. A full step by step description is left as an exercise.

Firstly if $x = 0$ then halt with output 0. Otherwise copy the input x by writing a block of x consecutive 1's to the right of the input leaving a blank square between the input and the copy.

Use the input-1 as a counter to keep track of how many times x has been multiplied by itself. While x has been multiplied by itself less than $x - 1$ times (that is, there are still some 1's on the counter) delete a 1 from the counter then use the copy of x to add another x copies of the answer computed so far (written to the right of the copy of x on the tape). Initially, the algorithm computes $1x$ by simply copying the copy of x once. Then next time thorough the procedure the algorithm produces x copies of $1x$ which is a string of x^2 1's. The third time we produce x copies of x^2 giving a string of x^3 many 1's. And so on. . . ..

Finally the tape is tidied up leaving just the output of x^x many 1's.

(iii) $f(x) = 1$ if x is prime and $f(x) = 0$ otherwise.

There are many different algorithms for determining whether a given number is prime. One solution is sketched as follows:

Firstly if $x = 0$ or $x = 1$ then output 0 since x is not prime.

Otherwise we use a counter to test whether n divides x for all $2 \leqslant n \leqslant x$.

Initially the counter is set as a string of two 1's.

By moving between the counter and the input, turn a 1 in the counter to a 0 and then look for a 1 in the input to turn it to a 0. Repeat this until either there are no more 1's in the input or no more 1's in the output.

If no more 1's can be found in the input check to see if there are any more 1's in the counter. If the counter is onl;y 0's then the current value of the ocunter, n say, has divided x. in this case compare x with n by replacing a 0 from n with a blank square followed by replacing a 0 from x with a blank square. When the counter runs out of 0's check if the input has totally disappeared as well. If it has then x is prime as the only number $n \geqslant 2$ dividing x was x itself. In this case clear the tape and output a 1. Otherwise x is not prime since it is divisible by some number $2 \leqslant n < x$. In this case output 0.

If no more 1's can be found in the input but there is at least a single 1 remaining in the counter, then the counter n has not divided x. In this case increment the counter by 1 by restoring the 0's to 1's and adding one 1. Also restore the input to 1's. Repeat the algorithm for the new value of the counter $n + 1$.

If no more 1's can be found on the counter but there is at least a single 1 remaining in the input then we restore the counter to a string of n many 1's and repeat the division process again.

In this way x is tested for division by numbers $2 \leqslant n \leqslant x$ until a divisor is found, determining x prime if and only if the divisor is less than x.

Exercise 3.2.6 This machine starts moving right until it hits a blank, remembering how many 1's it has seen modulo 3. 1. is for 0, 3. for 1 and 5. for 2.

    1. If B go to 12
2. R
3. if B go to 8.
4. R
5. If B go to 8.
6. R
7. Go to 1.
8. L
9. If B go to 12. 10. Print B.
11. Go to 8.
12. Stop.

Exercise 3.2.7 Quadruples needed for a Turing machine to run on a planar tape are as follows:

$\langle q_i, x_i, A_{i,j}, q_{i,j} \rangle$ where $q_i, q_{i,j} \in \{q_0, \ldots, q_m\}$ with q_0 the initial state; $x_i \in \Sigma$; $A_{i,j} \in \Sigma \cup \{N, S, E, W\}$.

A quadruple is interpreted as: if I am in state q_i reading symbol x_i then perform action $A_{i,j}$ by writing a symbol in Σ or moving in one of the directions north, south, east or west, and then go to state $q_{i,j}$.

The following list of quadruples is a program for covering an empty planar tape with the symbol A. It works by moving around in a clockwise spiral.

$\{\langle q_0, B, A, q_0 \rangle, \langle q_0, A, S, q_1 \rangle, \langle q_1, A, N, q_2 \rangle, \langle q_1, B, A, q_3 \rangle, \langle q_2, A, E, q_0 \rangle, \langle q_3, A, W, q_4 \rangle,$
$\langle q_4, A, E, q_5 \rangle, \langle q_4, B, A, q_6 \rangle, \langle q_5, A, S, q_5 \rangle, \langle q_5, B, A, q_3 \rangle, \langle q_6, A, N, q_7 \rangle, \langle q_7, A, S, q_8 \rangle,$
$\langle q_7, B, A, q_9 \rangle, \langle q_8, A, W, q_8 \rangle, \langle q_8, B, A, q_6 \rangle, \langle q_9, A, E, q_{10} \rangle, \langle q_{10}, A, W, q_{11} \rangle,$
$\langle q_{10}, B, A, q_0 \rangle, \langle q_{11}, A, N, q_{11} \rangle, \langle q_{11}, B, A, q_9 \rangle\}$

Quadruples 1 to 5 move east writing the symbol A in blank squares until finding a blank square to the south. Similarly, quadruples 6 to 10 move south writing A's until finding a blank square to the west. quadruples 11 to 15 move west writing A's until detecting a blank row to the north, and quadruples 16 to 20 write A's north until finding a blank square to the east. The process then repeats forever.

Exercise 3.3.1

(i) Use recursion. $f(a, 0) = 1 = P(S(Z(a)))$. $f(a, b+1) = f(a, b) \cdot b$. (We have already shown that multiplication is primitive recursive.)
Use induction to prove that it works. For the induction step, if we assume $f(a, b) = a^b$, then $f(a, b+1) = f(a, b) \cdot a = a^b \cdot a = a^{b+1}$, as required.

(ii) Use recursion again. $f(0) = 1$, $f(a+1) = (a+1) \cdot f(a)$, etc.

Exercise 4.1.3 These are all similar. You can convert the regular languages into deterministic finite automata. Then, for example, asking if $L(M) = \emptyset$ is asking whether there is a string σ accepted by the machine. Since there are only n states for fixed M, if M accepts anything it will accept something of length $\leqslant n$, as the longest path without repetitions will have length n. Thus, at worst try all strings of length $\leqslant n$.

Exercise 4.1.5 Let $f(x) = \varphi_x(x) + 1$ if $\varphi_x(x) \downarrow$. Then f is partial computable, by the universal Turing Machine. Suppose that g is a computable function extending f. Then there is some z with $g = \varphi_z$. But then as g is total, $g(z) \downarrow$. But by definition of f, $f(z) = \varphi_z(z) + 1$. However, since g extends f, $\varphi_z(z) = g(z) = f(z) = \varphi_z(z) + 1$, a contradiction.

Exercise 4.2.6 A Minsky machine to compute the function $f(x) = x^2 + 1$ is:

(Here I am using the shorthand $R_i -$ p, q to mean "If the content of R_i is > 0 go to instruction p, and if $= 0$, go to instruction labeled q." This separates the "take 1" option, into two instructions, which happens at instruction p.)

0. $R_1 -$	2, 11	7. $R_4 +$	5
1. $R_1 -$	2, 4	8. $R_4 -$	9, 4
2. $R_2 +$	3	9. $R_2 +$	8
3. $R_3 +$	1	10. $R_2 -$	10, 11
4. $R_3 -$	5, 10	11. $R_1 +$	12
5. $R_2 -$	6, 8	12. Halt.	
6. $R_1 +$	7		

A vector game derived from the above Minsky Machine is:

$$1 \ \langle 0,0,0,0 \mid 1,-1 \rangle \qquad 11 \ \langle 0,-1,0,0 \mid -5,6 \rangle$$
$$2 \ \langle 0,0,0,0 \mid -12,0 \rangle \qquad 12 \ \langle 0,0,0,0 \mid -5,8 \rangle$$
$$3 \ \langle 1,0,0,0 \mid -11,12 \rangle \qquad 13 \ \langle 0,0,-1,0 \mid -4,5 \rangle$$
$$4 \ \langle 0,-1,0,0, \mid -10,10 \rangle \ 14 \ \langle 0,0,0,0 \mid -4,10 \rangle$$
$$5 \ \langle 0,0,0,0 \mid -10,11 \rangle \qquad 15 \ \langle 0,0,1,0 \mid -3,1 \rangle$$
$$6 \ \langle 0,1,0,0 \mid -9,8 \rangle \qquad 16 \ \langle 0,1,0,0 \mid -2,3 \rangle$$
$$7 \ \langle 0,0,0,-1 \mid -8,9 \rangle \qquad 17 \ \langle -1,0,0,0 \mid -1,2 \rangle$$
$$8 \ \langle 0,0,0,0 \mid -8,4 \rangle \qquad 18 \ \langle 0,0,0,0 \mid -1,4 \rangle$$
$$9 \ \langle 0,0,0,1 \mid -7,5 \rangle \qquad 19 \ \langle -1,0,0,0 \mid 0,2 \rangle$$
$$10 \ \langle 1,0,0,0 \mid -6,7 \rangle \qquad 20 \ \langle 0,0,0,0 \mid 0,11 \rangle$$

A rational game for $f(x) = x^2 + 1$ derived from the above vector game is:

$$1 \ r_1 = 2^0 \cdot 3^0 \cdot 5^0 \cdot 7^0 \cdot 11^1 \cdot 13^{-1} = 11/13 \qquad 11 \ r_{11} = 13^6/(3 \cdot 11^5)$$
$$2 \ r_2 = 2^0 \cdot 3^0 \cdot 5^0 \cdot 7^0 \cdot 11^{-12} \cdot 13^0 = 1/11^{12} \quad 12 \ r_{12} = 13^8/11^5$$
$$3 \ r_3 = 2^1 \cdot 11^{-11} \cdot 13^{12} = (2 \cdot 13^{12})/11^{11} \qquad 13 \ r_{13} = 13^5/(5 \cdot 11^4)$$
$$4 \ r_4 = 13^{10}/(3 \cdot 11^{10}) \qquad 15 \ r_{15} = (5 \cdot 13)/11^3$$
$$5 \ r_5 = 13^{11}/11^{10} \qquad 16 \ r_{16} = (3 \cdot 13^3)/11^2$$
$$6 \ r_6 = (3 \cdot 13^8)/11^9 \qquad 17 \ r_{17} = 13^2/(2 \cdot 11)$$
$$7 \ r_7 = 13^9/(7 \cdot 11^8) \qquad 18 \ r_{18} = 13^4/11$$
$$8 \ r_8 = 13^4/11^8 \qquad 18 \ r_{18} = 13^4/11$$
$$9 \ r_9 = (7 \cdot 13^5)/11^7 a \qquad 19 \ r_{19} = 13^2/2$$
$$10 \ r_{10} = (2 \cdot 13^7)/11^6 \qquad 20 \ r_{20} = 13^{11}$$

Finally, the relevant Collatz function is the following. (To simplify the calculation, I will use the least common multiple of the denominators instead of their product as the base p.)

Let $p = 2 \cdot 3 \cdot 5 \cdot 7 \cdot 11^{12} \cdot 13$. The value of $f(n)$ is

1. $\frac{11}{13}n$ if $n = 13k \bmod p$ for some k.
2. $\frac{1}{11^{12}}n$ if $n = 11^{12}k \bmod p$ for some k and the previous case does not apply.
3. $\frac{2 \cdot 13^{12}}{11^{11}}n$ if $n = 11^{11}k \bmod p$ for some k and the previous cases do not apply.
4. $\frac{13^{10}}{3 \cdot 11^{10}}n$ if $n = (3 \cdot 11^{10})k \bmod p$ for some k and the previous cases do not apply.
5. $\frac{13^{11}}{11^{10}}n$ if $n = 11^{10}k \bmod p$ for some k and the previous cases do not apply.
6. $\frac{3 \cdot 13^8}{11^9}n$ if $n = 11^9k \bmod p$ for some k and the previous cases do not apply.
7. $\frac{13^9}{7 \cdot 11^8}n$ if $n = (7 \cdot 11^8)k \bmod p$ for some k and the previous cases do not apply.
8. $\frac{13^4}{11^8}n$ if $n = 11^8k \bmod p$ for some k and the previous cases do not apply.
9. $\frac{7 \cdot 13^5}{11^7}n$ if $n = 11^7k \bmod p$ for some k and the previous cases do not apply.
10. $\frac{2 \cdot 13^7}{11^6}n$ if $n = 11^6k \bmod p$ for some k and the previous cases do not apply.
11. $\frac{13^6}{3 \cdot 11^5}n$ if $n = (3 \cdot 11^5)k \bmod p$ for some k and the previous cases do not apply.
12. $\frac{13^8}{11^5}n$ if $n = 11^5k \bmod p$ for some k and the previous cases do not apply.
13. $\frac{13^5}{5 \cdot 11^4}n$ if $n = (5 \cdot 11^4)k \bmod p$ for some k and the previous cases do not apply.
14. $\frac{13^{10}}{11^4}n$ if $n = 11^4k \bmod p$ for some k and the previous cases do not apply.
15. $\frac{5 \cdot 13}{11^3}n$ if $n = 11^3k \bmod p$ for some k and the previous cases do not apply.
16. $\frac{3 \cdot 13^3}{11^2}n$ if $n = 11^2k \bmod p$ for some k and the previous cases do not apply.
17. $\frac{13^2}{2 \cdot 11}n$ if $n = (2 \cdot 11)k \bmod p$ for some k and the previous cases do not apply.
18. $\frac{13^4}{11}n$ if $n = 11k \bmod p$ for some k and the previous cases do not apply.
19. $\frac{13^2}{2}n$ if $n = 2k \bmod p$ for some k and the previous cases do not apply.
20. $13^{11}n$ if none of the precious cases apply.

Exercise 4.3.11 $\sqrt{a}$ is rational iff $\exists x \exists y ((x+1)^2 - a(y+1)^2 = 0)$.

Exercise 4.3.12 Suppose S is Diophantine. $a \in S \iff (\exists \overline{x})[p(a, \overline{x}) = 0]$. Define $q(a, \overline{x}) = (a+1)(1 - 2p(a, \overline{x})^2) - 1$. Then if $p(a, \overline{x}) = 0$ then $q(a, \overline{x}) = (a + 1(1 - 2 \cdot 0) - 1 = a$. On the other hand, if $p(a, \overline{x}) \neq 0$, $1 - 2p(a, \overline{x})^2 < 0$, so $q(a, \overline{x}) < 0$ (unless $a < 0$). Therefore $S = \mathrm{ra}(q) \cap \mathbb{N}$

Exercise 4.3.14 $\ell = 6, x = 2, s = 7$ so that $Q = 32$.

```
2 2 1 1 0 0 0 R1
1 0 0 0 0 1 0 L0
0 1 0 0 0 0 0 L1
0 0 1 0 0 0 0 L2
0 0 0 1 0 1 0 L3
0 0 0 0 1 0 0 L4
0 0 0 0 0 0 0 L5
1 0 0 0 0 0 1 L6
```

11.4 Chapter 5

Exercise 5.1.3 Suppose that A is infinite and c.e. Then there is a computable f with $f(\mathbb{N}) = A$. We can "thin f down" as we *know* that A is infinite. Thus we can define a new function g with $y(0) = f(0)$ and $g(n+1) = x$ where $(x \notin \{g(0), \ldots, g(n)\} \wedge \exists s(s$ least with $f(s) = x))$. Then g is 1-1.

Exercise 5.1.4 1. Suppose that $A = \mathrm{ra}\ (f)$ with f computable and increasing. Then to decide if $x \in A$, that is $\chi_A(x)$, compute $\{f(0), \ldots, f(x+1)\}$. Then $x \in A$ iff $x \in \{f(0), \ldots, f(x+1)\}$.

2. Let A be c.e. and infinite, with $A = \mathrm{ra}\ g$ and g computable. By waiting, we can have a computable function f with $f(0) = g(0)$, $f(n_1) = g(m)$ where m is least with $f(m) > g(n)$. Then define $B = \mathrm{ra}\ f$.

Exercise 5.3.4 Actually, the easiest proof is to take $f(x) = x^2$ and to take an index e with $\varphi_e = f$. then $\varphi_e(e) = e^2$. The original proof I had was the following, and a similar proof would show that there is an index e with $\varphi_e(e) = e^2$ and $\varphi(x) \uparrow$ for $x \neq e$ using $g(x, y) = y^2$ if $x = y$ and $\uparrow$, else. Define $g(x, y) = y^2$ for all x. Use the s-m-n theorem to get computable s with $g(x, \cdot) = \varphi_{s(x)}(\cdot)$. Now apply the Recursion Theorem to get a fixed point, $\varphi_{s(z)} = \varphi_z$. Let $e = s(z)$. Then $g(e, e) = \varphi_{s(e)}(e) = \varphi_e(e) = e^2$.

Exercise 5.6.3. (Sketch) Take the proof of the Limit Lemma and observe that if Φ is a wtt functional, then it can change its mind on x at most $2^{\varphi(x)+1}$ many times. For the other direction, use the same proof but observe that once you know the computable mind change bound, then you can bound the search in B. Since B is c.e., $B \leq_m K_0$, so we can bound the use in the halting problem represented by K_0.

Exercise 5.6.4 (Sketch) The normal proofs from a course in calculus can easily be made to be computable. This is because the relationship between ε and δ are explicit functions.

Exercise 5.7.2 Build $A = \lim_s A_s$ by finite extension to meet

$$R_e : \exists x [\Phi_e^{A \setminus \{x\}}(x) \neq A(x)].$$

Having met R_j for $j < e$ and given A_s, we pick a large fresh number x and ask $\emptyset'$ whether there is a string σ extending A_s with

$$\Phi_{s+1}^{\widehat{\sigma}}(x) = 1?$$

Here $\widehat{\sigma}$ is the same as σ except $\widehat{\sigma}(x) = 0$. If the answer is yes, let $A_{s+1} = A_s\widehat{\sigma}$. If the answer is no, let A_{s+1} be the extension of A_s of length x which is all 0 except at length x where we have $A_{s+1}(x) = 1$. In either case we meet R_{s+1}.

Exercise 5.7.3 (Sketch) (i) Let V_e be the e-th c.e. set of strings. Meet a requirement which meets or avoids V_e. at stage s see if there is a σ with $A_s\sigma \in V_e$ to meet R_{s+1}.

(ii) If A was c.e. then as it is infinite it has a computable subset $B = \{b_0 < b_1 < \ldots \}$. Consider the collection of strings $V = \{\tau_j : |\tau| = b_j \wedge \tau(b_j) = 0, j \in \mathbb{N}^+\}$. Then for all $\sigma \prec A$, there is a $\tau_j \in V$ with $\sigma \preccurlyeq \tau_j$, but for no j is $\tau_j \prec A$.

Exercise 5.7.10 1. In the same was as we turned Kleene-Post into a priority argument, turn the solution of Exercise 5.7.2 into a priority argument.

2. Suppose that $A = A_1 \sqcup A_2$ with $\Gamma^{A_1} = A_2, \Delta^{A_2} = A_1$. We build an autoreduction Φ for A. Run the enumerations of $A_{i,s}$ so that at every stage $A_s = A_{1,s} \sqcup A_{2,s} \restriction \xi(s)$. Here $\xi(s) = \max\{j \leqslant s : \gamma(j)[s], \delta(j)[s]\}$. Now to compute if $x \in A$, Φ computes the least stage s such that $(A_s \setminus \{x\}) \restriction \xi(x)[s] = (A \setminus \{(x)\} \restriction \xi[s]$. Then $x \in A$ iff $x \in A_s$. The point is that if x enters A after stage s, it must enter one of A_1 and A_2, but *also* the other A_i must change as well, and this cannot use x. Therefore $(A_s \setminus \{x\}) \restriction \xi(x)[s]) \neq (A \setminus \{(x)\} \restriction \xi[s]$.

The reasoning more or less reverses. If $\Phi^{A\setminus\{(x\}} = A(x)$ is an autoreduction, then whould x enter $A_{s+1} \setminus A_s$, then some *other* number must enter $A_{s+1} \setminus A_s$, once we speed up the computations so that at each stage s the length of agreement is at least s. For the least x entering $A_{s+1} \setminus A_s$ we put the least number into $A_{1,s+1}$ and all the others into $A_{2,s+1}$. Then it is not hard to show that $A_1 \equiv_T A_2$.

11.5 Chapter 6

Exercise 6.1.2

(i) Let $\Gamma_e(x)$ be a measure such that $\Gamma_e(x) = s$ means that $\varphi_e(x)$ uses exactly s space. $\varphi_e(x) \downarrow$ implies that $\varphi_e(x)$ halts in some space s, and hence $\Gamma_e(x) \downarrow = s$. Conversely, if $\Gamma_e(x) = s$ then $\varphi_e(x)$ halts in exactly space s, which must mean that $\varphi_e(x) \downarrow$. Hence, $\varphi_e(x) \downarrow$ if and only if $\Gamma_e(x) \downarrow$, so Blum's first axiom is satisfied for Γ.

To show the question $\Gamma_e(x) = s$? is computable we consider four cases: we begin running $\varphi_e(x)$ and keep track step by step of how much space is used and whether we have seen a configuration at step t occur at some earlier step.

    a. If $\varphi_e(x)$ reaches a stage when it uses more than s space then we answer the above question *NO*.

    b. If $\varphi_e(x) \downarrow$ at some stage and has used space strictly less than s then we answer *NO*.

    c. If $\varphi_e(x) \downarrow$ at some stage having used exactly space s then we answer *YES*.

    d. If we see a configuration at some stage that we have already seen at an earlier stage, and strictly less than s space has been used, then we know that the computation has entered an infinite loop of repeating configurations, and so we can answer *NO*.

Hence Γ satisfies Blum's second axiom, and so Γ is a complexity measure.

(ii) To show Blum's axioms are independent we give two examples, (there being many others, of course) one for which axiom 1 holds but axiom 2 does not hold, and vice-versa, one for which axiom 2 holds but axiom 1 does not hold: Let

$$\Gamma_e(x) = \begin{cases} 1 & \text{if } \varphi_e(x) \downarrow \\ \uparrow & \text{otherwise} \end{cases}$$

then clearly $\varphi_e(x) \downarrow$ if and only if $\Gamma_e(x) \downarrow$ and so axiom 1 holds. But the question $\Gamma_e(x) = s$? cannot be computable because then the Halting problem would be computable, a contradiction.

Now let

$$\Gamma_e(x) \downarrow = 1$$

for all e and x. Then clearly Blum's second axiom holds since to answer the question $\Gamma_e(x) = s$? we just say *YES* if $s = 1$ and *NO* otherwise. Also, axiom 1 does not hold since $\Gamma_e(x)$ is always defined, even if $\varphi_e(x)$ is undefined.

Exercise 6.1.3 The simulation of each step of the 2-tape machine M requires finding and recording the symbols under the two heads but using one tape and one head. The alphabet for the one tape machine M' uses symbols to encode what is being read by each tape of M together with whether the tape heads for M are reading a particular square or not. If L is accepted in time $f(n)$ by M then M' never needs to have more than $2f(n)$ tape squares in use. Searching for squares of M' that represent the square of each tape in M that is being read by each head can therefore only take time $\leqslant 4f(n)$. Then to perform the required change of state and tape writing action will be at most another $2f(n)$ time steps, so $\leqslant 6f(n)$ in total (moving to the opposite end of the tape) and we ignore some constant number of steps that are needed to perform the action. So each step of M takes $O(f(n))$ steps by M'. Therefore the at most $f(n)$ steps that M performs can be done by M' in $O(f(n)^2)$ steps.

Exercise 6.1.7 We first show that $P \subseteq \mathrm{DTIME}(n^{\log n})$.

Recall that $P = \bigcup_d \mathrm{DTIME}(n^d)$, where n is the length of the input x, and $\mathrm{DTIME}(f(n)) = \{L : $ there is some machine M that accepts almost all $x \in L$ in time $\leqslant f(n)\}$. *Almost all* means all but finitely many, that is there exists n_0 such that for all $n \geqslant n_0$, M accepts x in time less than $f(n)$ where n is the length of the input x.

We show that given $L \in P$ (so $L \in \mathrm{DTIME}(n^d)$, for some d) there is an n_0 such that for all strings x of length $\geqslant n_0$, $n^{\log n} > n^d$.

Choose $n_0 = 3^d$. Then $e^d < 3^d$ and so $e^d < n_0$. Therefore $e^d < e^{\log n_0}$ and so $d < \log n_0$. It follows that $n_0^d < n_0^{\log n_0}$ as required.

Therefore $P \subseteq \mathrm{DTIME}(n^{\log n})$.

Now we show the inclusion is strict, that is $P \neq \mathrm{DTIME}(n^{\log n})$. Suppose $f \in \mathrm{DTIME}(n^{\log n})$ and $g \in P$, that is $g \in \mathrm{DTIME}(n^k)$ for some $k \in \mathbf{N}$. Then by the hierarchy theorem, since

$$\lim_{n \to \infty} \frac{g(n)}{f(n)} = \lim_{n \to \infty} \frac{n^k}{n^{\log n}} = \lim_{n \to \infty} \frac{1}{n^{\log n - k}} = 0,$$

we have that $P \neq \mathrm{DTIME}(n^{\log n})$.

11.6 Chapter 7

Exercise 7.2.2 Suppose $L \in NP$. Then there is some nondeterministic Turing machine M and a polynomial q such that for almost all $x \in L$, there is an accepting computation of x by M in time $\leqslant q(n)$, where $n = |x|$, the length of the input x. That is there is some k such that M accepts almost all $x \in L$ is time $\leqslant |x|^k$.

We give a polynomial time computable language B : for the alphabet for B we extend the alphabet Σ of M as follows. Suppose M has q states, then we add q many new symbols to Σ to get the alphabet for B. Since M accepts x in at most $\leqslant |x|^k$ steps then M uses at most $|x|^k$ tape squares. Then using the alphabet for B we can encode a configuration of M that accepts x in $|x|^k + 1$ symbols. Now suppose for input x M goes through the configurations $c_1, c_2, \ldots, c_m$ and $m \leqslant |x|^k$. Add $|x|^k - m$ copies of c_m to the end of this list and so to code the whole computation requires $|x|^k$ configurations, and so requiring a string of $|x|^k(|x|^k + 1) = p(|x|)$ symbols. Define

$$B = \{\langle x, z \rangle \mid z \text{ encodes an accepting computation for } x\}.$$

Then B is a polynomial time computable language. Now $x \in L$ if and only if $\exists z[|z| = p(|x|) \,\&\, \langle x, z \rangle \in B]$, as required.

Conversely, suppose $x \in L$ if and only if $\exists z[|z| = p(|x|) \,\&\, \langle x, z \rangle \in B$ for some polynomial p and computable language B. Then $L \in NP$ since to nondeterministically compute if $x \in L$ in polynomial time we perform the following nondeterministic algorithm:

1. Guess a string z of length $p(|x|)$.
2. Compute whether $\langle x, z \rangle \in B$.
3. Accept x if $\langle x, z \rangle \in B$.

Since B is a polynomial time computable language, the above algorithm can be performed in polynomial time.

Exercise 7.2.3 Suppose that $\widehat{L} \leqslant_m^P L$ via f. Using Exercise 7.2.2, let $x \in L$ iff $\exists z[|z| = p(|x|) \wedge \langle x, z \rangle \in B]$, with $B \in P$. then $u \in \widehat{L}$ iff $f(u) \in L$ iff $\exists z[|z| = p(|f(u)|) \wedge \langle x, z \rangle \in B]$.

Exercise 7.2.10 Given G, ask if G has a Hamilton sycle. If not then say no. If yes, start at any vertex v and look at any edge uv. Consider $G_1 = G \setminus \{uv\}$. If G_1 has a Hamilton cycle, try to remove anoth edge from v. If G_1 has no Hamilton cycle, the uv is essential to any cycle. Continue this way removing edges until only a cycle remains.

Exercise 7.2.11 (Sketch) Consider $k = 3$. Let X be an instance of 3-SAT. The only problem might be that X has one of more clauses C of size < 3. We use local replacement to replace them with clauses of size 3. For example if C has size 2, say $m \vee n$ for literals m, n, then we replace C by $(m \vee n \vee z) \wedge (m \vee n \vee \neg z)$ for some fresh variable z, etc.

Exercise 7.2.17 The solution is a case analysis. As the graph is symmetric, it is enough to consider the opposite corners on the horizontal axis.

Suppose that x has colour 1. We need to show y has colour 1. Then give s colour 2. Hence h has colour 3.

Case 1. r has colour 3. Then ℓ has colour 1 and q has colour 2 and k has colour 3. Hence t has colour 1, p has colour 2, m has colour 3, v has colour 2 meaning that y must have colour 3.

Case 2. r has colour 2. Then ℓ has colour 1 and hence q and k have colour 3. Thus v has colour 2, and n has colour 1, meaning that y must be 3.

Exercise 7.2.20 The problem is in NP by guess and check. To establish NP-hardness, we reduce from VERTEX COVER. Let (G, k) be an instance of VERTEX COVER.

We construct a new graph H such that H has a dominating set of size k iff G has a vertex cover of size k. To make H, begin with G, and for each edge $xy \in E(G)$, add a new vertex v_{xy} to G. We then add edges $v_{xy}x$ and $v_{xy}y$ to specify H as G together with the new vertices and edges.

Now suppose H has a dominating set of size k. Since every vertex of H must be dominated, all the v_{xy} must be dominated, so for each edge $xy \in E(G)$, we must choose x, y, or v_{xy}, since they are the only vertices connected to v_{xy}. In particular, one vertex corresponding to each *edge* of G must be chosen, and hence there must be k vertices which are a vertex cover in G. Conversely, if G has a vertex cover of size k, say $u_1, \ldots, u_k$, then this is also a dominating set in H.

Exercise 7.2.21 This is NP-complete. Let (G, k) be an instance of DOMINATING SET. Now form a new graph H by adding a single vertex v, joined to every vertex of G. H has a dominating set of size 1, namely $\{v\}$. If we later remove v (and all its edges) from H then solving DOMINATING SET REPAIR is the same as finding a dominating set in G.

Exercise 7.2.22 (i) First see if the graph is 2-colourable. If it is not then evidently no schedule exists. If it is 2-colourable, then we need to decide, for each component, whether to schedule the red vertices at time 0, an the blue ones at time 1 or vice versa, so that there are at most m vertices scheduled at any one time. This can be done in polynomial time as can be seen by induction on the number of components.

(ii) The problem is equivalent to the following. Given a graph G can the vertices of G be properly coloured into t colours and each monochromatic set has cardinality at most m? This solves t-COL by taking $m = n$, and hence is NP-complete.

Exercise 7.2.23 (i) We reduce from SAT. Suppose that we have an instance φ of SAT with a variable x occurring $k > 3$ times. Then we will replace all but one of these occurrences by new variables, $\{s_1, \ldots, s_{k-1}\}$ (chosen to be distinct for each x). We then add the new clauses saying that each s_i is equivalent to x. This is enforced by $x \to s_1 \to s_2 \to \ldots s_{k-1} \to x$, which is equivalent to the conjunction:

$$(\neg x \lor s_1)(\neg s_1 \lor s_2) \land \ldots (\neg s_{k_1} \lor x).$$

(ii) If the variable x occurs only positively then we can make all clauses containing x true by making it true, and eliminate them, similarly if all occurrences are negative. Therefore we can reduce to the case where every variable occurs both positively and negatively, and, since it can occur at most twice, they occur *exactly* once positively and once negatively. If those two clauses contain no other variable, you would get $x \land \neg x$ so no satisfying assignment. Then so solve the question, we can always apply a "*resolution rule*" that we can combine two clauses, one containing x and one containing $\neg x$, by throwing x away. For example, if we have $(x \lor y \lor z)$ and $(\neg x \lor p \lor q \lor r)$ they combine to give $(y \lor z)(p \lor q \lor r)$ and the new clauses are satisfiable iff the old ones were. Continue in this way, until all variables are eliminated (then it is satisfiable) or we get a contradiction.

Exercise 7.2.31 First the problem is in NP as you can guess the solution and then check it. We reduce HAMILTON CYCLE to LONG PATH. Take G and suppose that G has n vertices. We make a new graph $\widehat{G}$ by taking a vertex x of G and adding an edge in its place of uv of G. Then we add two new vertices $\widehat{u}$ and $\widehat{v}$, and declaring $\widehat{u}u$ and $\widehat{v}v$ to be new edges; so that $\widehat{u}$ and $\widehat{v}$ have degree 1 in $\widehat{G}$. Then $\widehat{G}$ has $n+3$ many vertices and a long path will have $n + 2$ edges. If $\widehat{G}$ has a long path then it must start at $\widehat{u}$ and finish at $\widehat{v}$, up to reversal. This is because they have degree 1. Moreover, since vertices are not repeated, u and v are used exactly once. But then there is, in G a path from u to v consisting of only old edges, and then contracting uv in this path makes a Hamilton cycle in G. Conversely if G has a Hamilton cycle, then it has one beginning and ending at x. Then reversing the reasoning above, we can turn this into a long path.

Exercise 7.2.32 Reduce from HAMILTON CYCLE. Let G be an instance with n vertices, then the cycle of n vertices is a topological minor iff G has a Hamilton cycle.

Exercise 7.4.5 (Sketch) Use the following widget, where the indicated crossing is replaced by the crossing widget.

Fig. 11.2: Crossing

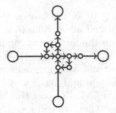

Fig. 11.3: Crossing Widget

11.7 Chapter 8

Exercise 8.1.4 $B = \{\langle e, x, 1^n \rangle :$ some computation path of Φ_e^B length $\leqslant n$ accepts $x\}$.
We need to prove that $\mathrm{NP}^B = \mathrm{P}^B$. The point is that on input $z = \langle e, x, n \rangle$, to determine
is z is put into B, we look at computation paths of length n, and $n = |1^n| < |\langle e, x, 1^n \rangle|$.
So whether or not we put z into B is determined by strings of length less than $|z|$. Thus
B is a well-defined and computable language. By the proof of Theorem 7.2.1, relativized,
we see that B is NP^B-complete, and certainly $B \in \mathrm{P}^B$, so $\mathrm{NP}^B = \mathrm{P}^B$.

11.8 Chapter 9

Exercise 9.4.11 (i) (2) implies (1). It suffices to prove that $E(C) \cap Y \neq \emptyset$ for any odd
cycle C. Let $E(C) \cap X = \{u_0 v_0, u_1 v_1, \ldots, u_q v_q\}$, which can be shown to be odd, and
numbered with vertices $u_i v_{(i+1) \bmod q}$ being connected by a path in $G - X$. Since Φ is
a valid colouring, we see that $\Phi(v_i) \neq \Phi(u_i)$ for all $0 \leqslant i \leqslant q$. As q is odd, there must
be some pair $v_i u_{(i+1) \bmod q}$ with $\Phi(v_i) \neq \Phi(u_{(i+1) \bmod q}$. But removal of Y destroys all
paths between black and white vertices, and hence $E(C) \cap Y \neq \emptyset$.

(1) implies (2). Let $C_X : V \to \{B, W\}$ be a two-colouring of the bipartite graph $G - X$,
and similarly G_Y a two-colouring of $G - Y$. Now define $\Phi : V \to \{B, W\}$ with $\Phi(v) = B$ by
having if $C_X(v) = C_Y(v)$ and $\Phi(v) = W$, otherwise. The claim is that given the restriction
of Φ to $V(X)$, $\widehat{\Phi}$ is a valid colouring with Y an edge cut in $G - X$ between the white and
the black vertices of $\widehat{\Phi}$.

(ii)The method is quite similar to that used for VERTEX COVER. Let $E(G) = \{e_1, \ldots, e_m\}$.
At step i we consider $G[\{e_1, \ldots, e_i\}] := G_i$, and either construct from X_{i-1} a minimal
bipartization set X_i or return a "no" answer. $X_1 = \emptyset$. Consider step i, and suppose that
$|X_{i-1}| \leqslant k$. If X_{i-1} is a bipartization set for G_i, we are done, as we can keep $X_{i-1} = X_i$,
and move on to step $i + 1$. Else, we consider $X_{i-1} \cup \{e_i\}$, which will clearly be a minimal
edge bipartization set for G_i. If $|X_{i-1} \cup \{e_i\}| \leqslant k$ then set $X_i = X_{i-1} \cup \{e_i\}$ and move
on to step $i + 1$. If $|X_{i-1} \cup \{e_i\}| = k + 1$, we seek an X_i that will have $\leqslant k$ edges, or we
report "no." The plan is to use Lemma 9.4.1 with "$X = X_{i-1}$" and "$Y = X_i$" to seek new
bipartizations. This lemma needs Y to be disjoint from X. This is achieved by a simple
reduction at this step. For each $uv \in X_{i-1}$, we delete this edge, and replace it in G with
3 edges, $uw_1, w_1 w_2, w_2 v$ (w_1, w_2 are distinct for each pair u, v). Then in X_{i-1} we include
one of these edges for each uv, for instance $w_1 w_2$. Notice that if uv is necessary for a
minimal edge bipartization before this step, then we can use either of the other two in its
stead for G_i. Now we proceed analogously to VERTEX COVER. Namely, we enumerate all
valid colourings Φ of $V(X_{i-1})$, and determine a minimum edge cut between the white and
black vertices of size $\leqslant k$. Each of the minimum cut problems can be solved using bipartite

matching and hence in time $O((k+1)i)$, as this has $k+1$ rounds. If no such cut is found in any of the partitions we say "no". If we find a Y, set $G_i = Y$, and move to step $i+1$. The total running time is thus $O(\sum_{i=1}^{m} 2^{k+1}ki) = O(2^k km^2)$.

Exercise 9.4.13 Let N denote the number of distinct coordinates that appear in the r-tuples of M, and assume that these coordinates are $\{1, \ldots, N\}$. Let $K = kr$. Let n denote the total size of the description of M.

Define a *solution schema* S for the problem to be a set of k r-tuples with all coordinates distinct, and with the coordinates chosen from $J(K)$. If α is an r-tuple in M, $\alpha = (x_1, \ldots, x_r)$, let $h(\alpha)$ denote the image r-tuple, $h(\alpha) = (h(x_1), \ldots, h(x_r))$.

Algorithm 11.8.1 1. For each $h \in \mathcal{H}(N, K)$ and each solution schema S:
  2. Compute $h(\alpha)$ for each $\alpha \in M$.
  3. Determine whether the solution schema S is *realized*, meaning that every r-tuple in S occurs as the image $h(\alpha)$ of some $\alpha \in M$. If so, answer "yes."
  4. If no hash function h realizes a solution schema, then answer "no."

The correctness of the algorithm is nearly immediate. If M has a k-matching, then this involves $K = kr$ distinct coordinates, and by the lemma, there must be some $h \in \mathcal{H}(N, K)$ that maps these invectively to $J(K)$, and the image under h is the solution schema that causes the algorithm to output "yes. " Conversely, if there is an h that realizes a solution schema S, then choosing one r-tuple in each of the k preimages yields a matching.

The running time of the algorithm is bounded by $O(K!K^{3K+1}n(\log n)^6)$.

11.9 Chapter 10

Exercise 10.2.1 This uses a simple gadget. Suppose that we had such an approximation algorithm and we can approximate to within k. Let G be a graph and with an instance of classical INDEPENDENT SET. To G add an independent set with k elements to make H. Then G has an independent set of size t iff H has one of size $t + k$.

Exercise 10.2.9 Suppose that $(\mathcal{I}_1, \mathcal{S}_1) \leqslant_{AP} (\mathcal{I}_2, \mathcal{S}_2)$, with f, g and d as above. Let A be an approximation algorithm for $(\mathcal{I}_2, \mathcal{S}_2)$. Then the algorithm $g(I, A(f(I, r)), r)$ is an approximation algorithm for $(\mathcal{I}_1, \mathcal{S}_1)$ with ratio $1 + d(r - 1)$.

Exercise 10.2.15 This is proven by induction on the width k, and for each k we recursively construct an algorithm A_k. If $k = 1$ then G is what is called a caterpillar graph being one that is like a path except it can have places which look like small stars with spikes of length 1; and we can use greedy minimization which will use at most 3 colours. So suppose $k > 1$, and let G_n have vertices $\{v_1, \ldots, v_n\}$. The computable algorithm A_k will have computed a computable partition of G, which we denote by $\{D_y \mid y < k\}$. We refer to the D_y as *layers*. Consider v_{n+1}. If the pathwidth of $G_{n+1} = G_n \cup \{v_{n+1}\}$ is $< k$, colour v_{n+1} by A_{k-1}, and put into one of the cells D_y, for $y < k - 1$ recursively. (In the case of pathwidth 1, this will all go into D_0.) We will be colouring using using the set of colours $\{1, \ldots, 3k - 2\}$.

If the pathwidth of G_{n+1} is k, consider H_{n+1}, the induced subgraph of G_{n+1} generated by $G_{n+1} \setminus D_k$. If the pathwidth of H_{n+1} is $< k$, then again colour v_{n+1} by A_{k-1}, and put into one of the cells D_y, for $y < k - 1$, recursively, and colour using the set of colours $\{1, \ldots, 3k - 2\}$. If the pathwidth of H_{n+1} is k, then we put v_{n+1} into D_{k-1}. In this case, that is in D_{k-1}, we will use first fit using colours $3k - 2 < j \leqslant 3k + 1$.

The validity of this method follows from the fact that the maximum degree of vertices restricted to D_{k-1} is 2, and induction on k. Assume that A_{k-1} is correct and colours the subgraph of G_n induced by the layers $\{D_y \mid y < k - 1\}$ using colours $\{1, \ldots, 3k - 2\}$.

Note that the construction ensures that the pathwidth of this subgraph H_k is at most $k - 1$. Moreover, induction on n ensures that A_{k-1} would colour the vertices of the subgraph of G_n induced by $\{D_y \mid y < k-1\}$ using colours $\{1, \dots, 3k-2\}$ assuming the vertices of G_n in D_{k-1} did not exist, the same as A_k colours them. To see this, assuming that it is true up to step n, then if v_{n+1} is added to D_{k-1} then there is nothing to prove, and if it is added to $\cup_{j<k-1} D_j$, then this step will invoke A_{k-1} and since the colours of D_{k-1} will exceed $3k-2$, they have no effect on the colouring of the subgraph of G_{n+1} induced by $\cup_{j<k-1} D_j \cup \{v_{n+1}\}$.

Suppose that that v_{n+1}'s addition to G_n has pathwidth k. Now consider a path decomposition $B_1, \dots, B_q$ of G_{n+1}. Suppose, for a contradiction, that the degree of $v = v_{n+1}$ in D_k is $\geqslant 3$. Thus there are x, y, and z in D_k which are each connected to v. Without loss of generality, let's suppose that that they were added at stages $s_x < s_y < s_z \leqslant n$. Since each is in D_k, when we added them to D_k, we could not have added them to D_y for $y < k-1$. Since they were not added to such D_y it follows that at the stages they were added, they made the pathwidth of the relevant H_s ($s \in \{s_x, s_y, s_z\}$ to be k. Consider s_x. As the pathwidth of H_{s_x} was k, there must be some bag in any path decomposition of G_{s_x}, consisting of only members of G_{s_x} which has size $k+1$, and containing x. For $t > s_x$, this must still hold, that is, x must be in, at stage t, a bag of size $k+1$ consisting only of elements of G_{s_x}. For suppose this was not true at stage t. The pathwidth of G_t is k, and has bags $P_1, \dots, P_v$, say. Now delete all of the elements of $G_t \setminus G_{s_x}$ from the bags forming bags $P_1', \dots, P_v'$. This is a path decomposition of G_{s_x}, and hence must have pathwidth k, so there must be one of size $k+1$ containing x, and it only consists of elements of G_{s_x}.

Consider s_y. Since the pathwidth of H_{s_y} is k, it follows that s_y must be in a bag of size k in the path decomposition of H_{s_y} containing none of D_{k-1}. In particular, in any path decomposition of G_{s_y}, x and y must appear in bags Q_x and Q_y, respectively, of size k with $x \notin Q_y$ and $y \notin Q_x$,

Using the same reasoning, as above, this fact must hold also for each stage $t > s_y$. So we can conclude, using the same reasoning, that at stage $n+1$, x, y, z, and v are all in bags of size k, B_x, B_y, B_z, B_v, where $x \notin B_y \cup B_z \cup B_v$, and similarly for y, z and v.

Now consider B_v. Since xv is an edge, x and v lie together in some bag B_{xv}. If B_{xv} is left of B_v but B_{xv} is right of B_v we get a contradiction, since this would put x into B_v, by the interpolation property of pathwidth. So B_{xv} and B_x both lie, without loss of generality left of B_v. Similarly B_{yv} and B_y must lie on the same side, and this must be right. For if there were both left of B_v, then the interpolation property would make either B_x or B_y contain y of x respectively (considering the relevant orientations of B_x and B_y). But now we get a contradiction, since B_z cannot be either right or left of B_v without one of the B_x, B_y, or B_z containing a forbidden element. Thus, within D_{k_1} the degree of v is at most 2.

References

AB09.   Sanjeev Arora and Boaz Borak. *Computational Complexity: A Modern Approach*. Princeton University Press, 2009. vi

AB12.   Jeremy Avigad and Vasco Brattka. Computability and analysis: the legacy of Alan Turing. In Rodney Downey, editor, *Turing's Legacy*, pages 1–46. Springer-Verlag, 2012. xiv

ACG+99.   Giorgio Ausiello, Perlugio Crescenzi, Gorgio Gambosi, Viggo Kann, Alberto Marchetti-Spaccemela, and Marco Protasi. *Complexity and Approximation*. Springer-Verlag, 1999. vi, xix, 286, 287

Ack28.   Wilhelm Ackermann. Zum hilbertschen aufbau der reellen zahlen. *Mathematische Annalen*, 99:118–133, 1928. 60

AD22.   Matthew Askes and Rodney Downey. Online, computable, and punctual structure theory. *Logic Journal of the IGPL*, jzac065:44 pages, 2022. 293

ADF93.   Karl Abrahamson, Rodney Downey, and Michael Fellows. Fixed parameter tractability and completeness iv: On completeness for $W[P]$ and PSPACE analogs. *Annals of Pure and Applied Logic*, 73:235–276, 1993. 252

Adi55.   Sergei Adian. Algorithmic unsolvability of problems of recognition of certain properties of groups. *Doklady Akademii Nauk SSSR*, 103:533–535, 1955. 96

AHK77.   Kenneth Appel, Wolfgang Haken, and John Koch. Every planar map is four colorable. ii. reducibility. *Illinois Journal of Mathematics*, 21(3):491–567, 1977. 193, 282

AIK84.   Akeo Adachi, Shigeki Iwata, and Takumi Kasai. Some combinatorial game problems require $\omega(n^k)$ time. *Journal of the ACM*, 31(2), 1984. 209

AK00.   Chris Ash and Julia Knight. *Computable structures and the hyperarithmetical hierarchy*, volume 144 of *Studies in Logic and the Foundations of Mathematics*. North-Holland Publishing Co., Amsterdam, 2000. vi, 150

AKCF+04. Faisal Abu-Khzam, Rebecca Collins, Michael Fellows, Michael Langston, Henry Suters, and Christopher Symons. Kernelization algorithms for the vertex cover problem: theory and experiments. In R. Sedgewick L. Arge, G. Italiano, editor, *ALENEX/ANALC, Proceedings of the Sixth Workshop on Algorithm Engineering and Experiments and the First Workshop on Analytic Algorithmics and Combinatorics, New Orleans, LA, USA*, pages 62–69. SIAM, 2004. 263

AKLSS06. Faisal Abu-Khzam, Michael Langston, Puskar Shanbhag, and Christopher Symons. Scalable parallel algorithms for FPT problems. *Algorithmica*, 45:269–284, 2006. 260, 262, 263

AKS04.   Manindra Agrawal, Neeraj Kayal, and Nitin Saxena. PRIMES in P. *Annals of Mathematics*, 160:781–793, 2004. 181

© The Editor(s) (if applicable) and The Author(s), under exclusive license
to Springer Nature Switzerland AG 2024
R. Downey, *Computability and Complexity*, Undergraduate Topics
in Computer Science, https://doi.org/10.1007/978-3-031-53744-8

AL17.   Dennis Amelunxen and Martin Lotz. Average-case complexity without the
             black swans. *Journal of Complexity*, 41:82–101, 2017. 300, 309

ALM⁺98.   Sanjeev Arora, Carsten Lund, Rajeev Motwani, Madhu Sudan, and Mario
             Szegedy. Proof verification and the hardness of approximation problems. *Jour-
             nal of the ACM*, 45(3):501–555, 1998. 201, 224, 253

Aro96.   Sanjeev Arora. Polynomial time approximation schemes for Euclidean TSP
             and other geometric problems. In *Proceedings of the 37th IEEE Symposium
             on Foundations of Computer Science*, 1996. 253, 254

Aro97.   Sanjeev Arora. Nearly linear time approximation schemes for Euclidean TSP
             and other geometric problems. In *Proc. 38th Annual IEEE Symposium on
             the Foundations of Computing (FOCS'97)*, pages 554–563. IEEE Press, 1997.
             254, 255

AW09.   Scott Aaronson and Avi Wigderson. Algebraization: A new barrier in com-
             plexity theory. In *Symposium on the Theory of Computing (STOC)*, 2009.
             224

AYZ94.   Noga Alon, Raphy Yuster, and Uri Zwick. Color-coding: A new method for
             finding simple paths, cycles and other small subgraphs within large graphs. In
             Proc. Symp. Theory of Computing (STOC), pages 326–335. ACM, 1994. 267,
             268

Bab90.   Lázló Babai. E-mail and the unexpected power of interaction. In *Proceedings
             Fifth Annual Structure in Complexity Theory Conference*, pages 30–44, 1990.
             224

Bab16.   Lázló Babai. Graph isomorphism in quasipolynomial time [extended abstract].
             In *STOC '16: Proceedings of the forty-eighth annual ACM symposium on
             Theory of Computing*, pages 684–697, 2016. 228

Baz95.   Christina Bazgan. Schémas d'approximation et complexité paramétrée. *Rap-
             port de stage de DEA d'Informatique à Orsay*, 1995. 255

BBM03.   Cyril Banderier, René Beier, and Kurt Mehlhorn. Smoothed analysis of three
             combinatorial problems. In Branislav Rovan and Peter Vojtáš, editors, *Math-
             ematical Foundations of Computer Science 2003*, pages 198–207, Berlin, Hei-
             delberg, 2003. Springer Berlin Heidelberg. 309

BD20.   Laurent Bienvenu and Rodney Downey. On low for speed oracles. *Journal of
             Computing and System Sciences*, 108:49–63, 2020. 174

BDFH09.   Hans Bodlaender, Rodney Downey, Michael Fellows, and Danny Hermelin.
             On problems without polynomial kernels. *Journal of Computing and System
             Sciences*, 75(8):423–434, 2009. 270, 273, 274, 275

BDG90.   José Balcázar, Josep Díaz, and Joaquim Gabarró. *Structural Complexity I
             and II*. Number 11 and 22 in EATCS Monographs on Theoretical Computer
             Science. Springer Verlag, 1988 and 1990. 170, 211

Bee85.   Michael J. Beeson. *Foundations of Constructive Mathematics: Metamathe-
             matical Studies*. Springer-Verlag, 1985. 72

BG81.   Charles Bennett and John Gill. Relative to a random oracle A, $P^A \neq NP^A \neq$co-
             NP^A with probability 1. *SIAM Journal on Computing*, 10:96–113, 1981. 222

BGG97.   Egon Börger, Erich Grädel, and Yuri Gurevich. *The Classical Decision Prob-
             lem*. Springer-Verlag, 1997. 107

BGS75.   Theodore Baker, John Gill, and Robert Solovay. Relativizations of the P=?NP
             question. *SIAM Journal on Computing*, 4:431–442, 1975. 218, 219, 220, 221,
             222

BH08.   Harry Buhrman and John Hitchcock. NP-hard sets are exponentially dense
             unless coNP ⊆ NP/poly. In *Proceedings of the 23rd Annual IEEE Conference
             on Computational Complexity*, pages 1–7. IEEE, 2008. 272, 273

BHPS61.   Yehoshua Bar-Hillel, Micha Perles, and Eli Shamir. On the formal properties
             of the simple phrase structure grammars. *Z. Phon. Sprachwiss Kummun.
             Forsch.*, 14(2):143–172, 1961. 21, 39

Blu67.   Manuel Blum. A machine-independent theory of the complexity of recursive
            functions. *Journal of the Association for Computing Machinery*, 14(2):322–
            336, 1967. 173

Bod96.   Hans Bodlaender. A linear-time algorithm for finding tree-decompositions of
            small treewidth. *SIAM J. Comput.*, 25(6):1305–1317, 1996. 291

Bod11.   Hans Bodlaender. A tutorial on kernelization. In D. Marx and P. Rossmanith,
            editors, *Parameterized and Exact Computation 6th International Symposium,
            IPEC '11, Saarbrücken, Germany*, 2011. 264

Boo59.   William Boone. The word problem. *Annals of Math*, 70:207–265, 1959. 96

Boo72.   Ron Book. On languages accepted in polynomial time. *SIAM Journal on
            Computing*, 1:281–287, 1972. 210

Bor12.   Émil Borel. Le calcul des intégral défines. *Journal de Mathématiques Pures
            and Appliquées*, 8:159–210, 1912. xiv, 137

Buc65.   Bruno Buchberger. *An Algorithm for Finding the Basis Elements of the
            Residue Class Ring of a Zero Dimensional Polynomial Ideal*. PhD thesis,
            University of Innsbruck, 1965. xiii

BV04.   Rene Beier and Berthold Vöcking. Random knapsack in expected polynomial
            time. *Journal of Computer and System Sciences*, 69(3):306–329, 2004. Special
            Issue on STOC 2003. 309

BYE85.   Reuven Bar-Yehuda and Shimon Even. A local-ration theorem for approxi-
            mating the weighted vertex cover problem. In G. Ausiello and M. Lucertini,
            editors, *Analysis and Desighn of Algorithms for Combinatorial Problems*, vol-
            ume 109 of *Annals of Discrete Mathematics*, pages 27–46 Elsevier Science,
            1985. 285

Cai96.   Leizhen Cai. Fixed-parameter tractability of graph modification problems for
            hereditary properties. *Information Processing Letters*, 58(4):171–176, 1996.
            261

Can74.   Georg Cantor. Über eine eigenschaft des inbegriffes aller reellen algebraischen
            zahlen. *Journal für die Reine und Angewandte Mathematik*, 77:258–262, 1874.
            3, 6, 7, 13

Can78.   Georg Cantor. Ein beitrag zur mannigfaltigkeitslehre. *Journal für die Reine
            und Angewandte Mathematik*, 84:242–258, 1878. 3, 6, 7

Can79.   Georg Cantor. Über unendliche, lineare punktmannichfaltigkeiten. 1. *Mathe-
            matische Annalen*, 15:1–7, 1879. 3, 10, 11, 120

CBSW17.   Jack Copeland, Jonathan Bowen, Mark Sprevak, and Robin Wilson. *The
            Turing Guide*. Oxford University Press, 2017. 54

CC97.   Liming Cai and Jianer Chen. On fixed-parameter tractability and approxima-
            bility of NP-hard optimization problems. *Journal of Computer and Systems
            Sciences*, 54:465–474, 1997. 255

CCDF97.   Liming Cai, Jianer Chen, Rodney Downey, and Michael Fellows. The pa-
            rameterized complexity of short computation and factorization. *Archive for
            Mathematical Logic*, 36(4/5):321–338, 1997. 237, 247

CCF+05.   Jianer Chen, Benny Chor, Michael Fellows, Xiushen Huang, David Juedes,
            Iyad Kanj, and Ge Xia. Tight lower bounds for certain parameterized np-
            hard problems. *Information and Computation*, 201:216–231, 2005. 279

CCG+94.   Richard Chang, Benny Chor, Oded Goldreich, Juris Hartmanis, Johan Hastad,
            Desh Ranjan, and Pankaj Rohatgi. The random oracle hypothesis is false.
            Journal of Computing and System Sciences, 49(1):24–39, 1994. 222

CDF96.   Kevin Cattell, Michael Dinneen, and Michael Fellows. A simple linear-time
            algorithm for finding path-decompositions of small width. *Inform. Process.
            Lett.*, 57:197–203, 1996. 291

CFG04.   Yijia Chen, Joerg Flum, and Martin Grohe. On miniaturized problems in pa-
            rameterized complexity theory. In *International Workshop in Parameterized
            Complexity and Exact Computation*, pages 108–120, 2004. 277

CFK+16.   Marek Cygan, Fedor Fomin, Lukasz Kowalik, Dniel Lokshtanov, Dániel Marx,
            Daniel Pilipczuk, Michaa Pilipczuk, and Saket Saurabh. *Parameterized Algo-
            rithms*. Springer-Verlag, 2016. vi, xviii, 260, 267, 269
CI97.   Marco Cesati and Miriam Di Ianni. Computation models for parameterized
            complexity. *MLQ Math. Log. Q.*, 43:179–202, 1997. 251, 275
CJ01.   Liming Cai and David Juedes. Subexponential parameterized algorithms col-
            lapse the W-hierarchy. In *28th International Colloquium on Automata, Lan-
            guages and Programming*, pages 273–284, 2001. 278, 279
CJ03.   Liming Cai and David Juedes. On the existence of subexponential parameter-
            ized algorithms. *Journal of Computing and System Sciences*, 67(4):789–807,
            2003. 277, 278, 279
CK00.   Chandra Chekuri and Sanjee Khanna. A ptas for the multiple knapsack prob-
            lem. In *Proceedings of the ACM-SIAM Symposium on Discrete Algorithms
            (SODA 2000)*, pages 213–222, 2000. 253, 254
CKX10.   Jianer Chen, Iyad Kanj, and Ge Xia. Improved upper bounds for vertex cover.
            Theoretical Computer Science A, 411:3736–3756, 2010. 233, 259
CL19.   Yijia Chen and Bingkai Lin. The constant inapproximability of the parame-
            terized dominating set problem. *SIAM Journal on Computing*, 48(2):513–533,
            2019. 291
CM99.   Jianer Chen and Antonio Miranda. A polynomial-time approximation scheme
            for general multiprocessor scheduling. In *Proc. ACM Symposium on Theory
            of Computing (STOC '99)*, pages 418–427. ACM Press, 1999. 253, 254
Coh63.   Paul Cohen. The independence of the continuum hypothesis, [part 1]. *Pro-
            ceedings of the National Academy of Sciences of the United States of America*,
            50(6):1143–1148, 1963. 14
Coh64.   Paul Cohen. The independence of the continuum hypothesis, [part 2]. *Pro-
            ceedings of the National Academy of Sciences of the United States of America*,
            51(1):105–110, 1964. 14
Con72.   John Conway. Unpredictable iterations. In *1972 Number Theory Conference,
            University of Colorado, Boulder*, pages 49–52. Springer-Verlag, 1972. xv, 77,
            79, 83
Coo71.   Stephen Cook. The complexity of theorem proving procedures. In *Proceedings
            of the Third Annual ACM Symposium on Theory of Computing*, pages 151–
            158, 1971. xviii, 182
CS88.   Marek Chrobak and Maciej Ślusarek. On some packing problem related to
            dynamic storage allocation. *RAIRO Inform. Théor. Appl.*, 22(4):487–499,
            1988. 293
CT97.   Marco Cesati and Luca Trevisan. On the efficiency of polynomial time ap-
            proximation schemes. *Information Processing Letters*, pages 165–171, 1997.
            258
DASMar.   Rodney Downey, Klaus Ambos-Spies, and Martin Monath. Notes on Sacks'
            Splitting Theorem. *Journal of Symbolic Logic*, to appear. 155
Dav77.   Martin Davis. Unsolvable problems. In Jon Barwise, editor, *Handbook of
            Mathematical Logic*, volume 90 of *Studies in Logic and the Foundations of
            Mathematics*, pages 567–594. North-Holland Publishing Co., 1977. 107
DECF+03.   Rodney Downey, Vladimir Estivill-Castro, Michael R. Fellows, Elena Prieto-
            Rodriguez, and Frances A. Rosamond. Cutting up is hard to do: the pa-
            rameterized complexity of k-cut and related problems. *Electronic Notes in
            Theoretical Computer Science*, 78:205–218, 2003. 277
Deh11.   Max Dehn. Über unendliche diskontinuierliche gruppen. *Math. Ann.*,
            71(1):116–144, 1911. xiv, 85, 94, 95
DF92.   Rodney Downey and Michael Fellows. Fixed parameter tractability and com-
            pleteness. *Congressus Numerantium*, 87:161–187, 1992. 231, 235, 237, 238,
            239, 250, 251

DF93.   Rodney Downey and Michael Fellows. Fixed parameter tractability and com-
             pleteness iii: Some structural aspects of the W-hierarchy. In K Ambos-Spies,
             S. Homer, , and E. Schöning, editors, *Complexity Theory: Current Research*,
             pages 166–191. Cambridge Univ. Press, 1993. 231, 238
DF95a.   Rodney Downey and Michael Fellows. Fixed parameter tractability and com-
             pleteness i: Basic theory. *SIAM Journal on Computing*, 24:873–921, 1995.
             231, 238, 239, 250, 251
DF95b.   Rodney Downey and Michael Fellows. Fixed parameter tractability and com-
             pleteness ii: Completeness for $W[1]$. *Theoretical Computer Science A*, 141:109–
             131, 1995. 231, 237, 238, 239, 241, 245, 246
DF95c.   Rodney Downey and Michael Fellows. Parameterized computational feasibil-
             ity. In P. Clote and J. Remmel, editors, *Proceedings of Feasible Mathematics
             II*, pages 219–244. Birkhauser, 1995. 260
DF98.   Rodney Downey and Michael Fellows. *Parameterized Complexity*. Springer-
             Verlag, 1998. 234, 262, 269
DF13.   Rodney Downey and Michael Fellows. *Fundamentals of Parameterized Com-
             plexity*. Springer-Verlag, 2013. vi, xviii, 40, 241, 251, 252, 264, 267, 269, 270,
             279, 291, 292
DFM06.   Rodney Downey, Michael Fellows, and Catherine McCartin. Parameterized
             approximation problems. In Hans Bodlaender and Michael Langston, editors,
             *Parameterized and Exact Computation. Second International Workshop, IW-
             PEC '06. Zürich, Switzerland, September 13–15, 2006. Proceedings*, LNCS
             4169, pages 121–129. Springer, 2006. 292
DGH24.   Eric Demaine, William Gasarch, and Mohammad Hajiaghayi. *Computational
             Intractability: A Guide to Algorithmic Lower Bounds*. MIT Press, 2024. xvii,
             xviii
DH10.   Rodney Downey and Denis Hirschfeldt. *Algorithmic randomness and complex-
             ity*. Theory and Applications of Computability. Springer, New York, 2010. vi,
             147, 150, 162, 209
DHNS03.   Rodney Downey, Denis Hirschfeldt, André Nies, and Frank Stephan. Trivial
             reals. In *Proceedings of the 7th and 8th Asian Logic Conferences*, pages 103–
             131, Singapore, 2003. Singapore Univ. Press. 150
Die05.   Reinhard Dietstel. *Graph Theory, 3rd Edition*. Springer-Verlag, 2005. 291
Din07.   Irit Dinur. The PCP Theorem by gap amplification. *Journal of the ACM*,
             54(3):12–es, 2007. 201
Dir89.   G. P. Lejeune Dirichlet. Über die darstellung ganz willkürlicher funktionen
             durch sinus- und cosinusreihen (1837). In *Gesammelte Werke*, pages 135–160.
             Bd. I. Berlin, 1889. 12
DJW12.   Vida Dujmovic, Gwenael Joret, and David Wood. An improved bound for
             first-fit on posets without two long incomparable chains. *SIAM J. Discrete
             Math*, 26:1068–1075, 2012. 293
DK92.   Rodney Downey and Julia Knight. Orderings with αth jump degree $\mathbf{0}^{(\alpha)}$.
             Proc. Amer. Math. Soc., 114(2):545–552, 1992. 138
DLP+12.   David Doty, Jack Lutz, Matthew Patitz, Robert Schweller, Scott Summers,
             and Damien Woods. The tile assembly model is intrinsically universal. In
             *Proceedings of the Fifty-third Annual IEEE Symposium on Foundations of
             Computer Science (FOCS 2012, New Brunswick, NJ, October 20-23)*, pages
             302–310. IEEE, 2012. 92
DM40.   Ben Dushnik and Evan Miller. Concerning similarity transformations of lin-
             early ordered sets. *Bull. Amer. Math. Soc.*, 46:322–326, 1940. 153
DMar.   Rodney Downey and Alexander Melnikov. *Computable Structure Theory:
             A Unified Approach*. Springer-Verlag, to appear. vi, 137, 150, 151
Dow03.   Rodney Downey. Parameterized complexity for the skeptic. In *Computational
             Complexity, 18th Annual Conference*, pages 147–169. IEEE, 2003. 253, 254

Dow14.   Rodney Downey. *Turing's Legacy*. Cambridge University Press, 2014. 54, 67

DPR61.   Martin Davis, Hillary Putnam, and Julia Robinson. The decision problem for exponential diophantine equations. *Annals of Mathematics*, 74:425–436, 1961. 99

Dru12.   Andrew Drucker. New limits to classical and quantum instance compression. In *Foundations of Computer Science, FOCS 2012*, pages 609–618, 2012. 276

DTH86.   Phan Dinh Dieu, Le Cong Thanh, and Le Tuan Hoa. Average polynomial time complexity of some NP-complete problems. *Theoretical Computer Science*, 46:219–327, 1986. 296, 297

Edm65.   Jack Edmonds. Paths, trees, and flowers. *Canadian Journal of Mathematics*, 17:449–467, 1965. xvii, 159, 236

EJS01.   Thomas Erlebach, Klaus Jansen, and Eike Seidel. Polynomial time approximation schemes for geometric graphs. In *Proc. ACM Symposium on Discrete Algorithms (SODA'01)*, pages 671–679, 2001. 254

Epp95.   David Eppstein. Subgraph isomorphism in planar graphs and related problems. In *Proceedings of the Sixth Annual ACM-SIAM Symposium on Discrete Algorithms, 22–24 January 1995. San Francisco, California*, pages 632–640, 1995. 275

ERV14.   Matthias Englert, Berthold Röglin, and Heiko Vöcking. Worst case and probabilistic analysis of the 2-opt algorithm for the TSP. *Algorithmica*, 68(1):190–264, 2014. 307, 309

Eul36.   Leonard Euler. Solutio problematis ad geometriam situs pertinentis. *Comment. Academiae Sci. I. Petropolitanae*, 8:128–140, 1736. xvii

Fei68.   Lawrence Feiner. *Orderings and boolean algebras not isomorphic to recursive ones*. ProQuest LLC, Ann Arbor, MI, 1968. Thesis (Ph.D.)–Massachusetts Institute of Technology. 142

Fel02.   Michael Fellows. Parameterized complexity: the main ideas and connections to practical computing. In *Experimental Algorithmics*. Apringer-Verlag, LNCS 2547, 2002. 256, 257

FG06.   Jorg Flum and Martin Grohe. *Parameterized Complexity Theory*. Springer-Verlag, 2006. vi, xviii, 235, 251, 269

FHRV09.   Michael Fellows, Danny Hermelin, Frances Rosamond, and Stéphanie Vialette. On the parameterized complexity of multiple-interval graph problems. *Theoretical Computer Science A*, 410:53–61, 2009. 249

FL88.   Michael Fellows and Michael Langston. Nonconstructive proofs of polynomial-time complexity. *Information Processing Letters*, 26:157–162, 1987/88. 269

FLSZ19.   Fedor Fomin, Daniel Lokshtanov, Saket Saurabh, and Meirav Zehavi. *Kernelization: Theory of Parameterized Preprocessing*. Cambridge University Press, 2019. vi, xviii, 263, 270

FR98.   David Fowler and Eleanor Robson. Square root approximations in old babylonian mathematics: YBC 7289 in context. *Historia Math*, 25(4):366–378, 1998. xii

Fri57.   Richard Friedberg. Two recursively enumerable sets of incomparable degrees of unsolvability. *Proceedings of the National Academy of Sciences of the United States of America*, 43:236–238, 1957. 143, 147, 148

FS56.   A. Fröhlich and J. Shepherdson. Effective procedures in field theory. *Philos. Trans. Roy. Soc. London. Ser. A.*, 248:407–432, 1956. 122, 123

FS11.   Lance Fortnow and Rahul Santhanam. Infeasibility of instance compression and succinct PCPs for NP. *Journal of Computing and System Sciences*, 77(1):91–106, 2011. 270, 271

FSU83.   Aviezri Fraenkel, Edward Scheinerman, and Daniel Ullman. Undirected edge geography. *Theoretical Computer Science*, 112(2):371–381, 1983. 208

FV11.   Fedor Fomin and Yngve Villanger. Subexponential parameterized algorithm for minimum fill-in. *CoRR, abs/1104.2230*, 2011. 261

GJ79.   Michael R. Garey and David S. Johnson. *Computers and Intractability*. W.
             H. Freeman and Co., San Francisco, Calif., 1979. A guide to the theory of
             NP-completeness, A Series of Books in the Mathematical Sciences. xviii, 200,
             253, 264

GMMU07.   Robert Gilman, Alexei Miasnikov, Alexander Myasnikov, and Alexander
             Ushakov. Notes on generic case complexity (37 pages, dated august 25, 2007).
             Personal communication to the author, 2007. 302, 303

GNR01.   Jens Gramm, Rolf Niedermeier, and Peter Rossmanith. Exact solutions for
             closest string and related problems. In *Proc. 12th ISAAC*, pages 441–453.
             Springer-Verlag, 2001. 260

Goe40.   Kurt Goedel. *The Consistency of the Continuum-Hypothesis*. Princeton Uni-
             versity Press, 1940. 14

GT04.   Ben Green and Terry Tao. The primes contain arbitrarily long arithmetic
             progressions. *Annals of Mathematics*, 167(2):481–547, 2004. 22

GV21.   Vijay Ganesh and Moshe Y. Vardi. On the unreasonable effectiveness of sat
             solvers. In TimEditor Roughgarden, editor, *Beyond the Worst-Case Analysis
             of Algorithms*, pages 547–566. Cambridge University Press, 2021. 309

HA77.   Wolfgang Haken and Kenneth Appel. Every planar map is four colorable. i.
             discharging. *Illinois Journal of Mathematics*, 21(3):429–490, 1977. 193, 282

Has87.   Johan Hastad. *Computational Limitations of Small Depth Circuits*. PhD
             thesis, Massachusetts Institute of Technology, 1987. 224

Her26.   Grete Herrmann. Die frage der endlich vielen schritte in der theorie der poly-
             nomideale. *Math. Ann.*, 95:736–788, 1926. xiv

Her95.   Rolf Herken. *The Universal Turing Machine: A Half Century Survey*.
             Springer-Verlag, 1995. 54, 67

Hie73.   Carl Hierholzer. Über die möglichkeit, einen linienzug ohne wiederholung und
             ohne unterbrechung zu umfahren. *Mathematische Annalen*, 4:30–32, 1873.
             xvii

Hig61.   Graham Higman. Subgroups of finitely presented groups. *Proc. Roy. Soc. Ser.
             A*, 262:455–475, 1961. 96

Hil90.   David Hilbert. Über die theorie der algebraischen formen. *Mathematische
             Annalen*, 36(4), 1890. xiii

Hil12.   David Hilbert. Mathematical problems. *Bulletin of the American Mathemat-
             ical Society*, 8(10):437–479, 1912. xiii, 98

HM07.   Michael Hallett and Catherine McCartin. A faster FPT algorithm for the
             maximum agreement forest problem. *Theory of Computing Systems*, 41, 2007.
             260

HN06.   Danny Harnik and Moni Naor. On the compressibility of NP instances and
             cryptographic applications. In *47th Annual IEEE Symposium on Foundations
             of Computer Science, FOCS 2006*, pages 719–728, 2006. 270

HS65.   Juris Hartmanis and Richard Sterns. On the computational complexity of
             algorithms. *Transactions of the American Mathematical Society*, pages 285–
             306, 1965. xvii, 159, 170

HS66.   Fred Hennie and Richard Sterns. Two-tape simulation of multitape turing
             machines. *Journal of the ACM*, 13(4):533–546, 1966. 163, 170

HS11.   Steven Homer and Alan Selman. *Computability and Complexity, 2nd Edition*.
             Springer-Verlag, 2011. vi

HU79.   John Hopcroft and J Ullman. *Introduction to Automata Theory, Languages
             and Computation*. Addison-Wesley Reading, 1979. 26

Huf54.   David Huffman. The synthesis of sequential switching circuits. *Journal of the
             Franklin Institute*, 257:161–190, 275–303, 1954. 43

Imp95.   Russell Impagliazzo. A personal view of average-case complexity. In *Pro-
             ceedings of Structure in Complexity Theory. Tenth Annual IEEE Conference*,
             pages 134–147, 1995. 309

IPZ01.   Russell Impagliazzo, Ramamoham Paturi, and Francis Zane. Which problems
            have strongly exponential complexity? *Journal of Computing and System
            Sciences*, 63(4):512–530, 2001. 277, 278

JM79.   James Jones and Yuri Matijacevič. Diophantine representation of enumerable
            sets. *Journal of Symbolic Logic*, 49(3):181–829, 1979. 99, 100, 101, 102, 103,
            105

Joh73.   David Johnson. *Near-Optimal Bin Packing Algorithms*. PhD thesis, Mas-
            sachusetts Institute of Technology, 1973. 284

JP78.   Carl Jockusch and David Posner. Double jumps of minimal degrees. *The
            Journal of Symbolic Logic*, 43:715–724, 1978. 145

Kar73.   Richard Karp. Reducibility among combinatorial problems. In Raymond
            Miller and James Thatcher, editors, *Complexity of Computer Computations*,
            pages 85–103. Plenum Press, 1973. 70, 189, 191, 192, 193, 195, 196, 197, 198,
            200, 238

Kar84.   Narendra Karmarkar. A new polynomial-time algorithm for linear program-
            ming. *Combinatorica*, 4(4):373–395, 1984. 198

Kar11.   Richard Karp. Heuristic algorithms in computational molecular biology. *Jour-
            nal of Computing and System Sciences*, 77(1):122–128, 2011. 265

Kha79.   Leonid Khachiyan. A polynomial algorithm for linear programming. *Doklady
            Akademii Nauk SSSR.*, 224(5):1093–1096, 1979. 198

Kho02.   Subhash Khot. On the power of unique 2-prover 1-round games. In *Proceedings
            of the thirty-fourth annual ACM symposium on Theory of computing*, pages
            767–775, 2002. 286

KL80.   Richard Karp and Richard Lipton. Some connections between nonuniform
            and uniform complexity classes. In *Proceedings of the 12th Symposium on the
            Theory of Computing*, pages 302–309, 1980. 210, 211

Kle36.   Stephen Kleene. λ-definability and recursiveness. *Duke Mathematical Journal*,
            2(2):340–352, 1936. 63

Kle38.   Stephen Kleene. On notation for ordinal numbers. *The Journal of Symbolic
            Logic*, 3:150–155, 1938. 116, 117

Kle56.   Stephen Kleene. Representation of events in nerve nets and finite automata.
            In C. Shannon and J. McCarthy, editors, *Automata Studies*, Annals of Math-
            ematics Studies, pages 3–42. Princeton University Press, 1956. 18, 26, 33,
            36

KM96.   Sanjeev Khanna and Rajeev Motwani. Towards a syntactic characterization
            of PTAS. In *Proceedings of Symposium on the Theory of Computing*, pages
            329–337, 1996. 255

KMSS03.   Ilya Kapovich, Alexander Myasnikov, Paul Schupp, and Vladimir Shpilrain.
            Generic case complexity, decision problems in group theory and random walks.
            Journal of Algebra, 264:665–694, 2003. xix, 302, 304, 305

Kob78.   G. N. Kobzev. On tt-degrees of r.e. t-degrees. *Mat. Sb.*, 106:507–514, 1978.
            225

Koz06.   Dexter Kozen. *Theory of Computation*. Springer-Verlag, 2006. vi, 210, 215

KP54.   Stephen Kleene and Emil Post. The upper semi-lattice of degrees of recursive
            unsolvability. *Annals of Mathematics. Second Series*, 59:379–407, 1954. 143

KQ95.   Hal Kierstead and Jun Qin. Coloring interval graphs with First-Fit. *Discrete
            Math.*, 144(1–3):47–57, 1995. Combinatorics of ordered sets (Oberwolfach,
            1991). 293

Kro82.   Leonid Kronecker. Grundzüge einer arithmetischen theorie der algebraischen
            grossen. *J. Reine Angew. Math.*, 92:1–123, 1882. xiv

KS81.   Jussi Ketonen and Robert Solovay. Rapidly growing Ramsey functions. *Annals
            of Mathematics*, 113(2):267–314, 1981. 60

KST93.   Johannes Köbler, Uwe Schöning, and Jacobo Toran. *The Graph Isomorphism
            Problem: Its Structural Complexity*. Springer-Verlag, 1993. 229

KST94.   Hiam Kaplan, Ron Shamir, and Robert Tarjan. Tractability of parameterized completion problems on chordal and interval graphs: Minimum fill-in and DNA physical mapping (extended abstract). In *35th Ann. Proc. of the Foundations of Computer Science (FOCS '94)*, pages 780–891, 1994. 260, 261

KST16.   Hal Kierstead, David Smith, and William Trotter. First-fit coloring on interval graphs has performance ratio at least 5. *European J. Combin.*, 51:236–254, 2016. 293

KT99.   Antonin Kučera and Sebastiaan Terwijn. Lowness for the class of random sets. *The Journal of Symbolic Logic*, 64:1396–1402, 1999. 293

Kuč86.   Antonin Kučera. An alternative priority-free solution to Post's problem. In J. Gruska, B. Rovan, and J. Wiederman, editors, *Proceedings, Mathematical Foundations of Computer Science*, Lecture Notes in Computer Science 233, pages 493–500. Springer, Berlin, 1986. 150

Kun11.   Kenneth Kunen. *Set Theory, Revised Edition*. Studies in Logic: Mathematical Logic and Foundations. College Publications, 2011. 3, 14

Kur83.   Stuart Kurtz. On the random oracle hypothesis. *Information Processing Letters*, 57:40–47, 1983. 222

Lac76.   Alastair Lachlan. A recursively enumerable degree which will not split over all lesser ones. *Annals of Mathematical Logic*, 9(4):307–365, 1976. 148

Lad73.   Richard Ladner. Mitotic recursively enumerable sets. *Journal of Symbolic Logic*, 38(2):199–211, 1973. 153

Lad75.   Richard Ladner. On the structure of polynomial time reducibilities. *Journal of computing and System Sciences*, 22:155–171, 1975. 226, 227, 228

Lau83.   Clemens Lautemann. BPP and the polynomial hierarchy. *Information Processing Letters*, 17(4):215–217, 1983. 214

Lei81.   Ernst Leiss. The complexity of restricted regular expressions and the synthesis problem for finite automata. *Journal of Computing and Systems Sciences*, 23(3):348–354, 1981. 30

Ler81.   Manuel Lerman. On recursive linear orderings. In *Logic Year 1979–80 (Proc. Seminars and Conf. Math. Logic, Univ. Connecticut, Storrs, Conn., 1979/80)*, volume 859 of *Lecture Notes in Math.*, pages 132–142. Springer, Berlin, 1981. 142

Ler83.   Manuel Lerman. *Degrees of Unsolvability*. Perspectives in Mathematical Logic. Springer, Berlin, 1983. 143

Lev73.   Leonid Levin. Universal search problems. *Problems of Information Transmission*, 9(3):115–116, 1973. xviii, 182, 188

Lev86.   Leonid Levin. Average case complete problems. *SIAM Journal on Computing*, 15:285–286, 1986. xix, 293, 298

LFKN92. Carsten Lund, Lance Fortnow, Howard Karloff, and Noam NIsan. Algebraic methods for interactive proof systems. *Journal of the ACM*, 39:859–868, 1992. 222, 223, 224

Liv10.   Noam Livne. All natural NP-complete problems have average case complete versions. *Computational Complexity*, 19:477–499, 2010. 298

Lob51.   Nikolai Lobachevsky. On the vanishing of trigonometric series (1834). In *Collected Works*, pages 31–80. Moscow-Leningrad, 1951. 12

LPS+08. Michael Langston, Andy Perkins, Arnold Saxton, Scharff Jon, and Brynn Voy. Innovative computational methods for transcriptomic data analysis: A case study in the use of FPT for practical algorithm design and implementation. *The Computer Journal*, 51:26–38, 2008. 234

LS80.   David Lichtenstein and Michael Sipser. Go is polynomial-space hard. *Journal of the ACM.*, 27(2):393–401, 1980. 206, 210

LS01.   Roger C. Lyndon and Paul E. Schupp. *Combinatorial group theory*. Classics in Mathematics. Springer-Verlag, Berlin, 2001. Reprint of the 1977 edition. 97, 304

Luc78.   Édouard Lucas. Théorie des fonctions numériques simplement périodiques. *American Journal of Mathematics*, 1:184–240, 1878. 101

Luk82.   Eugene Luks. Isomorphism of graphs of bounded valence can be tested in polynomial time. *Journal of Computer and System Sciences*, 25:42–65, 1982. 228

LV93.   Ming Li and Paul Vitanyi. *An introduction to Kolmogorov Complexity and its Applications*. Texts and Monographs in Computer Science. Springer-Verlag, 1993. vi, 162

Mar05a.   Dániel Marx. The closest substring problem with small distances. In *Annual IEEE Symposium on Foundations of Computer Science*, pages 63–72, 2005. 258

Mar05b.   Dániel Marx. Efficient approximation schemes for geometric problems? In *Proceedings of 13th Annual European Symposium on Algorithms*, pages 448–459, 2005. 258

Mar08.   Dániel Marx. Parameterized complexity and approximation algorithms. *The Computer Journal*, 51(1):60–78, 2008. 256, 258

Mar10.   Dániel Marx. Completely inapproximable monotone and antimonotone parameterized problems. In *Proceedings of the 25th Annual IEEE Conference on Computational Complexity, CCC 2010. Cambridge, Massachusetts, June 9–12, 2010*, pages 181–187. IEEE Computer Society Press, 2010. 292

Mat70.   Yuri Matijasević. The diophantineness of enumerable sets. *Doklady Akademii Nauk SSSR (N.S.)*, 191:279–282, 1970. 98, 99

Mat93.   Yuri Matijasević. *Hilbert's tenth problem*. Foundations of Computing Series. MIT Press, Cambridge, MA, 1993. Translated from the 1993 Russian original by the author, With a foreword by Martin Davis. 106

MG92.   Jayadev Misra and David Gries. A constructive proof of Vizing's Theorem. *Information Processing Letters*, 41(3):131–133, 1992. 282

MM69.   Albert Meyer and Edward McCreight. Classes of computable functions defined by bounds of computations. *Proc. Symb. Th. Comp.*, pages 79–88, 1969. 175

MN82.   George Metakides and Anil Nerode. The introduction of nonrecursive methods into mathematics. In *The L. E. J. Brouwer Centenary Symposium*, pages 319–335. North-Holland, Amsterdam, 1982. xiii

Moh84.   Jeanleah Mohrherr. Density of a final segment of the truth-table degrees. *Pacific Journal of Mathematics*, 115(2):409–419, 1984. 225

Mon19.   Antonio Montalbán. Martin's conjecture: A classification of the naturally occurring Turing degrees. *Notices of the American Mathematical Society*, 66, 2019. 143

Moo56.   Edward Moore. Gedanken experiments on sequential machines. In *Automata Studies: Annals of Mathematics Studies 34*, pages 129–153. Princeton University Press, 1956. 43

Mor98.   Bernard Moret. *The Theory of Computation*. Addison-Wesley, 1998. 283

Mos09.   Yiannis Moschovakis. *Descriptive set theory*, volume 155 of *Mathematical Surveys and Monographs*. American Mathematical Society, Providence, RI, second edition, 2009. 150

MP43.   W. McCulloch and W. Pitts. A logical calculus of ideas imminent in nervous activity. *Bull. Math. Biophys.*, 5:115–133, 1943. 26

MS72.   Albert Meyer and Larry Stockmeyer. The equivalence problem for regular expressions with squaring requires exponential space. In *SWAT (now known as FOCS)*, pages 125–129. IEEE Computer Society, 1972. 178, 205, 209

Muc56.   Albert Muchnik. On the unsolvability of the problem of reducibility in the theory of algorithms. *Doklady Akademii Nauk SSSR (N.S.)*, 108:194–197, 1956. 143, 147, 148

MY60.   Robert McNaughton and Hiroaki Yamada. Regular expressions and state graphs for automata. *IEEE Trans. Electron. Comput.*, ED-9:39–47, 1960. 30, 36

Myh57.   John Myhill. Finite automata and representation of events. *WADD TR-57-624, Wright-Patterson AFB, Ohio*, pages 112–137, 1957. 36, 43

Ner58.   Anil Nerode. Linear automaton transformations. *Proceedings of the American Math. Soc.*, 9:541–544, 1958. 36, 43

Nie06.   Rolf Niedermeier. *Invitation to Fixed-Parameter Algorithms*. Oxford University Press, 2006. vi, xviii, 259, 262, 264, 266, 269

Nie09.   André Nies. *Computability and Randomness*, volume 51 of *Oxford Logic Guides*. Oxford University Press, Oxford, 2009. vi

Nov55.   Pyotr Novikov. On the algorithmic unsolvability of the word problem in group theory. *Trudy Mat. Inst. Steklov*, 44:1–143, 1955. 96

NR00.   Rolf Niedermeier and Peter Rossmanith. A general method to speed up fixed-parameter-tractable algorithms. *Inform. Process. Lett.*, 73(3–4):125–129, 2000. 262

NSB08.   N. S. Narayanaswamy and R. Subhash Babu. A note on first-fit coloring of interval graphs. *Order*, 25(1):49–53, 2008. 293

NT75.   George Nemhauser and Leslie Trotter. Vertex packings: Structural properties and algorithms. *Math. Program.*, 8:232–248, 1975. 263, 284, 288

Odi90.   Piogeorgio Odifreddi. *Classical Recursion Theory. The theory of functions and sets of natural numbers*. Number 125 in Studies in Logic and the Foundations of Mathematics. North-Holland Publishing Company, Amsterdam, 1990. vi

Owi73.   James Owings, Jr. Diagonalization and the recursion theorem. *Notre Dame Journal of Formal Logic*, 14:95–99, 1973. 117

Pap94.   Christos Papadimitriou. *Computational Complexity*. Addison–Wesley, 1994. 204

PCW05.   Iman Poernomo, John Crossley, and Martin Wirsing. *Adapting Proofs-as-Programs: The Curry-Howard Protocol*. Monographs in Computer Science. Springer-Verlag, 2005. 72

PER89.   Marion Pour-El and Ian Richards. *Computability in Analysis and Physics*. Perspectives in Mathematical Logic. Springer Verlag, Berlin, 1989. vi, 137

Pip79.   Nick Pippenger. On simultaneous resource bounds. In *20th Annual Symposium on Foundations of Computer Science, FOCS 1979, San Juan, Puerto Rico, 29–31 October 1979. Proceedings*, pages 307–311. IEEE Computer Society, 1979. 211

Poc12.   Henry Pocklington. The determination of the exponent to which a number belongs, the practical solution of certain congruences, and the law of quadratic reciprocity. *Proceedings of the Cambridge Philosophical Society*, 16:1–5, 1912. xvii

Poo14.   Bjorn Poonen. Undecidable problems: a sampler. In J. Kennedy, editor, *Interpreting Gödel*, pages 211–241. Cambridge Univ. Press, 2014. 107

Pos44.   Emil Post. Recursively enumerable sets of positive integers and their decision problems. *Bulletin of the American Mathematical Society*, 50:284–316, 1944. 143

Pos47.   Emil Post. Recursive unsolvability of a problem of Thue. *Journal of Symbolic Logic*, 12:1–11, 1947. 86, 91, 121, 122

Pre08.   Charles Pretzgold. *The Annotated Turing*. Wiley, 2008. 54

Rab58.   Michael Rabin. Recursive unsolvability of group theoretic problems. *Annals of Mathematics*, 67:172–194, 1958. 96

Raz85.   Alexander Razborov. Some lower bounds for the monotone complexity of some boolean functions. *Soviet Math. Dokl.*, 31:354–357, 1985. 224

Ric53.   Henry Rice. Classes of recursively enumerable sets and their decision problems. *Transactions of the American Mathematical Society*, 74:358–366, 1953. 115

Rob52.   Julia Robinson. Existential definability in arithmetic. *Transactions of the American Mathematical Society*, 72:437–449, 1952. 101

Rog87.   Hartley Rogers. *Theory of recursive functions and effective computability*. MIT Press, Cambridge, MA, second edition, 1987. vi, 130, 143

Rot65.   Joseph Rotman. *The Theory of Groups*. Allyn and Bacon, 1965. 96
RR97.   Alexander Razborov and Stephen Rudich. Natural proofs. *Journal of Com-
            puter and System Sciences*, 55, 1997. 224
RS69.   Michael O. Rabin and Dana Scott. Finite automata and their decision prob-
            lems. *IBM J. Res. Dev.*, 3:114–125, 1969. 28, 30, 32
RSV04.   Bruce Reed, Kaleigh Smith, and Adrian Vetta. Finding odd cycle transversals.
            Operations Research Letters, 32:299–301, 2004. 265, 266
Sac63.   Gerald Sacks. On the degrees less than $\mathbf{0}'$. *Annals of Mathematics. Second
            Series*, 77:211–231, 1963. 153, 155
Sac64.   Gerald Sacks. The recursively enumerable degrees are dense. *Annals of Math-
            ematics Second Series*, 80:300–312, 1964. 155
Sav70.   Walter Savitch. Relationships between nondeterministic and deterministic
            tape complexities. *Journal of Computer and System Sciences*, 4(2):177–192,
            1970. 203
Sch78.   Thomas Schaefer. On the complexity of some two-person perfect-information
            games. *Journal of Computer and System Sciences*, 16(2):185–225, 1878. 206
Sch88.   Uwe Schöning. Graph isomorphism is in the low hierarchy. *Journal of Com-
            puting and System Sciences*, 37(3):312–323, 1988. 229
See14.   Abigail See. Smoothed analysis with applications in machine learning. Tripos
            Part III, Essay, Cambridge University, 2014. 309
SG76.   Sartaj Sahni and Teofilo Gonzales. P-complete approximation problems. *Jour-
            nal of the Association for Computing Machinery*, 23, 1976. 287
Sha92.   Adi Shamir. IP=PSPACE. *Journal of the ACM*, 39:869–877, 1992. 222
Sho59.   Joseph Shoenfield. On degrees of unsolvability. *Annals of Mathematics. Sec-
            ond Series*, 69:644–653, 1959. 136
Sho79.   Richard Shore. The homogeneity conjecture. *Proceedings of the National
            Academy of Science, USA*, 76(9):4218–4219, 1979. 218
Sip83.   Michael Sipser. A complexity theoretic approach to randomness. In *15th
            Symposium on the Theory of Computing*, pages 330–335, 1983. 214
Sma83.   Steve Smale. On the average number of steps of the simplex method of linear
            programming. *Mathematical Programming*, 27:241–262, 1983. 306
Soa87.   Robert I. Soare. *Recursively enumerable sets and degrees*. Perspectives in
            Mathematical Logic. Springer-Verlag, Berlin, 1987. A study of computable
            functions and computably generated sets. vi, 117, 143, 148
Soa16.   Robert I. Soare. *Turing Computability*. Theory and Applications of Com-
            putability. Springer-Verlag, Berlin, 2016. Theory and applications. vi, 143
SST06.   Arvind Sankar, Daniel A. Spielman, and Shang-Hua Teng. Smoothed analysis
            of the condition numbers and growth factors of matrices. *SIAM Journal on
            Matrix Analysis and Applications*, 28(2):446–476, 2006. 309
ST98.   Ron Shamir and Dekel Tzur. The maximum subforest problem: Approxima-
            tion and exact algorithms. In *Proc. ACM Symposium on Discrete Algorithms
            (SODA'98)*, pages 394–399. ACM Press, 1998. 253, 254
ST01.   Daniel Spielman and Shang-Hua Teng. Smoothed analysis of algorithms: why
            the simplex algorithm usually takes polynomial time. In *Proceedings of the
            Thirty-Third Annual ACM Symposium on Theory of Computing*, pages 296–
            305. ACM, 2001. xix, 306
ST09.   Daniel Spielman and Shang-Hua Teng. Smoothed analysis: an attempt to
            explain the behavior of algorithms in practice. *Communications of the ACM*,
            52(10):76–84, 2009. 306
Tao22.   Terry Tao. Almost all orbits of the Collatz map attain almost bounded values.
            Forum of Mathematics, Pi, 10(e12), 2022. 76
Tho68.   Ken Thompson. Regular expression search algorithm. *Communications of the
            ACM*, 11(6):419–422, 1968. 29, 32
Tho98.   Robin Thomas. An update on the four-color theorem. *Notices of the American
            Mathematical Society*, 45(7):848–859, 1998. 282

TMAS77.   Shuji Tsukiyama, Ide Mikio, Hiromu Ariyoshi, and Isao Shirikawa. A new algorithm for generating all the maximal independent sets. *SIAM Journal on Computing*, 6:506–517, 1977. 297

Tra84.   Boris Trakhtenbrot. A survey of Russian approach to perebor (brute force) algorithms. *Annals of the History of Computing*, 6:384–400, 1984. xvi

Tur36.   Alan Turing. On computable numbers with an application to the Entscheidungsproblem. *Proceedings of the London Mathematical Society*, 42:230–265, 1936. correction in *Proceedings of the London Mathematical Society*, vol. 43 (1937), pp. 544–546. xiv, 46, 53, 74, 126

Tur37a.   Alan Turing. Computability and λ-definability. *Journal of Symbolic Logic*, 2(4):153–163, 1937. 63

Tur37b.   Alan M. Turing. On Computable Numbers, with an Application to the Entscheidungsproblem. A Correction. *Proceedings of the London Mathematical Society*, 43:544–546, 1937. 126

Tur39.   Alan Turing. Systems of logic based on ordinals. *Proceedings of the London Math. Society*, 45:154–222, 1939. 129

Tur52.   Alan Turing. The chemical basis of morphogenesis. *Philosophical Transactions of the Royal Society of London B*, 237:37–72, 1952. vii

Var82.   Moshe Vardi. The complexity of relational query languages. In *Proceedings STOC '82*. ACM, 1982. 208

Vaz01.   Vijay Vazirani. *Approximation Algorithms*. Springer-Verlag, 2001. vi, 286, 287

Viz64.   Vadim Vizing. On an estimate of the chromatic class of a p-graph. *Diskret. Analiz.*, 3:25–30, 1964. 282

VV86.   Leslie Valiant and Vijay Vazirani. NP is as easy as detecting unique solutions. *Theoretical Computer Science*, 47:85–93, 1986. 215

Wan65.   Hao Wang. Games, logic and computers. *Scientific American*, 213(5):98–106, 1965. 92

Wll85.   Herbert Wilf. Some examples of combinatorial averaging. *American Math. Monthly*, 92:250–261, 1985. 294

Win98.   Erik Winfree. *Algorithmic Self Assembly of DNA*. PhD thesis, California Institute of Technology, 1998. 92

WXXZ23.   Virginia Vassilevska Williams, Yinzhan Xu, Zixuan Xu, and Renfei Zhou. New bounds for matrix multiplication: from alpha to omega, 2023. 166

Yan81.   Michael Yannakakis. Computing the minimum fill-in is NP-complete. *SIAM J. Algebr. Discrete Methods*, 2, 1981. 261

Yap83.   Chee Yap. Some consequences of non-uniform conditions on uniform classes. *Theoretical Computer Science*, 26:287–300, 1983. 211, 212

Index

Printed in the United States
by Baker & Taylor Publisher Services

Printed in the United States
by Baker & Taylor Publisher Services